DARWIN'S NEMESIS

Phillip Johnson and the
Intelligent Design Movement

FORWORD BY
SENATOR RICK SANTORUM

EDITED BY
WILLIAM A. DEMBSKI

IVP Academic

An imprint of InterVarsity Press
Downers Grove, Illinois

Inter-Varsity Press
Leicester, England

InterVarsity Press, USA
P.O. Box 1400, Downers Grove, IL 60515-1426, USA
World Wide Web: www.ivpress.com
Email: mail@ivpress.com

Inter-Varsity Press, England
38 De Montfort Street, Leicester LE1 7GP, England
Website: www.ivpbooks.com
Email: ivp@ivp-editorial.co.uk

InterVarsity Press®, USA, is the book-publishing division of InterVarsity Christian Fellowship/USA®, a student movement active on campus at hundreds of universities, colleges and schools of nursing in the United States of America, and a member movement of the International Fellowship of Evangelical Students. For information about local and regional activities, write Public Relations Dept., InterVarsity Christian Fellowship/USA, 6400 Schroeder Rd., P.O. Box 7895, Madison, WI 53707-7895, or visit the IVCF website at <www.intervarsity.org>.

Inter-Varsity Press, England, is the publishing division of the Universities and Colleges Christian Fellowship (formerly the Inter-Varsity Fellowship), a student movement linking Christian Unions in universities and colleges throughout Great Britain, and a member movement of the International Fellowship of Evangelical Students. For information about local and national activities write to UCCF, 38 De Montfort Street, Leicester LE1 7GP, email us at email@uccf.org.uk, or visit the UCCF website at <www.uccf.org.uk>.

Stephen C. Meyer's contribution "The Origin of Biological Information and the Higher Taxonomic Categories" first appeared in Proceedings of the Biological Society of Washington 117, no. 2 (2004). Used by permission.

Sections of chapter 16, "A Taxonomy of Teleology," as well as figures 16.1 and 16.2 and table 16.1 first appeared in Marcus Ross, "Who Believes What? Clearing Up Confusion over Intelligent Design and Young-Earth Creationism," Journal of Geoscience Education 53, no. 3 (2005): 319-23. Used by permission.

Design: Cindy Kiple

Images: Charles Darwin (1809-1882) (photo), Cameron, Julia Margaret (1815-1879)/Private Collection, The Stapleton Collection/Bridgeman Art Library

USA ISBN-10: 0-8308-2836-2
    ISBN-13: 978-0-8308-2836-4
UK ISBN-10: 1-84474-128-1
    ISBN-13: 978-1-84474-128-1

Printed in the United States of America ∞

Library of Congress Cataloging-in-Publication Data
Darwin's nemesis: Philip Johnson and the intelligent design movement
/ edited by William A. Dembski.
    p. cm.
Includes bibliographical references and index.
ISBN-13: 978-0-8308-2836-4 (pbk.: alk. paper)
ISBN-10: 0-8308-2836-2 (pbk.: alk. paper)
1. Johnson, Phillip E., 1940- 2. Johnson, Phillip E., 1940- Wedge
of God. 3. Johnson, Phillip E., 1940- Darwin on trial. 4.
Apologetics. 5. Naturalism, 6. Intelligent design (Teleology) I.
Dembski, William A., 1960-
BT1220.D28 2006
576.8—dc22
                                                          2005033144

| **P** | 20 | 19 | 18 | 17 | 16 | 15 | 14 | 13 | 12 | 11 | 10 | 9 | 8 | 7 | 6 | 5 | 4 | 3 | 2 | 1 |
| **Y** | 23 | 22 | 21 | 20 | 19 | 18 | 17 | 16 | 15 | 14 | 13 | 12 | 11 | 10 | 09 | 08 | 07 | 06 |

CONTENTS

FOREWORD 9
   Senator Rick Santorum

PREFACE 11
   William A. Dembski

INTRODUCTION: A MYTHIC LIFE 25
   John Mark Reynolds

PART I: PORTRAITS OF THE MAN AND HIS WORK

1 YOUR WITNESS, MR. JOHNSON:
   A Retrospective Review of *Darwin on Trial* 33
   Stephen C. Meyer

2 FROM MUTTERING TO MAYHEM:
   How Phillip Johnson Got Me Moving 37
   Michael J. Behe

3 HOW PHIL JOHNSON CHANGED MY MIND 48
   Jay Wesley Richards

4 PUTTING DARWIN ON TRIAL:
   Phillip Johnson Transforms the Evolutionary Narrative 62
   Thomas Woodward

PART II: THE WEDGE AND ITS DESPISERS

5 DEALING WITH THE BACKLASH AGAINST INTELLIGENT DESIGN 81
William A. Dembski

6 IT'S THE EPISTEMOLOGY, STUPID!
Science, Public Schools and What Counts as Knowledge 105
Francis J. Beckwith

7 CUTTING BOTH WAYS:
The Challenge Posed by Intelligent Design
to Traditional Christian Education 117
Timothy G. Standish

PART III: TWO FRIENDLY CRITICS

8 TWO FABLES BY JORGE LUIS BORGES 135
David Berlinski

9 DARWINISM AND THE PROBLEM OF EVIL 139
Michael Ruse

PART IV: JOHNSON'S REVOLUTION IN BIOLOGY

10 THE WEDGE OF TRUTH VISITS THE LABORATORY 153
David Keller

11 COMMON ANCESTRY ON TRIAL 164
Jonathan Wells

12 THE ORIGIN OF BIOLOGICAL INFORMATION
AND THE HIGHER TAXONOMIC CATEGORIES 174
Stephen C. Meyer

13 GENETIC ANALYSIS OF COORDINATE FLAGELLAR AND
TYPE III REGULATORY CIRCUITS IN PATHOGENIC BACTERIA 214
Scott A. Minnich and Stephen C. Meyer

PART V: EVER-INCREASING SPHERES OF INFLUENCE

14 INTELLIGENT DESIGN AND THE DEFENSE OF REASON 227
   Nancy Pearcey

15 PHILLIP JOHNSON WAS RIGHT:
   The Rivalry of Naturalism and Natural Law 244
   J. Budziszewski

16 A TAXONOMY OF TELEOLOGY:
   Phillip Johnson, the Intelligent Design Community
   and Young-Earth Creationism. 261
   Marcus Ross and *Paul Nelson*

17 COMPLEXITY, CHAOS AND GOD. 276
   Wesley D. Allen and *Henry F. Schaefer III*

PART VI: EPILOGUE

18 PHILLIP JOHNSON AND THE INTELLIGENT DESIGN MOVEMENT:
   Looking Back and Looking Forward. 305
   Walter L. Bradley

19 THE FINAL WORD. 315
   Phillip E. Johnson

Notes 318

Contributors 348

FOREWORD

SENATOR RICK SANTORUM

This volume celebrates Phillip Johnson's leadership in the intelligent design (ID) movement. Scholars who have known Phil best and worked with him most closely assembled in April 2004 at Biola University to present him with a collection of papers in his honor. I wish I could have been there to offer my congratulations and thanks in person. Instead, I have the privilege of writing this brief foreword from Washington.

Since the publication of *Darwin on Trial* more than ten years ago, Phillip Johnson has provided extraordinary leadership for an extraordinary cause, namely, to rid science of false philosophy. The importance of the cause is clear: what could be more important than showing that only a shallow, partisan understanding of science supports the false philosophy of materialist reductionism, with its thoroughly unscientific denial of formal and final causes in nature and its repudiation of the first cause of all being? As the decline of true science has been a major factor in the decline of Western culture, so too the renewal of science will play a big part in cultural renewal.

Johnson's extraordinary leadership also is clear: rather than fall into the trap of building a cult of personality around himself and his own considerable talents, he has instead helped raise up and promote a whole group of intellectual leaders in the cause of scientific renewal. This kind of selfless Christian leadership is a shining example to us all, young and old.

Speaking of the young, I personally wish to commend Phil for the great help he has given me in my efforts to inject a renewed and unbiased understanding of science and its practice into the curricula of our public schools. There is much more for us to do, but working with Phil's colleagues at Seattle's Discovery Institute, we have begun the difficult fight for removing the stranglehold of philosophical materialism on textbook science.

Phil, I congratulate and praise you for your tireless work to return science to a sure philosophical grounding in the nature of things as they really are. Please know that during your Biola celebration, I was with you and your colleagues in spirit. As much as I was delighted when I first heard about this celebration in your honor, I am again delighted now that the proceedings from that celebration have appeared in book form.

Senator Rick Santorum

PREFACE

WILLIAM A. DEMBSKI

Early in the morning on April 22, 2004, Phillip Johnson's wife, Kathie, led him, unsuspecting, into a Biola University conference room where he was greeted with a standing ovation from people who were his friends from all over the world. One by one, these people thanked Phil for his pivotal role in their lives and went on to deliver speeches summarizing their written contributions to what now has become this book.

Later that evening, Phil was again surprised to be the guest of honor at a large banquet, where he was the inaugural recipient and namesake of The Phillip E. Johnson Award for Liberty and Truth, an award honoring the lifetime achievements of an individual who has radically expanded the scope of academic freedom and truth seeking. The banquet also kicked off a two-day conference at Biola University titled "Intelligent Design and the Future of Science" <www.arn.org/arnproducts/videos/v053.htm>. Hundreds of scientists, teachers and lay people attended this conference.

Those of us who organized this event decided that the best way to honor our friend would be to bring together Phil's closest colleagues and simply "turn them loose." It was a treat to sit back and listen to Phil's colleagues speak of his accomplishments and impact on their lives, and it was gratifying to witness their eloquence and passion in describing the master. Phil's close friend and Biola protégé, John Mark Reynolds, lauded his mentor as the "Wizard of Berkeley" and likened him to Tolkien's Gandalf leading his followers to destroy the ring of Scientific Naturalism. On the other side, Michael Ruse, Johnson's best "enemy," undressed down to a T-shirt emblazoned with Darwin's face and exclaimed that he was "Darwin-man," adding that whereas "God had put Phillip Johnson on this earth to do good and spread light," he had been put there to "make Phillip Johnson's life that much more uncomfortable."

The celebration in April 2004 was the happy conclusion to what all of us in the intelligent design (ID) community knew was long overdue, namely, a suitable tribute to the man who was the ID movement's chief architect and guiding light. In particular, during the years leading up to these festivities, colleagues and I conversed at length about what kind of lifetime achievement award or book publication would be worthy of honoring Phillip Johnson and his accomplishments. Through the tireless efforts of faculty, students and staff at Biola University, the financial support of the Servants Trust, and the enthusiasm of InterVarsity Press, Phil's closest colleagues were able to honor him with the inaugural Liberty and Truth award and now, with the publication of this festschrift volume, *Darwin's Nemesis*. Appropriately, all royalties from this volume go toward that award.

The title *Darwin's Nemesis* applies to Johnson's roles as fearless leader, trusted friend and far-seeing visionary. Professionally and publicly, Johnson stepped into the controversy over intelligent design at just the right time as the fledgling movement's field marshal. Later, eschewing authoritarianism and any desire to become a cult figure, he made room for a new generation of ID scholars, stepping back so that the movement could flourish, not through a top-down chain of command but through its own inner vitality. Personally and privately, Phil entered each of our lives at just the right time and in just the right way to help us channel our various talents in the right direction.

Strategist, Teacher, Prophet

I first heard of Phillip Johnson at the 1989 Spring Systematics Symposium at the Field Museum in Chicago. Some biologists from Calvin College in Grand Rapids, Michigan, had traveled to Chicago for this event, and one of them mentioned a Berkeley law professor who had written on evolution. This Calvin biologist subsequently mailed me a messy looking manuscript by this Berkeley professor. I'm afraid I gave the manuscript only a cursory look at the time; after all, what does a lawyer know about evolution? Johnson did have this going for himself, however: he was a graduate of the University of Chicago, where I had done my graduate studies in mathematics. I kept his name in mind.

Two years later that law professor created a sensation with the book that emerged from that manuscript—*Darwin on Trial.* It is hard to overestimate the impact of that book. Some of the chapters in this collection will describe its influence in detail, so I won't repeat what will be said there. In my view the book's primary importance is as the beachhead that finally put effective criticism of Darwinism on the map. Beachheads by themselves are not enough. They are the means toward greater ends. Johnson, as a master strategist, saw how to make this beachhead into the start of a comprehensive intellectual program for reforming science and revitalizing culture.

I had a conversation with Phil some years after the publication of *Darwin on Trial* in which I asked him how this book fit into his overall strategy for unseating Darwinian naturalism. He remarked that the book, though effective at laying out criticisms of Darwinism, would by itself never have been enough to unseat Darwinian control over science. As he noted, the opposition would either ridicule the book or, more likely, just ignore it—and, after some months or years of ignoring it, would dismiss it as "that discredited book that was refuted a long time ago." Phil's comment brought to mind that old Marx brothers movie *Duck Soup,* in which Groucho Marx, as president of Freedonia, presides over a meeting of his cabinet. The following exchange ensues between Groucho and one of Freedonia's ministers:

Groucho: "And now, members of the Cabinet, we'll take up old business."

Minister: "I wish to discuss the Tariff!"

Groucho: "Sit down, that's new business! No old business? Very well—then we'll take up new business"

Minister: "Now about that Tariff . . ."

Groucho: "Too late—that's old business already!"

It was not enough for Phil to write a great book (which he did and which, by the way, the scientific community did not ignore). Also required was a plan of action and the follow-through to bring the Darwinian edifice crashing down. Johnson had both the plan and the will. Trained in the law, he knew that contests are meant to be won, and he was not about to lose this one, especially since he had the better case.

The first glimmers of Johnson's grand designs could be seen in 1992, the year following the publication of *Darwin on Trial.* Several events that year and the next foreshadowed the juggernaut that Johnson would unleash. The highlight was a symposium organized by Jon Buell, Tom Woodward and Steve Sternberg, which took place at Southern Methodist University (SMU) in the spring of 1992, and had as its stated topic Johnson's critique of evolution. Phil Johnson and Michael Ruse were the star attractions and key antagonists at this symposium, but it also included other key people on both sides of this debate. It was at this event that I first met Phil, Mike Behe and Steve Meyer.

Why was this event so significant? Here, for the first time, a radical non-materialist critique of Darwinism and naturalistic evolutionary theories was put on the table for a high-level, reasoned, academic discussion without anyone promoting a religious or sectarian agenda. In particular, this symposium was not about reconciling science with the Bible or about showing that evolution is inconsistent with traditional views about religion or morality. It was about looking at the merits of design and Darwinism on their own terms.

This symposium legitimized the debate over evolution at the highest levels of the academy. With that legitimization came an openness to think new and hitherto forbidden thoughts. And with that openness came converts and volunteers to the cause of intelligent design. Within a year following that symposium, Phil had gathered a band of such converts and volunteers, mainly scientists and philosophers. The next step was to organize that band. This Phil did in 1993. Two moves on his part proved crucial for organizing the fledgling ID movement. The first was to organize a private several-day meeting of potential leaders in the ID movement at Pajaro Dunes, south of San Francisco. (This event is described in the opening segment of the video *Unlocking the Mystery of Life*.) The second, which followed that meeting, was to insist that the participants get on e-mail and be part of a listserv that he would run from Berkeley. The meeting at Pajaro Dunes was critical for getting all of us on the same page. The listserv was critical for keeping us networked and for building momentum.

Phil's listserv, which continues to this day under the auspices of the International Society for Complexity, Information and Design <www.iscid.org>, was in the early days the lifeblood of our movement. It still remains important, but with other centers and discussion boards that address ID, it is no longer as critical. Today, twelve years later, it may be difficult to appreciate just how important it was in those early days to get networked over the Internet. Today, everyone has e-mail. That was not the case in 1993. Moreover,

Netscape and Internet Explorer did not come into common use until two years later—surfing the Web was largely unknown at the time. But Phil saw what was coming and how the Internet would allow for the dissemination of knowledge that would make it increasingly difficult for secular elites to maintain control over what people think. He even offered to buy e-mail accounts for those of us who were without university appointments and thus without free access to university servers.

Phil's listserv brought us together, continually increased our talent pool and, most importantly, gave us a support network. One reason nothing like an ID movement critical of Darwinian naturalism had blossomed previously is that critics of Darwinism were typically isolated; thus the Darwinian establishment could go after them mercilessly without anyone coming to their aid. With Phil's virtual community that was no longer the case. To see this, contrast the case of Dean Kenyon with that of Percival Davis. Both were co-authors of the ID supplemental textbook *Of Pandas and People*. Davis, as it turned out, was also a coauthor of the bestselling basal biology textbook known simply as *Villee* (after Claude Villee). Someone complained to the publisher of *Villee* about Davis also being the author of an ID textbook. The publisher immediately dropped Davis as an author for subsequent editions of *Villee*. As a consequence, Davis now has little to do with ID and makes his living as a computer programmer.

Dean Kenyon likewise faced pressures for expressing views favorable to ID. But Kenyon, unlike Davis, had the benefit of Phil's support network. Kenyon's department chair tried to prevent him from teaching freshman biology at San Francisco State University even though Kenyon was a full professor in the biology department. Why? Because Kenyon had expressed doubts about Darwinism and a preference for intelligent design in his classes. In this case, however, the persecution backfired. Steve Meyer wrote an op-ed piece for the *Wall Street Journal* detailing the biology department's assault on Kenyon's academic freedom. Within days the university called the biology department on the carpet and reinstated Kenyon's right to teach freshman biology.

The SMU symposium and the Pajaro Dunes meeting were watershed events in the ID movement. There have been more visible events subsequently (for instance, the "Mere Creation" conference at Biola in 1996 and the "Nature of Nature" conference at Baylor in 2000). But these early events laid the foundation for all that followed. Around the same time as these wa-

tershed events, two lesser events happened that to this day characterize the
main challenges facing the ID movement.

The first is an encounter between Phillip Johnson and Howard Van Till
at Trinity Evangelical Divinity School in 1992. Phil had hoped that his message would be received warmly by evangelicals, and it largely was among
the rank and file. But among evangelicals teaching at Christian colleges and
universities, Phil found that he faced considerable opposition. A view of divine action in which God plays no evident role in nature had taken root in
Christian higher education. Thus, rather than being welcomed for critiquing
Darwinian naturalism, he found himself attacked for proposing an outdated
and ill-conceived theology of nature. Howard Van Till epitomized this position and strongly objected to *Darwin on Trial.* Phil in turn characterized Van
Till's position as "theistic naturalism." At the time Van Till rejected that designation. Nowadays he accepts it, having embraced the process theology of
David Ray Griffin.

If the Council for Christian Colleges and Universities was divided over
Johnson's message, the secular elites, notably in the academic and scientific
establishments, were not. As far as they were concerned, Johnson was a
threat to science and needed to be stopped. There's no dearth of examples
in this book about the sort of opposition that Phil has had to face from this
quarter, but an event in the spring of 1993 epitomizes for me the downright
shamelessness of this opposition. During the academic year 1992-1993, I
was a postdoctoral fellow in the history and philosophy of science at North-
western University. That spring, David Hull, one of the premier philosophers of biology, gave a seminar that I attended. One day, during the seminar, he mentioned *Darwin on Trial* and remarked that an editor at *Nature*
had contacted him about writing a negative review of that book. Hull, a convinced Darwinist, was only too happy to oblige, duly delivering the re-
quested hatchet job. Hull told the story with pride, as though he had done
a good deed. To this day I find many Darwinian naturalists exhibiting this
same means-justifies-the-ends mentality.

Phil's pivotal role in shaping the ID movement following those early days
is well known and discussed at length in this volume. Nevertheless, it is the
seeds that Phil sowed early on, emphasizing the need for a community of
thinkers willing to challenge Darwinism at the root, that made the success
of this movement possible. In the summer of 2000 I invited Michael Ruse to
speak at a seminar on ID that I was conducting at Calvin College. At that

seminar Ruse, though no friend of ID, remarked that the ID movement has not made one misstep. Ruse was exaggerating, but not by much. The ID movement has exhibited a remarkable cohesiveness, unity and focus.

Intellectual movements often fragment, with cults of personality developing around key figures and bitter rivalries ensuing. Nothing like this has happened in the ID movement (or, indeed, shows any sign of happening). Insofar as the credit here goes to any one person, it goes to Phil—and specifically in his capacity as teacher. Phil's role as a teacher is, of course, evident in his being a professor, a writer of books and a public intellectual. Yet, with respect to the health and vitality of the ID movement, Phil has done far more than merely teach us a set of propositions. He has taught by example. Accordingly, he has not taught us what or even how to think but rather a certain style of thinking or habit of thought. The issue for Phil has always been not whether you get the right answers and not even whether you pose the right questions (as important as these are, and notwithstanding the title of his last book), but rather the humility to consider all questions and not to reject any because our well-worn presuppositions tell us "you can't go there." In other words, Phil has taught us to put our presuppositions—all of them—on the table for examination, to ask the hard questions of them and to seek answers by following the evidence wherever it leads.

As a consequence, there are no secret handshakes or initiation rites needed to join Phil's club. The ID movement is a big tent and all are welcome. Even agnostics and atheists are not in principle excluded, provided they can adopt this open attitude of mind. In practice, however, most agnostics and atheists have their minds made up. Agnostics know that nothing is knowable about a transcendent reality. And atheists know that no transcendent reality exists, so again nothing is knowable about it. Accordingly, agnostics and atheists tend not to join the ID movement.

Johnson is a *radical skeptic,* insisting, in the best Socratic tradition, that everything be put on the table for examination. By contrast, most skeptics opposed to him are *selective skeptics,* applying their skepticism to the things they dislike (notably religion) and refusing to apply their skepticism to the things they do like (notably Darwinism). On two occasions I've urged Michael Shermer, publisher of *Skeptic* magazine, to put me on its editorial board as the resident skeptic of Darwinism. Though Shermer and I know each other and are quite friendly, he never got back to me about joining his editorial board.

Finally, I want to consider Phillip Johnson in his role as prophet. This may seem like a strange designation for Phil, but I contend that it fits. A popular stereotype conceives of prophets as wild-eyed individuals, living unconventionally, out of touch with the practicalities of life and offering enigmatic utterances about strange unseen worlds. But in fact, a prophet is anyone who holds up a mirror to the culture and forces it to acknowledge the idols it has constructed. Moreover, because we resent being told that our idols are idols, we tend to despise prophets. Only later, when we've constructed new idols, does it become clear that the old idols were indeed idols, at which point we no longer despise those prophets, now long gone, but instead venerate them. This is how hagiography is born.

What is an idol? Leslie Zeigler, in a book I edited with Jay Richards *(Unapologetic Apologetics)* offers the most perceptive analysis of idolatry that I know:

> This God who tells Moses "I am Who I am," who enters into contingent relationships with human beings at particular times and in particular places, who approves of certain actions and not of others, has always been, to say the least, hard to live with. Human beings have always preferred gods for whom they can write the job descriptions themselves.
>
> Scripture refers to these preferred gods as idols, and the author of Isaiah 44:9-20 gives us as clear a description as has ever been written of the idol maker and his idol. The craftsman cuts down a good, healthy tree, uses part of it for a fire to warm himself and to cook his dinner. Then from part of it he makes a graven image, to which he falls down and worships, praying—"Deliver me, for thou art my god!"
>
> It's only at this point that Isaiah delivers his punchline, a punchline which is all too often overlooked. He tells us of the awesome power of the idol. That piece of wood which the craftsman himself has formed has deceived him—has led him astray to the point that he no longer recognizes it as his own creation. He has been blinded—blinded by his own creativity—so that he no longer recognizes that he is worshipping a delusion and hence is no longer able to deliver himself. He is unable to ask himself, "Is there not a lie in my right hand?"
>
> Other Old Testament passages speak eloquently of the idols as being useless, unable to do anything, unable to support their people; instead, having to be carried around and being a burden to them. But Isaiah puts his finger on a far more dangerous characteristic: they have the power to delude and deceive their makers.

The work of those attempting to craft a god of their own making, the god *we* want, and for whom *we* can provide the job description, is rampant within the church today.

The trouble with Darwinian naturalism is that it turns nature into an idol, making brute material forces rather than the all-wise God into the source of creativity in nature. Moreover, it tries to justify this idolatry in the name of science. To the Darwinian naturalist, Johnson the prophet says, "Your idol cannot support itself because it is founded on a false philosophy and a biased construal of scientific evidence." Moreover, to the theistic evolutionist, Johnson the prophet says, "You have avoided turning nature into an idol, but at the cost of requiring God to act hiddenly in nature; yet what if nature reeks of design and our best scientific understanding confirms this, demonstrating that design is manifest in nature?" Like Francis Schaeffer a generation before him, Phillip Johnson has put his finger on the key place where our generation has forgotten God. For this generation it is the place of our origin. To a generation that regards God as increasingly distant, with nature as all there is and humans as mere appendages of nature, Johnson the prophet points us to the true God, the one in whose image we are made and to whom we must ultimately render an account.

Life After Dover

On December 20, 2005, as this book was going to press, Judge John E. Jones III rendered his verdict in the first court case involving intelligent design. In *Kitzmiller* v. *Dover*, also billed as Scopes II, Judge Jones not only struck down the Dover school board policy advocating intelligent design but also identified intelligent design as nonscientific and fundamentally religious. Accordingly, he concluded that the teaching of intelligent design in public school science curricula violates the Establishment Clause and therefore is unconstitutional.

It is hard to imagine that a court decision could have been formulated more negatively against intelligent design (for the actual decision, see <www.pamd.uscourts.gov/kitzmiller/kitzmiller_342.pdf>). In light of this decision, one may therefore wonder about the appropriateness of titling this book *Darwin's Nemesis*. To read Judge Jones's decision, one gets the impression that Darwin is alive and quite well. Even so, let me suggest that this decision is a bump in the road and that Phillip Johnson's program for dismantling Darwinism remains well in hand.

To see that Judge Jones's decision is not nearly the setback for intelligent design that its critics would like to imagine, let's start by considering what would have happened if the judge had ruled in favor of the Dover policy. Such a ruling would have emboldened school boards, legislators and grass-roots organizations to push for intelligent design in the public school science curricula across the nation. As a consequence, this case really would have been a Waterloo for the supporters of neo-Darwinian evolution (the form of evolution taught in all the textbooks).

Conversely, the actual ruling is not a Waterloo for the intelligent design side. Certainly it will put a damper on some school boards that would otherwise have been interested in promoting intelligent design. But this is not a Supreme Court decision. Nor is it likely this decision will be appealed since the Dover school board that instituted the controversial policy supporting intelligent design was voted out and replaced November 2005 with a new board that campaigned on the promise of overturning the policy.

Without an explicit Supreme Court decision against intelligent design, we can expect continued grass-roots pressure to promote intelligent design and undercut neo-Darwinian evolution in the public schools. Because of *Kitzmiller* v. *Dover*, school boards and state legislators may tread more cautiously, but tread on evolution they will—the culture war demands it!

It is therefore naive to think that this case threatens to derail intelligent design. Intelligent design is rapidly gaining an international following. It is also crossing metaphysical and theological boundaries. I now correspond with ID proponents from every continent (save Antarctica). Moreover, I've seen intelligent design embraced by Jews, Muslims, Hindus, Buddhists, agnostics and even atheists. The idea that intelligent design is purely an "American thing" or an "evangelical Christian thing" can therefore no longer be maintained.

Even if the courts manage to censor intelligent design at the grade and high school levels (and with the Internet, censorship means nothing to the enterprising student), they remain powerless to censor intelligent design at the college and university levels. Intelligent design is quickly gaining momentum among college and graduate students. Three years ago, there was one IDEA Center at the University of California at San Diego (IDEA = Intelligent Design and Evolution Awareness—see <www.ideacenter.org>). Now there are thirty such centers at American colleges and universities, including the University of California at Berkeley and Cornell University. These centers are fiercely pro-intelligent design.

Ultimately, the significance of a court case like *Kitzmiller* v. *Dover* depends not on a judge's decision but on the cultural forces that serve as the backdrop against which the decision is made. Take the Scopes Trial. In most persons' minds, it represents a decisive victory for evolution. Yet, in the actual trial, the decision went against evolution (John Scopes was convicted of violating a Tennessee statute that forbade the teaching of evolution).

Judge Jones's decision may make life in the short term more difficult for ID proponents. But the work of intelligent design will continue. In fact, it is likely to continue more effectively than if the judge had ruled in favor of intelligent design, which would have encouraged complacency, suggesting that intelligent design had already won the day when in fact intelligent design still has much to accomplish in developing its scientific and intellectual program.

Instead of ruling narrowly on the actual Dover policy, Judge Jones saw his chance to enter the history books by assuming an activist role, ruling broadly and declaring intelligent design to be unconstitutional. Yet if he and his ruling are remembered at all, it will be not for valiantly defending science but for pandering to a failed reductionist way of doing science.

Just as a tree that has been "ringed" (i.e., had its bark completely cut through on all sides) is effectively dead even if it retains its leaves and appears alive, so Darwinism has met its match with the movement initiated by Phillip Johnson. Expect Darwinism's death throes, like Judge Jones's decision, to continue for some time. But don't mistake death throes for true vitality. Ironically, Judge Jones's decision is likely to prove a blessing for the intelligent design movement, spurring its proponents to greater heights and thereby fostering its intellectual vitality and ultimate success.

A Celebration Volume

This volume takes its place alongside other intelligent-design compilations, serving as a handy reference for various ID research topics and personal stories. Yet, as a tribute to our well-loved friend, this book also plays a special role in the ID literature. As a festschrift, or celebration volume, whose contributors number twenty-one of Phil's closest colleagues, this book memorializes Phil's achievements in heading the ID movement and thanks him for his influence not only on our lives but also on the wider community of ID proponents and friendly critics. It is therefore especially appropriate that the publication of this volume marks the year that Phil turns sixty-five.

The opening of this book includes a foreword by Senator Rick Santorum and an introduction by John Mark Reynolds. Both of these individuals have played a special role in Phil's life. Senator Santorum, in formulating his now-famous amendment to the "No Child Left Behind" federal education act, provides perhaps the clearest example of how Phil's ideas are reaching the public square. That amendment, as it finally appeared in the report language that interprets the federal education act, included the following statement: "Where topics are taught that may generate controversy (such as biological evolution), the curriculum should help students to understand the full range of scientific views that exist, why such topics may generate controversy, and how scientific discoveries can profoundly affect society." This is all the opening needed to bring down Darwinian naturalism within public school science education.

Whereas Senator Santorum's amendment epitomizes Johnson's influence in the public square, John Mark Reynolds's Torrey Honors Institute epitomizes Johnson's vision for Christian higher education. The Torrey Honors Institute combines a traditional Great Books program with a vital engagement of contemporary culture and thought. Although John Mark is the most visible person associated with Torrey, the institute is really the joint brainchild of John Mark and Phil, who are close friends. (Their families were vacationing together when Phil had his first stroke.) If anyone has insight into Phil, it is John Mark.

The remainder of the book is divided into several parts. Part one describes first encounters with Phil and gives a window into Phil's appeal and persuasiveness. Steve Meyer gives us a vignette of first meeting Phil at a Greek restaurant in Cambridge just as Phil was beginning to take interest in the evolution controversy. Michael Behe and Jay Richards, on the other hand, describe meeting Phil first through reading *Darwin on Trial* and only later in person. Their chapters recount Phil's critical role in their own intellectual journey. Tom Woodward's chapter, adapted from his book *Doubts About Darwin* (which won *Christianity Today's* 2004 book award in the "Christianity and Culture" category), gives us a larger perspective on how *Darwin on Trial* transformed the origins debate.

Part two comprises three essays that address the controversy Phil's work has brought about. Phil has often referred to his work of intellectual renewal in terms of a wedge metaphor that splits off false philosophies and ideologies from genuine truth seeking. This part addresses the controversy arising from "the wedge." My chapter here describes the day-to-day dynamics of the debate, what to expect and what to watch out for. Frank Beckwith then turns to the teaching of science in public schools and how careful we must be not to cripple ourselves with false assumptions. Then Tim Standish takes up the challenge that intelligent design raises for Christian education.

Part three focuses on two of Johnson's most friendly critics. In fact, only one is a pure critic, Michael Ruse. Despite his close friendship with Phil, Michael remains an ardent Darwinist. Much of Michael's thinking these last several years has focused on the implications of Darwinism for religion, and not just for religion in general but specifically for traditional Christian belief. Michael's chapter takes on the most difficult problem in this area, the problem of evil. If Michael Ruse is a pure critic, then David Berlinski is a mixed critic. He admires Phil for having, as he once put it to me in an e-mail, done "amazingly well in creating a political movement." But David has reservations about ID's intellectual and scientific program. At the same time, he finds Darwinism's grand claims of being able to account for biological complexity as absurd. In a recent op-ed, he remarked, "The suggestion that Darwin's theory of evolution is like theories in the serious sciences—quantum electrodynamics, say—is grotesque. Quantum electrodynamics is accurate to thirteen unyielding decimal places. Darwin's theory makes no tight quantitative predictions at all." Berlinski's two fables call for skepticism of both Darwinism and intelligent design.

Part four examines Johnson's impact in biology. The first two chapters, one by David Keller and the other by Jonathan Wells, are more personal, describing how Johnson has influenced their approach to biology, and what implications such an ID-friendly approach would have for biology if widely adopted. The next two chapters are very different. These are actual research papers in biology. The solo-authored paper by Steve Meyer created a firestorm when it appeared in the *Proceedings of the Biological Society of Washington* in 2004. Despite having properly gone through the peer-review process, once this paper appeared, the Biological Society of Washington issued a statement (see <www.biolsocwash.org/id_statement.html>) declaring that this article should never have been published in its journal, and that in the future ID research would not be permitted in its pages, thus signifying at once that ID has been published in the peer-reviewed biological literature and that it faces fierce resistance getting published there. The other paper,

by Scott Minnich and Steve Meyer, examines the design of the bacterial flagellum, a molecular machine dubbed by Darwinists as the "icon of intelligent design."

Part five examines the spheres of Johnson's influence. The first chapter is
by master worldview theorist Nancy Pearcey. Here she details how, given
the widely held fact-value dichotomy, intelligent design provides the best
mainstay for a Christian worldview. Next, J. Budziszewski shows how traditional natural-law theory, which is consistent with intelligent design, is utterly incompatible with attempts to recover human nature on the basis of
naturalistic evolutionary theories. The chapter by Marcus Ross and Paul Nelson follows. It provides a window into how young-earth creationists have
assimilated Johnson. And finally, there is the essay quite distinct from the
rest, by Fritz Schaefer and Wesley Allen, who explicate complexity and
chaos as scientific concepts and then consider several philosophical and
theological implications of these physical principles.

An epilogue with two chapters concludes this festschrift. Walter Bradley,
whose book with Charles Thaxton and Roger Olsen titled *The Mystery of
Life's Origin* (Philosophical Library, 1984) is now widely regarded as the first
book in the intelligent design movement, has been at this debate longer than
any of the other contributors to this volume. In his chapter he reviews where
the debate over evolution was before Johnson weighed in on it. Next he
considers how Johnson transformed this debate by initiating a comprehensive intellectual movement for overturning Darwinian naturalism and developing intelligent design as a scientific theory. To end his chapter Walter reflects on where the intelligent design movement is headed and what it must
do to win the day.

One last chapter remains, and that is by the great one himself, our friend
Phil. He rightly gets the final word.

INTRODUCTION

A Mythic Life

JOHN MARK REYNOLDS

———

There is a moment in *Republic* where Socrates leans forward and says to his young disciple, "Inquire for yourself." All academics give lip service to this goal. Phillip Johnson has embodied it. The best measure of whether this is true for any public intellectual are the foes he makes by taking intellectual risks. Anyone can safely discomfit the fools and the socially marginal. Johnson has made intellectual opponents of great and clever men, as this book reveals. He has dared to challenge mainstream opinion, not just where it is obviously false but where it is seems most plainly true.

It is impossible to measure the full significance of a man's thought while he is still thinking. However, it might be possible to take a small measure of the impact of that thought and the direction it is headed.

For the intelligent design movement, *Darwin on Trial* was the seminal book. Critics have complained that it contains too little that is new. Hadn't people argued most of it before? Even friends have wondered why this book at this time sparked such a revolution of thought and brought so many people together. Of course, this overlooks that most of the arguments had not been drawn together into one place in an intellectually respectable manner. This is not as easy as it might sound post-Johnson.

Most antievolution books tried to do too much, took on side issues or were too sectarian to appeal to a wide audience. It is easy enough now to say how easy it should have been to write *Darwin on Trial,* but no one had done it before Johnson. Putting the puzzle together seems automatic after the puzzle is done; after all, the pieces were all there. But *someone has to do it.* Johnson put all the nagging doubts about Darwinism in one slim volume for everyone to see. His critics routinely overlook how many well-educated persons share those doubts in the privacy of their department offices.

Johnson's impact as a writer cannot be understood apart from his personality. Johnson is a reasonable and friendly man. He knows he might lose his case, but he does not think so. He is that rarity: an optimistic realist. Assaults on naturalism had, up to Johnson, been largely bitter and pessimistic. They came from culturally marginal people who felt the sting of their isolation from the mainstream or who were not comfortable or at home in the modern academy. Johnson loves the university, in particular Berkeley, the town and the school. It is his home, and he understands its complexity. He also understands its great good and sorrows when it turns from truth as he sees it. This love is communicated to his audience.

He is not a hater, not even of his enemies. However, most people with this attitude are not simultaneously fighters. They love their enemies too little to oppose them. Johnson was willing to offend people with whom he would just as soon have shared a drink. This is why so many who disagree with him can still respect him. He did them the honor of taking their ideas in deadly earnest and attacking them.

Reason in the Balance still strikes me as Johnson's most important work (a view Johnson himself takes of that book). It is the manifesto of a movement. This movement has actually only just begun to form as the ideas in it, which unify the sciences and the humanities, are digested. In the future, Johnson may be seen as a person who got scientists and humanists talking. For too long the university has been deeply divided on this fault line, neither group speaking to the other. Scientists need reminding that other academics deal with truth; at the same time, persons in the humanities need to recall the necessity to account for science in their work. Johnson's most recent book, *The Right Questions,* is a splendid example of this interdisciplinary style of writing.

In the end, Johnson is best understood as a truth seeker bored with any orthodoxy, even his own. His critics might find that statement odd since Johnson is one of the preeminent defenders of religious orthodoxy. Like Plato, he has followed the dialectic to a point where he thinks he sees truth.

He holds this truth provisionally, but his questions have only made him confident that his answers are correct. That does not mean, however, that the questions end—only that they grow in intensity.

Johnson set out to find the truth and much to his shock, and perhaps dismay, found it. To be skeptical is acceptable in the academy; to be skeptical about endless skepticism regarding religion is very bad indeed. Religion is acceptable only as a private thing of feelings. Johnson's questions have led him to believe, however tentatively, the astounding notion that Christianity is true and necessary in understanding the real world.

Phillip Johnson is one of those rare individuals who is always the same person. He asks the same hard questions in Sunday school as he does in the Berkeley classroom. He has a unified personality. I have seen him in hundreds of different circumstances, and there is no split in his soul. He embodies a reunification of deep religious thinking and rational public discourse. He is without pretense. He suffers fools gladly, if he believes there is a chance they can advance his own understanding or that of the public.

His thinking is subtle, but he is ultimately the master of the big picture. His is the strategic vision. If the devil is in the details, Johnson is on the side of the angels. Other people—one thinks of Michael Behe and Bill Dembski—fill in the mechanisms. Johnson's job is to set out the framework for the ideas.

As a result, the key to understanding his thinking is not in works of philosophy or in legal theory or practice. Johnson is a man of story and of epic. His work retells the story of the late twentieth century and early twenty-first in a new way.

When he went to school, the *Fantasia* myth was all the rage. Here man, like the famous "evolution" sequence in the Disney masterpiece, slowly ascends from the primordial chaos. Evolution is the creator. Religion is at best a thing that inspires and creates artistic beauty, but it must be carefully screened from reality lest it harm itself and others. Christianity is medieval, and in the legend of Johnson's early education that is another way of saying from the Dark Ages. Johnson came to see this myth as unhelpful and essentially false.

He does not make the postmodern mistake of confusing just any moving story with reality, but he also does not discount the power of the right story to move humans. Put another way, he found a movement with some of the right lyrics, but he gave it the right tune. Of course, he cleaned up the lyrics as well! Truth told properly has the power to cut through rubbish no matter how socially acceptable.

How can a person I have described as a Socrates love stories? One forgets that Plato pictures Socrates in two ways. There is the early Socrates, less mature, who is only interested in the "right questions." He is almost a cynic, leaving his listeners in confusion as he tears down their facile worldviews. Then there is the later Socrates, the Socrates of *Republic* 10 and the participant in hearing the grand cosmology of the *Timaeus*. This is the person who in his quest for truth has learned to embed the truths he has found into an appealing pattern.

To understand Johnson as a seeker for the stories that move him is to understand why Johnson has led the intelligent design movement as he has. First, there is the great novel, perhaps the greatest of the twentieth century, *That Hideous Strength,* by C. S. Lewis. One can see many of the ideas from this work in Johnson's writings, often much improved. More importantly, one sees Johnson living out some of the strategies in the book. To cite but one, Johnson does not center any movement in Phillip E. Johnson or in a centralized school or ministry. Like Ransom in the novel, he gathers men and women together and gives them a vision of the way things should be. He allows those inspired people to exercise their gifts. Many of the people in this volume were helped by Johnson early on and have marveled to never see him ask for a return on the favor. Of all the leaders I have known, he is the least concerned with money, power or formal institutions. He is not a guru expecting homage but a giver intent on serving others. Like any good teacher, he delights when his students surpass him, and I have seen him sit at conferences and allow others to advance arguments essentially his own.

The great myth of Tolkien's *Lord of the Rings* captures Johnson's imagination. The battle against all odds won in almost a humorous and unexpected manner fits his temperament. If there ever were a Council of Elrond for Christianity, surely Johnson would have to be invited. Tolkien's epic, much reread by Johnson, has been a model for his strategic vision. He plays to win and counts on the reality of the divine to save his cause. He takes calculated risks, but always risks that lead to victory, not a marginal existence on the edges of academic culture. If right, then he wants to win. To paraphrase Johnson, "We will advance the Ring so close to Mount Doom that only a miracle can save us. But since we believe in miracles, all will be well, or at least we will know we were wrong."

Of course, Johnson does not view *people* as the enemy. This is why he can be collegial with those with whom he has serious disagreements. He battles against false ideas, sloppy thinking and wicked notions. People are to be won over. Bad ideas are the orcs he would slay without mercy.

Like Tolkien, Johnson looks for leaders in unusual places. He does not care for current academic standing or present social acceptability. He is an educator and takes what is before him and helps it become what it was meant to be. He looks everywhere for promise. If the Shire holds a doughty and promising folk, then he will look there even if the "wise of this age" disregard the Shire folk.

In doing this, he has been pivotal in transforming hundreds of young adult lives. As the founding intellect of the Torrey Honors Institute at Biola University, his right questions have stimulated hundreds of the brightest students of Christianity. Today those students are in the best graduate programs in the nation. They are asking thoughtful questions in places like Oxford and the University of Southern California. From the offices of the White House to big business, they are finding places of influence to be reasonable and winsome. This is Johnson's work, but it is not limited to this.

There are thousands of home-school mothers and fathers who have embraced his books and used them in training their children. I meet them everywhere I go; they ask what it is like to know the great Wizard of Berkeley. There are thousands more who are students in colleges and universities all over America, top tier and not, who have heard him speak and have come alive. My own second life in academia came at a discouraging moment in graduate school, when I heard Johnson quietly dismantling Eugenie Scott on National Public Radio. I had not read *Darwin on Trial* but grasped that here was a man who understood the salient issues and *had a strategy to win*. The subsequent course of my life is, at least in part, due to that moment and that man. My story could be repeated thousands of times. His speaking pierced the fog of war and has given hope to many, especially the young.

This inspiration will be Johnson's greatest legacy. There are thousands of young adults all over the nation preparing to enter the fray. What will happen when there are many children in many places carrying on this work?

Most of these Halflings are still too young to be represented in this work, but they are growing in graduate schools and in colleges. They have learned to ask the right questions and to tell the best stories. And who knows? Even now some former Torrey student, his spiritual godchild, may be hard at work in the lab or laboring over the early draft of the critical book that will destroy the Ring and bring on a new and better age.

In any case, Johnson's internal dialectic will lead him on into further frays. He will, I think, never stop to rest even if he is victorious. His calling is to afflict the comfortable and comfort the afflicted. Already, he has begun a massive second work to assault naturalism in the humanities that will be as important as his work in the sciences.

Johnson is convinced that "in the beginning was the Word" is no mere romantic fiction. It is true, as true in its own way as any fact of the world. It is "the myth that is true." This is no longer intellectually fashionable, but Johnson has never cared much for intellectual fashion.

Like Gandalf, the white wizard, Johnson has faced death and suffering in his personal life and has come back to continue to lead the battle of his time. For like it or not, Johnson sees the intellectual movement he has organized and the resulting intellectual conflict he has set into motion as more significant than the nitpicking and logic-chopping that consume so much time in the academy. He sees an epic struggle and he has chosen a side. If he is right, then he is not the commander; that job is already eternally taken. Instead, his is a calling to gather the troops; in his words, to "unite the divided" and "divide the united."

This particular battle for the soul of the twenty-first century is his battle. Due to his efforts, Darwinism and its concomitant ills are on the way out. To be sure, other evils will vie to take their place. But this is a battle to be won. It is not a battle that is over yet, so we can anticipate, even those who love him, that Phillip Johnson will continue to make the comfortable uncomfortable. Some of those who honor him now may even wonder what he is up to and where he is going.

Johnson will go on asking the right questions at seemingly inopportune times. He will keep gathering misfits together and giving them a vision for another way of living. From his Berkeley home, over e-mail and the Internet he will keep strategizing for victory while seeing the problems clearly. Of course, Johnson believes that there is indeed a King who is returning. But of that future no one living can speak.

PART I

PORTRAITS OF THE MAN AND HIS WORK

YOUR WITNESS, MR. JOHNSON

A Retrospective Review of *Darwin on Trial*

STEPHEN C. MEYER

I first met Phillip Johnson at a small Greek restaurant on Free School Lane next to the Old Cavendish Laboratory in Cambridge. It was the fall of 1987. The meeting had been arranged by a fellow graduate student who knew Phil from Berkeley. My friend had told me only that his friend was "a quirky but brilliant law professor" who "was on sabbatical studying torts," and he "had become obsessed with evolution." "Would you talk to him?" he asked. His description and the tone of his request led me to anticipate a very different figure than I encountered. Though my own skepticism about Darwinism had been well cemented by this time, I knew enough of the stereotypical evolution-basher to be skeptical that a late-in-career nonscientist could have stumbled unto an original critique of contemporary Darwinian theory.

I should have known better, but only later did I learn of Johnson's intellectual pedigree: Harvard B.A., top of class University of Chicago law-school graduate, law clerk for Chief Justice Earl Warren, leading constitutional scholar, occupant of a distinguished chair at University of California, Berkeley. In Johnson, I encountered a man of supple and prodigious intellect who seemed in short order to have found the pulse of the origins issue.

Johnson told me that his doubts about Darwinism had started with a visit to the British Natural History Museum, where he learned about the controversy that had raged there earlier in the 1980s. At that time, the museum paleontologists presented a display describing Darwin's theory as "one possible explanation" of origins. A furor ensued, resulting in the removal of the display when the editors of the prestigious *Nature* magazine and others in the scientific establishment denounced the museum for its ambivalence about accepted fact.

Intrigued by the response to such an (apparently) innocuous exhibit, Johnson decided to investigate further. He began to read whatever he could find on the issue: Stephen Jay Gould, Michael Ruse, Matt Ridley, Richard Dawkins and Michael Denton's *Evolution: A Theory in Crisis*. What he read made him more suspicious of evolutionary orthodoxy. "Something about the Darwinists' rhetorical style," he told me later, "made me think they had something to hide."

An extensive examination of evolutionary literature confirmed this suspicion. Darwinist polemic revealed a surprising reliance on arguments that seemed to assume rather than demonstrate that life had evolved via natural processes. Johnson also observed an interesting contrast between biologists' technical papers and their popular defenses of evolutionary theory. When writing in scientific journals, he discovered, leading biologists acknowledged many significant difficulties with both standard and newer evolutionary models. Yet, when defending basic Darwinist commitments (such as the common ancestry of all life and the creative power of the natural selection/ mutation mechanism) in popular books or textbooks, Darwinists employed an evasive and moralizing rhetorical style to minimize problems and belittle critics. Johnson began to wonder why, given mounting difficulties, Darwinists remained so confident that all organisms had evolved naturally from simpler forms.

In *Darwin on Trial,* Johnson argued that evolutionary biologists remain confident about neo-Darwinism not because empirical evidence generally supports the theory but instead because their perception of the rules of scientific procedure virtually prevent them from considering any alternative view. Johnson cited, among other things, a communiqué from the National Academy of Sciences (NAS) issued to the Supreme Court during the Louisiana "creation science" trial. The NAS insisted that "the most basic characteristic of science" is a "reliance upon naturalistic explanations."

While Johnson accepted this principle of "methodological naturalism" as an accurate description of how much of science operates, he argued that treating it as a normative rule when seeking to establish that natural processes alone produced life assumes the very point that neo-Darwinists are trying to establish. Johnson reminded readers that Darwinism does not just claim that evolution (in the sense of change) has occurred. Instead, it purports to establish that the major innovations in the history of life arose by purely *natural* mechanisms—that is, without intelligent direction or design. Thus Johnson distinguished the various meanings of the term *evolution* (such as change over time or common ancestry) from the central claim of Darwinism, namely, that a purely undirected and unguided process had produced the appearance of design in living organisms. Following Richard Dawkins, the staunch modern defender of Darwinism, Johnson called this latter idea "the blind watchmaker thesis" to make clear that Darwinism as a theory is incompatible with the design hypothesis. In any case, he argued, modern Darwinists refuse to consider the possibility of design because they think the rules of science forbid it.

Yet if the design hypothesis must be denied from the outset, and if, as the NAS also asserted, exclusively negative argumentation against evolutionary theory is "unscientific," then Johnson argued that "the rules of argument . . . make it impossible to question whether what we are being told about evolution is really true." Defining opposing positions out of existence "may be one way to win an argument," but, said Johnson, it scarcely suffices to demonstrate the superiority of a protected theory.

To establish that such philosophical gerrymandering lay behind the success of the neo-Darwinism, *Darwin on Trial* evaluated the scientific arguments that ostensibly establish the "fact of evolution." Johnson trained a considerable facility for analysis on the whole edifice of Darwinist argumentation. He found a panoply of euphemism and wishful thinking masquerading as evidence: the pattern of gaps and sudden appearance in the fossil record described as "rapid evolutionary branching," superficial variations in moths or fruit flies cited to substantiate the possibility of grand "macroevolutionary" changes, elaborate depictions of human ancestors based on scanty bone fragments, and biochemical observations laden with evolutionary assumptions used to justify evolutionary claims.

Along the way, *Darwin on Trial* asks a good many questions rarely asked in polite biological society. Given the fossil evidence, how do we know that hypothetical "transitional" organisms ever existed? How do we know that natural selection can create complex organs and organisms when genetics suggests the vast improbability of random mutations producing advantageous and novel structures? How do we know that the first cells did arrange themselves from simple chemicals if we haven't yet established that they could? In each case, Johnson argued that "we know because we have equated scientific method with a philosophy of strict naturalism and materialism." We know because the rules of science require that some form of naturalistic evolution *must* be true.

Johnson's attempt to reopen such questions angered many members of the biological establishment who had grown accustomed to offering the public what Johnson called "proof through confident assertion." His criticism of Darwinist orthodoxy initially earned him dismissive reviews in *Science, Nature* and *Scientific American,* the latter written by Stephen Jay Gould. Yet these reviews also helped publicize Johnson's thesis, which has since struck a responsive chord with many scientists, including many writing in this festschrift. For example, biochemist Michael Behe, who later authored *Darwin's Black Box,* a seminal case for intelligent design, first came to Johnson's attention after Behe wrote a letter defending *Darwin on Trial* in response to the *Nature* review.

Moreover, some prominent neo-Darwinists such as Arthur Shapiro of the University of California, Davis, Michael Ruse of the University of Guelph and William Provine of Cornell University welcomed the spirited challenge that Johnson provided to their views. Shapiro, Ruse and eight other scientists and philosophers (including both defenders and critics of modern Darwinism) joined Johnson at Southern Methodist University in the spring of 1992 to debate the central thesis of his book. A much-publicized debate between Johnson and Provine at Stanford in 1993 helped propel Phil's case to even greater prominence.

The success of those events led to many others like them and a growing community of scientists and scholars willing to examine the issues that *Darwin on Trial* first raised. *Darwin on Trail* reopened long dormant questions by challenging the evolutionary establishment's reliance on philosophically tendentious rules of method. In the process, it helped inspire an intellectual movement and a scientific research program that has begun to redefine our understanding of the origin of life and the nature of science.

FROM MUTTERING TO MAYHEM

How Phillip Johnson Got Me Moving

MICHAEL J. BEHE

—

I was born into a staunch and devout Roman Catholic family, the third of an eventual eight children. My dad and mom were from Altoona, Pennsylvania, where both my grandfathers worked in the railroad shops. Mom was one of ten children, Dad one of six. In grade school my dad played on the Cathedral parochial school football team; the coach was the bishop. After moving around Pennsylvania and nearby states as my dad was transferred from office to office by his employer, Household Finance Corporation, our family settled down in Harrisburg, Pennsylvania, when I was in fourth grade. Like all my brothers and sisters[1] I attended Saint Margaret Mary Alacoque grade school and went on to Bishop McDevitt High School.

Since it stayed open late those evenings, every Thursday night my dad would work till nine at the office. Afterward Dad would usually bring home a couple bottles of beer, and he and Mom would stand in our tiny kitchen talking until eleven or twelve at night. As we kids became teenagers, we would frequently join in the kitchen conversation. Talk often centered on religion and politics. Mom and Dad were lifelong Democrats, convinced that Democrats were the party of the little guy, but they were disturbed by what was going on in the country in the 1960s. Equally unnerving were changes sweeping through the Catholic Church in the wake of the Second Vatican Council. All of this was fair game for discussion, and many and various were the opinions that echoed throughout a Thursday night.

Yet I don't recall evolution ever being a topic of discussion. It just didn't register at the Behe house. After I wrote *Darwin's Black Box* my dad said in quite colorful language that he thought I was right about this evolution stuff (Dad, a no-nonsense WWII vet, is always emphatic in his opinions) and my mom collected newspaper clippings about where I was speaking and such. But neither had a whole lot to say about the substance of the book. I think their lack of interest faithfully reflects a laissez-faire attitude of Catholics toward the theory. Evolution never was the problem in the Catholic Church that it was in various Protestant denominations. I recall in seventh grade that when the nature of God's creation was discussed, Sister David Marie told us that God was the Creator and that he could create the world and life any way he pleased, either directly in a moment or indirectly in eons. What's more, since God created the universe and its laws, if he wanted to make planets or life through his laws, who were we to say he couldn't?

I remember being entirely satisfied with such reasoning, and thinking that it fit well with the way I thought about the world anyway (to the extent that a seventh grader thinks about the world). Looking back, the reason why my adolescent sensibilities resonated with Sister's is because my thinking probably had already been formed in accord with Catholic reasoning. For example, a few years ago my wife bought a copy of the 1909 edition of *The Catholic Encyclopedia* at a local library's used-book sale. The encyclopedia had a scholarly twenty-thousand-word article on evolution written by two Jesuits, one of whom was a professor of biology at St. Ignatius college in Valkenburg, Holland, complete with the imprimatur of Cardinal John Farley of New York, and published "under the auspices of the Knights of Columbus Catholic Truth Committee." Consider the following excerpt from the article:

> Attitude of Catholics Towards the Theory.—One of the most important questions for every educated Catholic of today is: What is to be thought of the theory of evolution? Is it to be rejected as unfounded and inimical to Christianity, or is it to be accepted as an established theory altogether compatible with the principles of a Christian conception of the universe? We must carefully distinguish between the different meanings of the words *theory of evolution* in order to give a clear and correct answer to this question. We must distinguish (1) between the theory of evolution as a scientific hypothesis and as a philosophical speculation; (2) between the theory of evolution as based on theistic principles and as based on a materialistic and atheistic foundation; (3) between the theory of evolution and Darwinism; (4) between the theory of evolution as applied to the vegetable and animal kingdoms and as applied to man.[2]

Sweet reason itself. Catholics, and thoughtful theists generally, should not reject evolution but examine it and make necessary distinctions about the many different concepts that often are all bundled together under the title *theory of evolution*. The principle distinctions we need to make, to rephrase what the article states, are (1) between evolution as a scientific hypothesis reasonably supported by the data and evolution as an assumption about the way the world must be, regardless of the availability of supporting data, (2) between evolution as an utterly random process, foreseen by no one, and evolution as the intended result of God's will, (3) between evolution understood simply as descent with modification, with the question of how such a process could have happened left open, and evolution as Darwin's specific theory of change driven by natural selection, (4) between the theory applied to body and the theory applied to mind. We should consider the evidence, make distinctions, keep what is good and solid, and toss out the rest.

Reading Denton

With an attitude like this by the Church, it isn't hard to understand why a number of Catholics see no difficulty whatsoever in accepting Darwinian evolution as the way God chose to create life. In fact, I was for a long time one of their number. My attitude then reminds me of the attitude now of Catholic Darwinists such as Ernan McMullin, emeritus professor of philosophy at Notre Dame and an ordained Catholic priest, and Kenneth Miller, professor of biology at Brown University and author of *Finding Darwin's God*. Like both of them, I saw no theological problem with Darwin's theory (properly understood)—and still don't. Like Miller, I pointed to astronomical or physical data such as the big bang and fine-tuning of the universe to support the contention that God created a universe designed to produce life—and still do. Like Miller, I was insufferably smug about the superiority—both scientific and theological—of this view compared to the views of religious Protestant skeptics. In fact, as a postdoc in my mid-twenties at the National Institutes of Health, I remember keeping company with a pretty evangelical Protestant lab technician who attended Fourth Presbyterian Church in Bethesda, Maryland. Occasionally she'd bring up evolution. She reported approvingly that the church's pastor, Richard Halverson (an impressive man who later served as chaplain of the United States Senate), had proclaimed something like "apples can't turn into oranges." I sniffed superciliously, "Sez who?" "Given enough time . . ." "Many small steps . . ." blah, blah. No wonder she soon found she could manage quite well without my patronizing.

Later on as a young assistant professor at Queens College of the City University of New York, I met the woman I was destined from all time to marry. Celeste had attended Our Lady of Mount Carmel school in the Bronx, where her seventh grade class was once told by their teacher, a Holy Cross brother, that "Evolution is true—get used to it." This he spoke as he held up definitive evidence—a copy of Haeckel's famous drawing of nearly identical vertebrate embryos, now known to be, at the very least, grossly exaggerated. Like myself, Celeste had no particular theological objections to evolution and little interest in the topic. Celeste and I married in 1984. In 1985, with baby Grace in tow, we moved to Pennsylvania, where I had gotten a position as an associate professor in Lehigh's chemistry department. Our new home provided the room we would need to raise what would hopefully be a large family. At Lehigh I busied myself with my teaching and research, and things went quite well both professionally and personally, as Benedict and Clare followed Grace into our house on Apple Street.[3] My career was going in the linear fashion I'd hoped, with publications and grants coming in a more or less orderly fashion.

But then I hit what, I realize in retrospect, was a major bump in the road. In a flyer listing the monthly offerings of a book club I belonged to, I saw an ad for *Evolution: A Theory in Crisis,* by Australian geneticist Michael Denton. I was puzzled. Although I knew some folks, such as my evangelical friend at the NIH, who disbelieved evolution for religious reasons, I had never heard of a scientist questioning it. Yet Denton was billed as an actual scientist who questioned evolution based on science. Intrigued, I ordered the book and finished reading it in two days. I was stunned. In all the years of my scientific training—as a chemistry major at Drexel University in Philadelphia, as a biochemistry major in graduate school at the University of Pennsylvania, as a postdoc at the National Institutes of Health—never had I read a serious, systematic, scientific critique of Darwinian evolution. In fact, as I put the book down, I couldn't remember having heard *any* criticisms of Darwinism, other than joking ones.

One slow afternoon in the late 1970s I was hanging out in my lab at the National Institutes of Health near Washington, D.C., where I worked as a postdoctoral researcher investigating aspects of DNA structure. A fellow postdoc, Joanne Nickol, and I were chewing the fat about the big questions: God, life, the universe—that sort of thing. She and I were both Roman Catholics (Joanne's brother was a priest) and so had the same general attitude toward many topics. That included an easy acceptance of the idea of evolution, that life unfolded over a long time under the governance of secondary causes, natural laws. Unlike some Protestant friends of mine who seemed obsessed by it, we Catholics were always cool about evolution, because we knew that God could make life any way he wanted to, including indirectly. Who were we to tell him differently? The critical point was that God was the Creator of life, no matter how he went about it.

The course of Joanne's and my conversation in the lab hit a little snag. Because we were taught biology well in parochial school, we both knew that the evidence for Darwinian evolution by natural selection was ultra strong. But when the topic turned to the origin of life she asked, "Well, what would you need to get the first cell?" "You'd need a membrane for sure," said I. "And metabolism." "Can't do without a genetic code," said she, "and proteins." At that point we stopped, looked at each other and, in unison, hollered "Naaaahh!" Then we laughed and went back to work.

From a distance of years I notice three things about my conversation with Joanne (who died about a decade ago). The first is that the notion, widely accepted among scientists, that undirected physical laws started life, struck both of us—both well-trained young scientists who would be happy to accept it—as preposterous because of the many complicated preconditions necessary just to get things underway. Second, we apparently hadn't given it much thought before then. And third, we both just shrugged it off and went back to work. I suppose we were thinking that even if we didn't know how life started by natural processes, surely somebody must know. Or that somebody would figure it out before long. Or eventually. Or that it wasn't important. Or something.

In retrospect, I think I understand why Denton's writing had such a big impact on me. It has been said that Darwin's work came as a shock to nineteenth-century society not so much because of the specific mechanism of natural selection that he proposed to explain life. Rather, his very attempt to explain life by natural law made the natural development of life thinkable. God's guiding role was no longer the utterly obvious, unarguable fact it once was. It had been reduced to one possible idea among other possible ideas. Denton's writing had a similar effect on me, but of course the context was 180 degrees opposite from the context in which Darwin published. Now, in the late twentieth century, it was the truth of Darwin's theory that, to a well-educated person, was as obvious as the nose on one's face. It was the de-

velopment of life by the outplaying of natural law that needed no justification. And it was that serene assumption that was shattered for me by Denton's book.

Life in Another World

When I laid the book down, I lived in a different world. Not only was Darwinian evolution now a dubious story where a few days before it had been
boringly self-evident, but now I also had a reliability problem. If Darwinian
evolution had obvious and unsolved problems, as Denton argued convincingly, and if those problems were not hard to see, as they weren't in retrospect, then I had been seriously misled throughout my education on a topic
that was a major pillar of my worldview. I had been taught to accept Darwinism unquestioningly not because the evidence for it was unassailable but
because that's what I was supposed to believe. Darwinism was "the way we
think today" rather than an explanation forced on us by the evidence. In
short, I was taught to believe it for sociological, not scientific, reasons. If that
were the case, then I could no longer trust the breezy, confident assertions
that fill textbooks and science journals, that we know—or someone
knows—that this or that phenomenally complex biological system evolved
by random mutation and natural selection. I had to root through the science
literature on each topic and see for myself whether the claims were supported or were so much bluster.

The next morning I went into the lab. Still charged up from Denton's
book, I tried to explain the importance of what I'd read to my graduate students. They listened politely for ten minutes or so and then went back to
work. A bit frustrated, I talked to my postdoc. She was a rather demure person, who listened politely and then went back to work. I was beginning to
see a pattern. The force of Denton's sustained argument over the course of
his book was not easy to convey in a brief and casual conversation, even
with people, such as my grad students and postdoc, who were theoretically
supposed to listen to me with a degree of deference.

Even for a person like me who did indeed feel the force of the argument,
it was not at all clear what one could do about it. After talking to a few more
stray students and faculty colleagues wandering around the halls, I was reduced to muttering rude things about evolution to innocent passersby whenever an opportunity arose—not a very effective strategy for persuasion.[4]

One opportunity in the early years to do something with my new interest in evolution was provided by Lehigh. Like some other universities, Lehigh had decided to require first-year students to take a three-credit freshman seminar. The low-enrollment seminars would be offered on topics chosen by volunteer faculty and would be intended to show students the excitement of the intellectual life before fraternity and sorority rush shifted their attention elsewhere. I volunteered to teach a seminar I called "Popular Arguments on Evolution." In the class we would read and discuss both pro- and antievolution books and articles, using Denton's *Evolution: A Theory in Crisis* and Richard Dawkins's *The Blind Watchmaker* as the anchor texts for the two sides. Teaching the course was a fascinating experience for me. I remember going in the first day to the 8 a.m. class and realizing it was the first class these freshmen were having in their college careers—so I'd better set the right tone! After some preliminary pleasantries, one of the first questions I asked them was, who here believes evolution is true? Every hand went up. Next question—what is evolution? Every hand went down.

I smiled to myself.[5] It was then fairly easy to make the point that sometimes we believe in things—even important, fundamental concepts such as evolution—without having a good idea of what they are or what evidence supports and opposes them because we learn them from authority figures. Throughout the semester we would read arguments, and then ask what evidence the author cites to support his argument, whether his conclusions are justified or a stretch. We would tease apart the many different meanings of the word *evolution* and ask whether evidence for evolution in one sense was evidence for evolution in other senses. We would ask how, as was often the case, two authors could point to the exact same examples, such as the structure of the protein hemoglobin or the structure of the vertebrate eye, and use them to argue to completely opposite conclusions, as Denton and Dawkins did. It was a blast. I had the wonderful opportunity to see students realize that what they thought was a boring and blindingly obvious idea was actually quite complex. An excellent start to a college career, I thought!

Enter Phillip Johnson

Teaching the freshman seminar was certainly a step up from muttering to myself, but discussing evolution every few years in a small class makes for slow progress in changing the mind of the public. The next step toward my involvement in a much wider conversation happened, as did many things in this area, serendipitously. In the early 1990s another advertisement from my book club caught my attention. A new book questioning evolution had been written by a rather unlikely author—a law professor at one of the nation's most prestigious academic institutions, the University of California, Berkeley. I ordered the book and devoured it. Although it addressed the skimpy science used to support Darwinian evolution, unlike Denton's book Phillip Johnson's *Darwin on Trial* focused on the logical framework supporting Darwinism. Johnson demonstrated that, although the truth of materialism entailed that something like Darwinism had to be true, if one did not accept materialism (as most people in the United States do not) then the evidence supporting random mutation/natural selection looks pretty paltry indeed. His reasoning neatly encapsulated what I had vaguely surmised since reading Denton, and I began to include some of Phil's writings in my class, in particular a booklet titled *Evolution as Dogma: The Establishment of Naturalism,* published by Jon Buell's Foundation for Thought and Ethics, which also contained responses from Darwinists, including William Provine.

A month or two after I read Phil's book, I was in our departmental office at Lehigh and picked up the latest issue of *Science* magazine, which was kept on a table for visitors. Scanning the table of contents I noticed a news item titled "Johnson vs. Darwin." I turned to the brief article and found that it was not the discussion of the book that I was expecting. Rather, it was a warning to scientists to keep students away from this book because it would confuse them! (The article is reproduced in this endnote.)[6] After quoting Phil's answers to leading questions and likening him to creationists, the article turned for a response to a person I had never heard of—one Eugenie Scott—from an obscure organization called the National Center for Science Education (NCSE). I remember being royally ticked off, for several reasons. First, the article never mentioned what Phil's arguments actually were, only that they were similar to ones made by the Institute for Creation Research. Guilt by association. Second, I wondered why the article asked some unknown science educator for a remark rather than an evolutionary biologist or scientist. Later I was to learn of course that Scott and the NCSE are the police who monitor public discussions, tracking down and attacking any hint of deviance from evolutionary orthodoxy.

The news item made me so mad that I wrote a letter to the editor of *Science,* which they published. (The letter is also reproduced in this endnote.)[7] Briefly, I wrote that this Johnson fellow appears from his book to be a rather intelligent layman, and that scientists would do much better to address the substance of his arguments than to rely on ad hominem remarks. About a week later I received a letter with a return address of Boalt Hall. Phil thanked me for my letter and hinted broadly that it sure would be nice if someone with my knack for letter writing would send some comments to *Nature*, which had just published a scathing review of *Darwin on Trial* by philosopher David Hull. I took the hint and sent *Nature* a really swell letter, but they declined to print it.

What I didn't realize quite yet was that I was now in the loop—I was within the circle of Phil Johnson's acquaintances and useful contacts.

Going Public

About six months later the phone rang in my office and a voice with a Texas twang inquired if I was Michael Behe. Jon Buell of the Foundation for Thought and Ethics asked me if I would be willing to come to Texas to participate in a debate on evolution. Immediately I envisioned a dispute in some church basement with people waving Bibles and shouting at the participants. I was about to say no when Jon continued. Yes, he said, the debate would be on the campus of Southern Methodist University, and it would be held to celebrate the publication of Phillip Johnson's new book. In fact, Phillip Johnson himself would participate, along with a fellow named Michael Ruse, who had been a star witness for the evolution side during the Arkansas creation trial in the 1980s. *Wow,* I thought, *this ain't no church basement.* So I signed up, went to Texas and met Phil in person, as well as Steve Meyer, Walter Bradley and Bill Dembski.[8] The event was a smash, not only because it was conducted in a thoroughly professional manner by Jon (as well as Tom Woodward of the C. S. Lewis Fellowship and Steve Sternberg of Dallas Christian Leadership) but because it showed us participants on the skeptical side that we had good arguments that were difficult for the other side to address—in short, we had something to say that could stand up to vigorous public scrutiny, and our confidence was pumped up.

The early 1990s were also the early years of the Internet, and Phil was quick to take advantage to set up an e-mail discussion group on skepticism about evolution. That was very beneficial for me, as I got to try out some of the arguments about irreducible complexity of biochemical systems that would later become the basis for *Darwin's Black Box*. (Many of the participants in the group were actually reflexively antagonistic to antievolution and design arguments.) I became increasingly convinced that I had something to say that no one else in the group had heard. About a year after the SMU symposium I got an e-mail from Phil saying that he had received funding from some friends (the Ahmanson family) to hold a retreat for academic contacts who shared his skepticism of evolutionary claims, and he asked me to participate. I was delighted. Not only did I get to talk with some brilliant thinkers in fields far removed from my own, I also got to go to California! I had never been there before. The meeting at Pajaro Dunes consisted of a series of short talks by the participants and long periods for discussion, either at the beach house or while walking along the beach.[9]

From interacting with Phil and the intellectual circle he had gathered around him, a new idea popped into my head. Several of these people had either written books or planned to. Maybe I could do that! As a rule, most working scientists don't write books. They write research articles or review articles, but writing books is frowned on. None of my advisors in science, neither my Ph.D. advisor nor my postdoctoral adviser, had ever written a book, and they were very prominent scientists. I had heard of the occasional scientist writing a textbook, but it was always regarded as a distraction from the main business of bench research and was often done by investigators whose research efforts hadn't panned out. So the idea of writing a book that laid out my doubts about Darwinism literally never crossed my mind until I came under Phil's tutelage. Here at Pajaro Dunes were philosophers and scientists who were writing books that mattered to the intellectual conversation of our culture. I decided that I wanted to do that too.

As fortune would have it, I had forgotten to pack my tooth brush. I asked Phil (one of the few folks there with a car) to drive me to the commissary to buy one. On the way I asked him how one goes about getting a book published. His firm advice—get a literary agent! I went home from the meeting at Pajaro Dunes with a new resolve, to write a book explaining in detail for the public why modern biochemistry is not only resistant to Darwinian explanations but points strongly and insistently to intelligent design. From the journalist Tom Bethell, whose acquaintance I had made on Phil's e-mail listserv, I got the name of a literary agent in New York, who quickly agreed to take on the manuscript I sent her. She polished up my writing and shopped the manuscript around to a number of publishing houses. Quite a few expressed interest, but Free Press was especially excited about it, and we went with them.

The publication of *Darwin's Black Box* in 1996 allowed me to enter the mayhem of the public debate surrounding Darwinism versus intelligent design,[10] which I had wanted to do ever since putting down Michael Denton's book years earlier. Yet, if it were not for the political savvy, the vision and the networking of Phil Johnson, I'd still be muttering ineffectively to random passersby in Bethlehem, Pennsylvania. What Phil did for me—take a floundering critic and turn him into an effective spokesman—he did for many others, and it is through his strong leadership that the intelligent design movement has gained a forceful public voice.

Bibliography

Behe, Michael J. 2004. "Irreducible Complexity: Obstacle to Darwinian Evolution." In *Debating Design: From Darwin to DNA,* edited by M. Ruse and W. A. Dembski. Cambridge: Cambridge University Press, 2004.

———. "Design in the Details: The Origin of Biomolecular Machines." In *Darwinism, Design & Public Education,* edited by J. A. Campbell and S. C. Meyer. East Lansing: Michigan State University Press, 2003.

———. "The Modern Intelligent Design Hypothesis: Breaking Rules." In *God and Design: The Teleological Argument and Modern Science,* edited by Neil Manson. New York: Routledge, 2003.

———. "Reply to My Critics: A Response to Reviews of Darwin's Black Box: The Biochemical Challenge to Evolution," *Biology and Philosophy* 16 (2001).

———. Self-Organization and Irreducibly Complex Systems: A Reply to Shanks and Joplin. *Philosophy of Science* 67 (2000): 155-62.

How Phil Johnson
Changed My Mind

Jay Wesley Richards

—

W ith a title like "How Phil Johnson Changed My Mind," you're probably expecting an indulgent piece of navel gazing rather than reflections on Phil Johnson himself. But don't fear. I'm taking an autobiographical turn only because I suspect that the way I thought in 1991 when *Darwin on Trial* was released was how many academically oriented Christians tend to think. In other words, I suspect my views were a representative sample. By describing how Phil Johnson changed my mind, then, I hope to describe how Phil changed the minds of many like me and became the catalyst of something new.

In 1991 I was a seminary student, interested in things philosophical and theological. I also had a lifelong if avocational interest in science, especially astronomy and economics. Although I had been raised in the church and adequately schooled in the Christian faith, it wasn't really until college, after many bumps and starts, that I became a serious, thinking Christian. Like many untold thousands, the proximate cause of this was reading C. S. Lewis's books *Mere Christianity, Miracles* and the like. Lewis convinced me that a robust Christian faith was a real intellectual possibility and not just mere wish-fulfillment that couldn't withstand critical scrutiny. Moreover, it was Lewis who convinced me that naturalism (or what I prefer to call materialism) was little more than a modern prejudice that rested on dubious philosophical grounds.

Lewis's arguments alerted me to the unmistakable effects of this ubiquitous but unarticulated prejudice in disciplines as different as psychology, political science and biblical studies. In many ways, Lewis inoculated me against the effects of materialism.

I also recall the liberating effects of reading Thomas Kuhn's *Structure of Scientific Revolutions,* which I read around the same time. I rejected Kuhn's relativism and antirealism, and eventually found Michael Polanyi's understanding of the nature and history of science much more adequate. Nevertheless, Kuhn's analysis of how assumptions shape scientific inquiry provided a powerful interpretive grid for diagnosing and detecting the role of materialism in various academic disciplines. This was probably not what Kuhn hoped readers would glean from his book. But it was, for me, its most important take-home lesson.

These convictions did not immediately translate into a struggle over evolution. It's not that I simply avoided the issue. I really did try to think through issues like evolution and the implications it could have for belief in God. I wondered how and if the Christian doctrine of creation was compatible with evolution. It was on the basis of such reflection that I came up with a way of reconciling Darwinian evolution, which I took to be the best scientific theory of life's diversity and complexity, with Christian belief.

I understood Darwinian evolution as the claim that all organisms share a common ancestor and that adaptive complexity was the result of natural selection and random genetic mutations. This is standard neo-Darwinism. My way of reconciling this bundle of ideas with Christian belief started with the seemingly innocuous question, Couldn't God have used evolution to create us? (Of course, my idea was not rigorously developed. But Del Ratzsch, in *Mere Creation,* and Michael Ruse, in *Can a Darwinian Be a Christian?* have explored well the half-formed intuitions that I nurtured then.)[1]

Here's the basic idea, described with the benefit of hindsight. I envisioned God creating a world with certain natural laws and lots of degrees of freedom. (I was aware of the so-called fine-tuning of the physical constants and already thought they provided evidence for design at the cosmic level.) So, given a large universe and the right sort of laws, there would be plenty of opportunities for life to evolve. (At the time, I didn't appreciate the many factors needed to produce a habitable planet.)

I didn't entertain nontraditional process or open theism. On the contrary. I imagined that God, in his foreknowledge, would create just the world where the structures, beings and persons he chose would evolve through natural selection and random variations. These variations, genetic mutations or whatever would be random in the sense that they weren't uniquely determined by physical laws, and God need not directly cause them for the functional advantage of organisms. Nevertheless, they would take place within a world created and chosen by God based on his detailed foreknowledge. So they would fulfill his will. I didn't see anything obviously absurd or contradictory in this idea. In fact, I still don't, as long as it's left quite general.

I was also unimpressed with most of the arguments against evolution. Unlike those urbane people who grow up in large cities in the Northeast and never meet anyone who disagrees with them on anything important, I had several friends and acquaintances who were staunch creationists and critics of evolution. So I encountered the arguments first hand. Nevertheless, all the arguments I heard from my creationist friends seemed flaky and far-fetched.

There were those famous Paluxy footprints, which were really a hit in south Texas, where I went to college. And there was the story about NASA's misguided fear that the moon landers in the Apollo missions would sink into feet of lunar dust. This fear was the supposed result of NASA scientists assuming that matter had been accreting on the moon's surface for billions of years. The lack of such dust, then, was supposed to prove that the solar system was just a few thousand years old. Then there were the sightings of dinosaurs in Central America, and the plesiosaur remains picked up in the Pacific by Japanese fishermen.

You might dismiss these as fringe ideas, but they were the arguments that I most often heard from critics of evolution, even at a secular liberal arts college. In fact, I still get asked about the moon dust when I speak in churches. The other more substantive arguments I heard all seemed to focus on nit-picky anomalies, while ignoring the preponderance of relevant evidence. Perhaps there were some cogent arguments mixed in that I missed. But one bad argument has a way of chasing out several good ones.

Then there were the exegetical and theological arguments. I recall one common argument against my conciliatory reasoning. Several friends argued that the Bible claims we are made in the image of God, whereas evolution says we're descended from apes—or something like that. So obviously the Christian doctrine of humanity's creation wasn't compatible with evolution. This didn't seem like a great argument. It still doesn't. After all, on a straight-forward reading of Genesis 2, God used the dust of the ground to form

Adam. If God, using dust, could form us in his image, then, I reasoned, he could surely use an ape.

My creationist friends insisted on what I thought was a wooden and unnecessary interpretation of Genesis 1, in which the days had to refer to twenty-four-hour earth days. I didn't think this was an obvious deliverance of the text. They also insisted that all animal death was the result of human sin, which I didn't think was either claimed in Scripture or especially tenable. Moreover, I didn't think these issues made much difference theologically. At the very least, I didn't think they were as significant as my creationist friends made them out to be. I thought that most of the trouble was the result of doubtful presuppositions held by young-earth creationists rather than anything intrinsically bad in Darwinian theory.

So I developed an intellectual strategy that reconciled evolution with Christian doctrine, and I was not persuaded by contrary arguments. As a result, I didn't struggle with this issue, even after I had become a serious Christian.

What Phil Did

But something happened. First, a friend I trusted, Jonathan Witt, cajoled me into reading *The Mystery of Life's Origins.*[2] This book persuaded me that there was no good materialistic explanation for the origin of life and that the evidence I had taken as decisive—the famous Urey-Miller experiment—didn't do the trick. In fact, it didn't even take the stage.

Then, somehow I came across *Darwin on Trial,* with its gold cover and picture of Charles Darwin. It didn't look like those books that my creationist fraternity brother had on his bookshelves. It was written by a Berkeley law professor. I wish I could say that such superficialities weren't relevant. But they were.

Of course, appearances alone didn't persuade me. They just kept me from prejudging the book before I had read it. So I read the first chapter and thought, *Oh, this is different from what I've seen before.* I read the book that evening and finished the next morning. It produced a little crack in my intellectual edifice. Within a few weeks, I knew that the edifice had to be rebuilt from the ground up. So what did Phil do that not only clarified the issue for me and many others but catalyzed a full-blown movement? While there are many ways to describe any historical event, I would boil Phil's achievement down to three points:

1. He evaluated the empirical evidence for neo-Darwinism fairly and sep-arated this clearly from the philosophical assumptions that allow the theory to win by default. As Phil puts it at the end of his first chapter: "My purpose is to examine the scientific evidence on its own terms, being careful to dis-tinguish the evidence itself from any religious or philosophical bias that might distort our interpretation of that evidence."[3]

This sunk in. I never should have been satisfied with my little reconcili-ation of God and evolution. To have an informed opinion on the subject, I should have examined the scientific evidence on its own terms. Phil's mod-est proposal was quite subversive, because Darwinism in its totalizing form is about much more than a modest inference from the evidence. Phil had framed the issue so that that central issue was crystal clear. He had asked the right question. A penetrating analysis was sure to follow.

Recently, Phil and I spent many hours on a train together traveling from Seattle to Salem, Oregon. It allowed us to have one of those rambling con-versations that erases your perception of the passing of time. At one point I told Phil about the question that had allowed me to avoid this issue for sev-eral years: Couldn't God have used evolution? He said, quite perceptively: "Questions about what God could have done are the most boring questions you can ask!" After all, unless what you're proposing is logically impossible or morally repugnant, the answer is always yes. The problem with this ques-tion is that it uses a vague term, *evolution,* and then substitutes the mere existence of a logically possible world for a serious inquiry into what the actual world is like. We should want to know what the world is actually like, that is, what God actually did, not rest satisfied with what he could have done.

By focusing on mere logical possibility, the question also bypasses the way ideas gain currency historically. Human beings are not logical deduc-tion machines. Darwinism did not overtake the world of ideas because Dar-win had deductively proved his theory. (In fact, his claims in the *Origin of Species* are closer to conjecture than theory.) Darwinism won't be retired to the history of ideas because someone has decisively refuted it. Ideas wax and wane more subtly. And it doesn't really matter if some version of evo-lutionary theory is logically compatible with Christian theism. Before we even consider the issue of compatibility, we should try to figure out what it is we really need to reconcile.

Darwin on Trial persuaded me that Darwin's theory, at least in its most bloated form, aptly called the "blind watchmaker thesis," was quite weak in the way of evidence. As a global explanation of biological complexity, its force lies primarily in the fact that its key alternative has been ruled out of bounds. As a result, it operates not as a hypothesis that is tested empirically but as a deductive first principle, for which the theorist need do little more than tell possible stories and find a confirming example or two.

I confirmed this thesis by reading several of the leading defenses of Darwinism by Ernst Mayr and Richard Dawkins. I even read an authoritative encyclopedia of evolution. What I realized is that practically any defense of Darwinism is likely to confirm Phil's argument. In 154 pages, Phil had boiled down to their essence the arguments of the staunchest Darwinian advocates. Nothing I've seen in the last thirteen years has changed my initial judgment.

His analysis manages to walk the razor's edge. On the one hand, Phil avoids the simplistic positivism and naive empiricism that so often plagues this debate on all sides. He recognized that a successful analysis of Darwinism could not simply say: Not enough evidence. Philosophical assumptions pop up everywhere and often do a great deal more work than the evidence itself.

On the other hand, Phil avoids describing science as a mere language game in which different groups, with incommensurable assumptions, vie for power and prestige. He retains a strong sense that science, properly understood, should be about the pursuit of truth and that good scientific theories really should be put in empirical harm's way (to quote Del Ratzsch's apt phrase).

Often science is defined so narrowly that only lab chemistry, with its discrete, repeatable experiments, qualifies as science. Other times the scientific enterprise is cast as little more than presuppositions in search of confirming evidence. Darwinists have their presuppositions. Creationists have theirs. The best we can do is to disclose our biases and come up with a description of nature that is consistent with our basic assumptions. I don't think either of these extremes is right. Phil kept the role of empirical evidence in a careful dialectical tension with theoretical assumptions that underlie scientific investigation. And he did so in a way that is not only descriptively accurate but also preserves what is essential and admirable about science at its best.

I found Phil's critique of the function of "methodological naturalism" (as we now call it) especially persuasive. Yes, it's a theoretical assumption. But it's unjustifiable. In my brief autobiographical sketch, you'll notice that I didn't say anything about it. That's because it played no role in my earlier, more conciliatory attitude toward Darwinism. I felt no temptation to abide by its rules. I just didn't know enough about the evidence for Darwinism to make an informed decision about the theory.

In fact, I never thought the arguments for methodological naturalism were the least bit persuasive. On the contrary, it always seemed to be an arbitrary restriction on what we could consider. Since I took a realist understanding of science and truth seeking for granted, I never thought it made sense to impose a definition on science that would restrict not only the types of questions that one could ask but even what one could discover about the world. Besides, we can't determine anything about the evidence or the validity of an argument with a definition of science. If I present a set of evidence drawn from, say, biochemistry, make it a premise in an argument, and then draw a conclusion, how is the truth of the argument affected by labeling it, say, art history or microeconomics or religion or pseudoscience? It's not.

Double Nobel Laureate Linus Pauling once said, "Science is the search for truth." Well, natural science is the search for truth about the natural world. You can't decide that truth in advance by defining certain possibilities as outside science. The search for truth is an *intrinsic* good. An arbitrary restriction on such a search is not. In fact, it's intellectually perverse.

The position you take on this one question may explain whether you find Phil Johnson's argument persuasive in *Darwin on Trial.* After I read *Darwin on Trial* I became zealous for other friends to read it. Many were as impressed as I was, but some remained skeptical. To a person, the remaining skeptics were those who had accepted methodological naturalism as the essence of science. As a result, something like Darwinism had to win by default, because design was not a possibility.

2. Phil insisted on keeping separate issues separate. Keeping separate issues separate seems easy, especially once it's been done. But it's very hard to get people on all sides of this debate to do it. The first step in doing it successfully, of course, is to get definitions straight. If you open up a copy of *Darwin on Trial,* you will discover on the first page of chapter one that Phil discerned that the word *evolution* means all sorts of different things: "The conflict requires careful explanation," he says, "because the terms are confusing."[4] He discerns that defenders of Darwinism, especially advocates of the blind watchmaker thesis, seem to avoid careful use of the term. In fact, few advocates of Darwinian evolution seem willing to stipulate a definition of the word and then stick with it consistently. Strangely, they sometimes accuse critics of equivocating when they say "Darwinism" or "Darwinian evolution."

My hasty reconciliation of Christian doctrine with evolution was facilitated by my vague idea of evolution.[5] For me the main issue was the question of whether human beings had apes as ancestors or, to put it more accurately, whether human beings and apes shared some common ancestor. Natural selection had an intuitive plausibility, so I didn't think much about it. And I had already found a way to domesticate random variations within a theistic context, so that they only *appeared* random from our limited perspective. Armed with a coherent little system, I didn't need to spend much time looking into the evidence for Darwinism. My biology professor explained things effortlessly in Darwinian terms, so I assumed it was basically a done deal.

Moreover, I didn't consider the degree to which factors other than evidence might have shaped both the reception and the subsequent defense of Darwin's theory. I had no sense of how important Darwinism was ideologically. My reasoning was essentially this: Most smart people believed that Darwinism was true. In my science classes, I didn't hear about any serious scientific problems with it. No one has time to look into everything. So, as we all do most of the time, I trusted the official story.

So a muddled way of thinking prevailed. But the fault did not lie simply with Darwinist propaganda. Quite the contrary! In this debate, mush and muddle have been the trends all around. I recently spent an hour and a half lecturing to intelligent graduate students. Even they had a hard time separating the concept of common ancestry from Darwin's mechanism of natural selection working on random variations. Some of them never got the distinction.

This isn't merely a problem with students. Much of the creationist literature also fails to keep separate issues separate. For example, about six years ago I read a book called *Bones of Contention: A Creationist Assessment of Human Fossils* by Marvin Lubenow.[6] The book has its problems early on. But Lubenow did persuade me that our understanding of the human fossil record is about ninety-nine parts imaginative reconstruction and one part empirical evidence.

But Lubenow isn't satisfied with this simple, defensible thesis. So toward the end of the book, he suddenly launches into a critique of big bang cosmology! Aside from the merits of his arguments—which are few—such lack of focus within a book on the human fossil record is the writer's equivalent of attention deficit disorder. It only serves to weaken whatever useful arguments the book contains.

In contrast, *Darwin on Trial* is lean and focused. And this focus had an ironic result. Once Phil helped me get the various issues straight in my head, I realized that much of what I thought about matters such as age of the universe was not based on careful study of the issues. Rather, it was a reaction against the unsophisticated arguments I had encountered. So, to be intellectually honest, I decided I had to revisit these issues. After all, it was possible that God had created the universe a few thousand years ago in six twenty-four days.

First, I studied the opening chapters of Genesis in detail. I had learned biblical Hebrew and exegesis, and knew how to read from a wide range of biblical commentaries, including the arguments of young-earth creationists. Here, I can only report my conclusion. The days of Genesis 1, it seems to me, are literally God's days, not *our* twenty-four-hour earth days. Our earth days are a dim reflection, an analogy, of God's preeminent creative week. Our day and night is separated by the presence and then absence of the light of the sun (which shows up on day four of God's work week). God's work week is delineated by the light and darkness (which isn't explained) that God separates on his day one. Days one through six each end with an evening and a morning. So, presumably, just as we work from morning to evening and then rest from evening to morning in our ordinary work week on earth, God does the same in his work week. And as we rest on the seventh day, God does as well. God's seventh day, however, doesn't end like the previous six days. This suggests that God is now still in his seventh day, in his rest, as the Psalms and Hebrews intimate.[7]

So I decided that the young-earth creationist exegesis was wrong. There is no direct biblical claim that the universe is just a few thousand years old and that God created it all within the span of a single earth week. All this meant was that we couldn't deduce that the universe was young from the biblical texts. It didn't follow that it wasn't. I thought that question should be answered by looking at the available evidence.

For about a year I dipped into the relevant young-earth creationist literature, attempting to get a representative sample. Then in the summer of 1992 I spent three months obsessively reading far and wide. I had a friend who had ten years of back issues of *Creation Ex Nihilo* magazine, including the

Creation Ex Nihilo Technical Journal. I read every issue. And I read widely on cosmic-age-related fields such as geology and cosmology.

I quickly learned that there were better arguments for young-earth creationism than those off-putting stories about Paluxy footprints and moon dust. I learned that young-earth creationists had made many of the scientific arguments Phil treated in *Darwin on Trial*. But I also found that the literature was filled with bad philosophy of science, with both extremes represented—positivism and presuppositionalism—sometimes in the same place.

Moreover, I noticed that the chaff and wheat were closely intertwined, making it hard to harvest the wheat and discard the chaff. After a few months of reading the literature, I began to get the strong impression that the issues varied both in their importance and in the strength of their evidence. Showing the inadequacies of Darwin's mechanism was one thing. Showing that the universe was no more than ten-thousand-years old was quite another.

So I developed an informal ranking of the various issues, reproduced below. At the top are the propositions that I think are the easiest to substantiate and most important scientifically, philosophically and theologically. At the bottom are the ones that I think are the most difficult to substantiate (in fact, I think several are false, but that's not crucial here). Consider this list a first shot, since I have never bothered to publish it until now:

- Natural selection working on random genetic mutations does not adequately explain all or most of the adaptive complexity we observe in biological organisms.

- There is little evidence that random genetic mutations, however significant, can produce major, new functional biological structures.

- Some, if not most, of the popular evidence cited in favor of Darwinian evolution dissipates on closer inspection (e.g., vertebrate embryo diagrams, peppered moths, Darwin's finches).

- Some, if not much, of the (valid) evidence cited in favor of Darwinian evolution establishes a much more modest proposition than the theory itself. For example, antibiotic resistance in bacteria at best illustrates that natural selection can work within an individual population of a single species in an artificial selective environment.

- There are a number of strong philosophical objections to scientific and philosophical naturalism.

- Methodological naturalism is very difficult to justify nonarbitrarily, and it contradicts the true spirit of science, which is to seek the truth about the natural world, no holds barred.

- Darwinists often use methodological principles, like methodological naturalism, to fill gaps in their case for Darwinism.

- Darwinism often functions as a "designer substitute," and as such, often functions as a creation myth for materialists.

- Inferring design is often rational and truth preserving. In principle, there is nothing to prevent us from inferring design when studying nature or natural structures.

- The universe and certain natural structures within the universe provide evidence of intelligent design. Such evidence can be found in cosmology, physics, astronomy, biology, the human sciences and elsewhere.

- Universal common ancestry often functions as an axiom rather than a tentative inference based on empirical evidence.

- There is some empirical evidence against universal common ancestry (and not merely a lack of evidence for it).

- Some geological structures typically explained by slow, gradual processes can be explained as well or better in terms of one or several catastrophic events.

- Radiometric dating methods, when not independently calibrated, often lead to unreliable results.

I think there is good evidence in favor of all of these propositions. And I think that they can be rationally defended, despite the fact that many of them challenge intellectual orthodoxies.

Now let's consider some issues that I think are much harder to defend. (I'm not going to say why I think that. I'm just asserting.)

- All dating methods are hopelessly subjective, and are based on circular reasoning and materialistic assumptions.

- The age of the universe is just as important theologically as whether there is evidence for design in the natural world.

- Since materialism requires an old universe, Christians should oppose any scientific claims that suggest the universe is old.

- Since materialists can incorporate big bang cosmology into their worldview, big bang cosmology is intrinsically atheistic.

- The only honest way to interpret the days of creation in Genesis 1 is as twenty-four-hour earth days.

- A single global flood, which occurred a few thousand years ago, explains many if not most of the features of the geological column.

- God created the universe so that it would look old, say ten to twenty billion years old. But we know from Scripture that it's really no more than ten thousand years old.

- An unbiased evaluation of the available scientific evidence in geology, astronomy, physics, biology and so on leads to the conclusion that the universe was created in six twenty-four-hour days between six and ten thousand years ago.

If this ranking is anywhere near the truth, it should be clear why Phil's arguments in *Darwin on Trial* were so much more effective than the creationist arguments against evolution.

First, many of the issues in this list are not logically related. You can evaluate the power of natural selection without considering evidence for a recent global flood. Second, the role of background assumptions varies dramatically between the two lists. It was that difference that led me to make separate lists in the first place. To put it technically, the claims near the bottom seem much more theory and presupposition laden. In particular, they seem to depend much more on beliefs derived from a certain interpretation of biblical texts. In fact, some of them look plainly like biblical interpretations in search of confirming evidence rather than inferences from the evidence itself.

This certainly became obvious to me in reading young-earth creationist literature. Many young-earth creationists admit this openly. I know one prominent young-earth creationist, for example, who argues that scientific facts are slaves to presuppositions. In fact, he has even admitted that without a young-earth creationist premise, one would be unlikely to infer that the universe is young. But this isn't a problem for creationists, he argues, since *everyone* is just reasoning from presuppositions. The materialist has one set of assumptions, the creationist another. *Tu quoque.* So the creationist's only obligation is to provide an interpretation of the data that coheres with his assumptions.

This antirealist attitude, really a *reductio ad absurdum* of presuppositionalism, contradicts both the letter and the spirit of *Darwin on Trial.* Phil argues that Darwinists have illegitimately allowed materialist assumptions to distort their assessment of the evidence for Darwinism. This means that contaminating evidence with unwarranted assumptions, while common, is an intellectual *vice,* not a necessity and certainly not a virtue. If it's bad for Darwinists, then it's bad for creationists and everyone else as well.

When someone has solid independent evidence for a proposition, they talk about evidence. Abstruse claims about the omnipotence of presuppositions should be an indication that the evidence is weak. The young-earth-creationist claims in my second list are much weaker evidentially than the more general claims in the first list. And even many who believe the things in list two concede this point. What this means is that as long as the propositions in the first list are mixed with the propositions in the second, probably none of them has much chance of gaining traction. I suspect that many intelligent design (ID) critics understand this quite well. That's why they refuse to keep these matters separate, refer to intelligent design (and even mere critiques of Darwinism) as "intelligent design creationism," and always bring up religious motivations and the imminent dangers of theocracy. They *want* the arguments against the power of the Darwinian mechanism and the cogency of materialism, and arguments for design, to suffer guilt by association with logically unrelated and much less persuasive arguments about the youthful age of the universe and the explanatory power of a single worldwide flood.

It should now be clear what Phil did in *Darwin on Trial.* He exercised *wisdom.* Part of wisdom consists in discernment—the ability to separate the chaff from the wheat, the noise from the signal, the plausible from the implausible, the central from the tangential. In *Darwin on Trial* Phil focused on the very issues that were not only the most empirically tractable and defensible but also the most philosophically important. This allowed me and many others to see what was really at stake.

3. Phil avoided the one-man show syndrome. This is a subtle point. But I think it's important nonetheless. *Darwin on Trial* appeared at a particular time, within a particular context. It was written in English, in the United States, in 1991. It had to work against the firmly entrenched Scopes Monkey Trial stereotype, which prevented rational, public debates about the merits of Darwinism. And it had a certain base of evidence. The evidence that was available in 1991 was quite different from the evidence available in, say, 1859, 1925 or 1953. If it had been written at one of these earlier periods, it would have had a completely different effect. As it was, the evidence against the grand Darwinian story had begun to stack up, little by little, more or less unannounced.

Moreover, when Phil wrote *Darwin on Trial*, there were other individuals outside the traditional creationist communities who were deeply dissatisfied not only with Darwinian orthodoxy but with scientific materialism more generally. (There were also a few within the creationist community who were looking for a different approach.) Some were already interested in design in disciplines other than biology. A brave few had already dared to publish some heretical ideas. Most important, I think, were *The Mystery of Life's Origins* by Charles Thaxton, Walter Bradley and Roger Olsen, and *Evolution: A Theory in Crisis* by Michael Denton. These had less public impact than *Darwin on Trial,* but they helped fertilize the soil.

Many others who were future advocates of intelligent design (ID) were in graduate school or teaching at universities in places like eastern Pennsylvania. They had begun to think impure thoughts. Many had only begun to realize that there were others, perhaps many others, who noticed the intellectual and evidential inadequacies of scientific materialism. Phil Johnson created rhetorical and intellectual territory for these individuals, territory that was risky and bound to receive heavy fire but still defensible.

And after the ID movement started to pick up steam, Phil did not sap its creative potential by seeking to institutionalize the ID movement around himself. Rather, he wanted to see a thousand flowers bloom and acted to further that vision. This was both strategically wise and spiritually mature. I'm convinced that that one unnoticed decision is partly responsible for the progress we have made in the last fourteen years.

Phil arrived on the scene at just the right time—or to put it differently, just in time—with the right message, framed in the right way, when the conditions were right for his message to resonate with others who needed a public voice. In a sense *Darwin on Trial* has much in common with that other book written some 130 years earlier in Victorian England, called *The Origin of Species.*

Putting Darwin on Trial

Phillip Johnson Transforms the Evolutionary Narrative

Thomas Woodward

—

As Phillip Johnson returned to Berkeley from England in the summer of 1988, he carried in his briefcase an "ugly duckling": an eighty-eight-page single-spaced manuscript, mounting a radical critique of Darwinian macro-evolution. This initial distillation of his intellectual doubts was destined to evolve (through several stages of intelligent mutation) into *Darwin on Trial*. Because of the strategic importance of this one book in the development of the intelligent design (ID) movement, it invites intense scrutiny by rhetoricians of science. Yet what is unique about the target of such a study is that *Darwin on Trial* itself bears telltale qualifying marks to be counted as a study within the communication field known as "rhetoric of science." That is, Johnson's work is itself an undisguised "rhetorical analysis" of the scientific claims of neo-Darwinism. As a member of the rhetoric of science research community and in the context of a doctoral case study on the rhetoric of the entire ID movement, I set out to analyze Johnson and his first book. What struck me early on (and totally unexpectedly) was the decisive role of narrative reasoning that empowered Johnson's critique, both within the book and in the broader historical context of how the study was provoked and how it was received by Darwinian critics before and after its publication. Thus my Johnsonian focus began to revolve around the role of stories, as seen in factual stories of all kinds and in other specialized stories containing a mixture of fact and faith, which rhetoricians call fantasy themes—a term I changed to "projection themes" in *Doubts About Darwin*.

As I pointed out throughout my study, both kinds of stories—the factual sort and the projection themes—are constantly being deployed by both ID theorists and their Darwinian opponents. The key questions I raised focused on *how* and *why* these stories were being used, and specifically how they functioned to pivot the entire controversy away from an "Inherit the Wind" stereotype toward the consciousness of a possible paradigm shift in biological science. My purpose in this chapter is far narrower and is twofold: First, I want to briefly review the role of the historical narrative of Johnson's earliest work, in the period that led up to the publication of *Darwin on Trial*. I intend to show how Johnson's interactions with key evolutionists both provoked and encouraged him in his writing project. Second, I turn to the book itself, reviewing its four theses, showing how Johnson faced key obstacles, and finally sketching the narrative strategies he used to advance his critique.

Encouragement, Correction, Hostility

The literary reshaping of Johnson's paper that he brought from England in 1988 was not done in an ivory-tower vacuum. Johnson repeatedly sought out response and criticism from others in academia, especially evolutionary biologists. By the time of his return to teaching duties in California, Colin Patterson of the British Museum had already read and commented on several drafts, checking for scientific accuracy, and both Michael Ruse and William Provine (researchers who were well-known for their vigorous defense of Darwinism) had read and panned the paper in their letters to Johnson.[1] Among his most important and vigorous responders in those early days were a group of twenty Berkeley professors who gathered in September 1988 to discuss the original paper in a colloquium format. This remarkable encounter—retold in detail in chapter four of *Doubts About Darwin*—served as a *public debut* of his major theses, which would prove to remain substantially constant during the book's evolution. It was also a decisive test of his supporting arguments. (This crucial element of persuasion is called *logos* in rhetorical theory.) These logos arguments had to be both robust and cumulatively impressive in order to advance the plausibility of Johnson's radical claim in the colloquium: Macroevolutionary change, generally assumed as overwhelmingly "factual" in the intellectual world, is actually an "illusion," a belief built upon a background certainty of metaphysical naturalism.

Johnson's use of what I call a stigma word, in this case *illusion,* followed a similar practice of Michael Denton in labeling macroevolution the "cosmogenic myth of the twentieth century."[2] In the Berkeley colloquium the opposition returned the stigma favor. As Johnson attacked the assumption that naturalistic explanations would ultimately be found for evolutionary puzzles, one of his critics observed that such a stance was one that embraced "magic." Johnson objected to this rhetorical twist (which implied he was being irrational) and pressed harder for the intellectual foundations of Darwinian confidence. The most revealing responses to his paper in the colloquium were (a) the repeated citing of microevolutionary evidence to make good on the claim of macroevolution, and (b) the assertion that naturalism is indeed the *only rational foundation* for biological investigations of the origin of life's diversity and complexity. One of the most remarkable outcomes of this meeting (besides the steady circulating of the paper to ever-wider circles in academia) was the ongoing correspondence between Johnson and two of the professors—Jeremy Waldron and Montgomery Slatkin—who attended the symposium.

In this early stage of circulating copies of his paper to wider circles of scientific critics, there looms one other huge milestone: the private 1989 meeting of a dozen scholars at the Campion Center in Boston. Again, for details I will refer the reader to chapter four of *Doubts About Darwin,* but one highlight stands out: an impromptu debate with Stephen Jay Gould. To appreciate the debate, one must grasp the surprising (though quiet) certification of empirical accuracy that immediately preceded it.

The unexpected defender of Johnson here, shortly before the debate erupted, was David Raup, a well-known evolutionary paleontologist with a reputation of brutal honesty about empirical gaps in the neo-Darwinian scenario. Raup had already read Johnson's original Berkeley paper and had used it in a graduate seminar at the University of Chicago. He and his students had found no factual errors as they reviewed the paper. As an open-minded scientist, he came to respect Johnson's scholarship, although he was not persuaded to abandon hope that evolutionary explanations would ultimately be found for the nagging anomalies. In fact, Raup briefly described his view of Johnson's critique in a phone conversation with me in the fall of 2000, while recalling the Campion events: "Johnson's work is very good scholarship, and of course, this is widely denied. He cannot be faulted; he did his homework, and he understands 99 percent of evolutionary biology."[3]

The details of Raup's appreciative comments to the assembled group that Saturday, as Johnson wound up a review of his research, will never be precisely known. Yet his receptive attitude was enough to signal to the group that Johnson's logos was virtually error free, and his ethos (perceived qualities and competencies of a rhetor) was solid, if not unassailable. Raup's interjection seems to have provoked his long-time friend Stephen Jay Gould to launch a verbally intense, argumentative assault on Johnson. His line of argument, in effect, retreated from his previous outspoken criticism of evolutionary gradualism, *into a defense of classic neo-Darwinism.* This surprised several in the audience. Yet Johnson was not surprised at the attack. Gould had told him that morning after shaking hands, "You're a creationist, and I've got to stop you."

This now-legendary attack turned into a debate before a mesmerized group; it was fast and furious; it seesawed and ultimately came to a draw, from which Johnson took consolation. The significance of Gould's attempted crushing of the upstart critic from Berkeley can be seen at several levels: (1) the importance of the perceived danger in Johnson and his critique, (2) the extreme contrast between Gould's claim of Johnson's ignorance and incompetence on the one hand, and Raup's assurance that at the level of empirical fact, Johnson could not be faulted, (3) further understanding of how the counterattack on Johnson's theses would be waged in the future, and (4) the satisfaction and encouragement in Johnson's holding one of the world's most prestigious evolutionists to a draw—and this after only two years of active research into the subject.

As Johnson continued to work on the book draft in the year that followed (1990), he took many more opportunities to expose his ideas and arguments to the strongest possible critique from unfriendly and friendly readers alike. These stories, unfolding one after another during the birthing of the original paper and the multiple baptisms in fiery criticism, constituted a large part of the book's ultimate aura of importance and legitimacy. In June of 1991 the long-awaited book appeared, and the pace of Johnson's advocacy suddenly jumped to a new level of energy. His "skeptic's story" shifted from hopeful expectancy to published reality.

The Nature, Theses and Strategy of *Darwin on Trial*

The pivotal place of *Darwin on Trial* in the history of intelligent design can hardly be overstated. This book was neither just Phillip Johnson's commentary on the flaws of contemporary biology nor merely a much-anticipated overhaul of his Berkeley paper. It was nothing less than an intellectually savage manifesto designed to overwhelm the opposition, to expose Darwinism as, in Johnson's stigma word, a *pseudoscience.*[4]

Another stigma word, placed like a stinger at the end of his book's purpose statement, hints at the severity of his critique:

> My purpose is to examine the scientific evidence on its own terms, being careful to distinguish the evidence itself from any religious or philosophical bias that might distort our interpretation of that evidence. I assume that the creation-scientists are biased by their precommitment to Biblical fundamentalism, and I will have very little to say about their position. The question I want to investigate is whether Darwinism is based upon a fair assessment of the scientific evidence, or *whether it is another kind of fundamentalism.* (emphasis added)[5]

By the end of the book Johnson's answer is clear. He has indeed symbolically consigned Darwinism to the stigma realm of fundamentalism. The new perspective—incongruous to say the least—was deliberately provocative, even shocking.[6] It pictures Johnson as standing between and above the two fundamentalist perspectives, equidistant and critically detached from both religious and Darwinian fundamentalists.[7] This implicit positioning of Johnson in ideological space, in turn, was absorbed into ID's self-image and became a significant part of its developing rhetorical vision.

Blurbs on a book cover often convey the vitality and energy of the book's message, and no reviewer better captured Johnson's aggressive persuasion than Michael Denton:

> *Darwin on Trial* is unquestionably the best critique of Darwinism I have ever read. Professor Johnson combines a broad knowledge of biology with the incisive logic of a leading legal scholar to deliver a brilliant and devastating attack on the whole edifice of Darwinian belief. There is no doubt that this book will prove a severe embarrassment to the Darwinian establishment.[8]

Perhaps no biologist felt the book's engaging energy and sensed the imminent danger more than Stephen Jay Gould, who lashed back a year after its publication in a vitriolic and lengthy rebuttal. Gould's four-page review, appearing in *Scientific American,* charged that Johnson had produced a "very bad book" that "hardly deserves to be called a book at all."[9] Clearly, Gould's purpose in this attack was not so much to review Johnson's book or even to defend macroevolution. His goal was primarily *personal*—to thoroughly demolish Johnson's credibility (ethos) as a competent commentator on Darwinism. In a verbal bludgeoning, which was rare for Gould, he strove to convey the exact opposite of David Raup's courageous endorsement of

Johnson's scientific accuracy and quality, calling *Darwin on Trial* a "clumsy, repetitious abstract argument with no weighing of evidence, no careful reading of literature on all sides, no full citation of sources."

Gould's eighth paragraph began with an assault of *dismissive contempt—* a pithy rhetorical blow calculated to scorch and flatten Johnson: "The book, in short, is full of errors, badly argued, based on false criteria, and abysmally written." Gould then launched a meandering excursus of over two thousand words—a series of mini-essays on Johnson's alleged errors and misunderstandings.[10] Gould's attack was followed by Johnson's unsuccessful attempt to persuade *Scientific American* to let him publish a reply. Despite this setback, Johnson saw the whole incident as a major step forward, and he quickly added an epilogue to *Darwin on Trial* (which first appeared in the 1993 revised edition by InterVarsity Press) that described the clash, and responded point by point to every one of Gould's criticisms. (He pleaded guilty to only one error in a footnote.)

Gould concluded his 1992 diatribe with rebukes reminiscent of a biblical prophet. Invoking powerful "Inherit the Wind" fantasies, he thundered, "Johnson's *grandiose claims,* backed by such poor support in fact and argument, recall a variety of phrases from a mutually favorite source: 'He that troubleth his own house shall inherit the wind' (Proverbs 11:29, and source for the famous play that dramatized the Scopes trial); 'They have sown the wind, and they shall reap the whirlwind' (Hosea 8:7)."[11]

What were the "grandiose claims" Gould sought to annihilate? The delineation and rhetorical analysis of these claims and their persuasive presentation by the frames and tools of narrative is the main purpose of this chapter. Even though Johnson's book was half the length of Denton's book *Evolution: A Theory in Crisis,* his themes, critique and rhetorical strategy were similar. In fact, Denton and Johnson had over two years of friendly correspondence by the time *Darwin on Trial* was published. At one point Denton visited Johnson in Berkeley for several days, and as they hiked along Johnson's favorite nature trails near San Francisco, Denton gave him advice on confronting the bitter rhetorical onslaught that awaited him.[12]

Like Denton, Johnson affirmed Darwinian *microevolution* as respectable science, but he relentlessly attacked *macroevolution* as a thoroughly counterfactual enterprise. Here Johnson and Denton share a common negative thesis—non-Darwinian (or anti-Darwinian) inferences spring naturally from the relevant scientific data. Where Johnson differs from Denton is in his incorporation and extensive development of *three additional theses*. Let me list all four:

1. Johnson's negative thesis (shared with Denton) can be restated: *Biological and paleontological evidences and other scientific data, with very few exceptions, tend to falsify the Darwinian story of macroevolution and its chemical origin-of-life prelude.* I call this thesis, focused on scientific evidence, T-1.

2. An additional thesis (not shared with Denton) asserts: *Darwinian macroevolution, as a comprehensive truth claim, is ultimately grounded on the philosophical assumption of naturalism.* Johnson describes naturalism (or materialism) as a philosophy that "assumes the entire realm of nature to be a closed system of material causes and effects, which cannot be influenced by anything from 'outside.' "[13] This thesis, focused on the philosophical base of Darwinism, I call T-2.

3. A second additional thesis, more in the vein of rhetorical analysis, is a charge of "shoddy rhetoric": *When Darwinism is brought into question, it is routinely protected by empty labels, semantic manipulations and faulty logic.* This is an underappreciated thrust of Johnson's persuasion, yet it is quite prominent throughout *Darwin on Trial*. I label this thesis T-3.

4. Johnson's final thesis is stunningly radical and also a major conclusion drawn from the other three: *Therefore, Darwinism functions as the central cosmological myth of modern culture—as the centerpiece of a quasi-religious system that is known to be true a priori rather than as a scientific hypothesis that must submit to rigorous testing.* The religious-mythological functions of Darwinism and its corollary of shying away from testing are major recurring themes of *Darwin on Trial*. I label this thesis T-4.

I will return to these four theses periodically as I flesh out the strategy of the book. But first I need to probe Johnson's overarching purpose, and to do that it will help to connect *Darwin on Trial* to its historical context. There had developed a quietly deteriorating rhetorical situation within evolutionary biology in the latter half of the twentieth century. A simmering tension arose within the evolutionary consensus. This was stirred by the annoying stream of questions, puzzles, anomalies and murmurs of dissent against the reigning orthodoxy. In 1985-1986 Denton's critique was published in Britain and the United States, and he symbolically captured much of this stream of dissent and portrayed its flow as a whitewater narrative of an incipient paradigm crisis. Denton's massive tributary was joined in the late 1980s by other rhetorical rivulets, especially Thaxton, Bradley and Olsen *(Mystery of Life's Origin)*, and Johnson himself *(Evolution as Dogma)*, although by this time many others had contributed (Norman Macbeth, Pierre Grassé, Hubert Yockey, Michael Pitman, Robert Augos and George Stanciu, Thomas Bethell, and Robert Shapiro).[14]

By 1991 this river of rhetoric had carved out a proactive nucleus of several dozen American dissenters, all of whom were skeptical of Darwinism (agreeing with T-1) but also were dissatisfied with the approach of scientific creationism. Under the leadership of Thaxton and Johnson, this group of skeptics wielded *Darwin on Trial* as their new weapon of choice as they enlisted in a new campaign to bring into question *the most powerful scientific paradigm of all time.*

I realize that my italicized description above can be contested. Some may say that Copernicus, Newton or Einstein produced paradigms of greater import. In terms of a paradigm's effect in shaping metaphysics, I would reply that other paradigms may have indirect materialistic or antitheistic implications (e.g., a decentralization of the earth or the introduction of mechanism as the core notion of reality), but these paradigms imply no frontal assault on the intelligent design of nature's most complex systems. Darwinism clearly *does.* I agree with Richard Dawkins in *The Blind Watchmaker* when he observes that objects in space such as stars or planets, being relatively simpler objects, do not logically suggest an intelligent-type explanation for their existence. On the other hand, biological entities *do suggest* such explanation due to their watchlike complexity. Darwin's solution to this old riddle is the *most dramatic answer* to the universe's seeming (even to Dawkins!) implicit suggestions of theism. I heartily endorse Dawkins's dictum: "Darwin made it possible to be an intellectually fulfilled atheist."[15]

Thus ID's rhetorical vision came to include a central theme of an ultimate paradigm replacement of design over Darwinism. I suggest that this dramatic vision, which fueled the skeptics' efforts, was driven by its two key qualities of *cultural centrality* and *dramatic supremacy.* The first idea—that "a reliable creation narrative is *culturally central*"—is not controversial. Most observers recognize the function of naturalistic macroevolution as a keystone of the Western worldview and its myriad secularized subcultures and socioeconomic practices. Implicit recognition of this centrality is revealed in Steven Jay Gould's vehement and trembling reaction to Johnson's thesis at Campion.

If this kingpin of truth were to erode or be dislodged, the impact on the world's intellectual and social history would be extensive and even unprecedented, compared with earlier paradigm shifts. This expectation is the essence of the second key quality, *dramatic supremacy*.[16] Part of the supreme drama is that such a change would almost certainly mark a major reversal in the intellectual (and hence institutional) marginalization of theism that has proceeded unchecked in the last two centuries of "scientific enlightenment." Recall Gould's pointed summation of his own motives when he first met Johnson in December 1989. He said to Johnson, "You are a creationist, and I've got to stop you." Clearly a great deal more is at stake than the objective truth about creation, evolution or design, and all of this comes to be implied or embodied in the notion of a paradigm shift and revolution.[17]

Related to dramatic supremacy and cultural centrality is ID's driving motivation—*reformative urgency*. Johnson wrote in 1990 to Notre Dame philosopher Alvin Plantinga, "My goal . . . is to legitimate the critique of Darwinism and its philosophical background so that others, particularly in the scientific world, can step forward and finish the job."[18] This is an early glimmer of what later became articulated and symbolized as Johnson's *wedge* strategy, in which he sees himself functioning as the "thin edge of the wedge" with other ID scholars following in his wake, extending and deepening the critique.[19] Yet Johnson has frequently said that his even bigger long-range goal is to help open up higher education *in all spheres* to the possibility that scientific materialism may not be true, and thereby to "legitimate a theistic perspective in the universities."[20] This global dimension of the critique is a key to understanding the expansion of the rhetorical vision of intelligent design during the 1990s.

Rhetorical gusts and gales whip briskly at exactly this point. Defenders of the academic status quo who read these words are surely tempted to protest what they perceive as the dark political motives of Johnson and his theistic cohorts to legitimate a theistic perspective. They may suspect an effort to take over university education, which they believe would stifle and even devastate the sciences.[21] Theistic advocates of intelligent design would reply that the issue is not the religious party *taking control* but rather being given a chair at the intellectual table after being shut out in a long exile.

Perhaps the common ground that fair-minded persons on both sides can press toward at this point is the establishing of *pluralistic forums for unfettered rhetoric*—occasions for reasoned discussion between opposing points of view with all worldview assumptions placed on the table for inspection. In fact, such rhetoric of metaphysical engagement and negotiation began in the wake of *Darwin on Trial,* as seen in several interdisciplinary conferences held at American universities to discuss Johnson's shocking theses. The stories of these conferences have enhanced ID's self-image and its positive image in the public and on university campuses. They have also shaped and matured its rhetorical vision.[22]

Because most change in America's science-education system (in terms of content—what shall be taught) trickles down from the universities rather than going the other direction, Johnson primarily targeted his manifesto at what he calls the "high university world."[23] In this specialized world a theistic perspective is generally seen as having very little if anything to contribute to intellectual discourse. Theism is viewed as a personal quirk or preference—tolerable as long as it stays a private matter.[24] In Johnson's view, most intellectuals in universities like Harvard and Yale (both of which have divinity schools) actually possess very little respect for theology as a legitimate academic field.

Furthermore, if a Christian professor working in a scientific field wishes to integrate his research with his Christian perspective, says Johnson, he typically tracks the secular thinking in his area and adds some "excusable god-talk"[25] but rarely challenges naturalistic assumptions. Of course, university professors in the arts and literature as well as in social sciences and hard sciences are free in the West to hold theistic beliefs and function as members of a religious community both on and off campus. However, according to intelligent-design theorists, *these same theists often do not have the freedom to connect their academic work with their deepest convictions about the existence of a deity who transcends and rules the universe.*[26]

In my judgment, this last statement not only is an accurate depiction of Johnson's view of academia, it also corresponds generally to the *social reality* of Western universities at the start of the twenty-first century. Theology, as a system of belief or of rational investigation, holds only the most tenuous and marginal place in Western universities. Primarily, its visibility in that arena is only as an object of study—a persistent sociocultural phenomenon to be analyzed. For example, I was amazed once to hear a brilliant rhetorician whom I respect very highly describe the issue of God's existence as a nonrhetorical issue, implying that it is a purely subjective (that is, nonrational) issue, one that cannot really be argued at all.[27] Furthermore, if this situation implies certain unwritten rules about epistemic constraints in academic discourse, then *Darwin on Trial* constituted Johnson's massive act of dissent against those rules and their underlying intellectual rationale. As a manifesto it went well beyond the point of "comprehensive critique" of the dominant paradigm. It constituted a brief for the rejection of the status quo, namely, *submitting to naturalism as an inviolable starting point of all research and teaching in academia.*

Johnson's new path for biology was not as clear. *Darwin on Trial* itself did not present a well-developed case for the consideration of intelligent causes. As I said earlier, this second stage—the explicit case for design— began in 1996 with the proliferation of the arguments of Michael Behe, William Dembski and many others.[28] *Darwin on Trial* prepared the way for those later discourses by plowing the ground and sowing doubt about the claims of Darwinism. It heightened the existing tension and encouraged readers to initiate their own reassessment.[29] With these strategic purposes in view, Johnson aimed at a target audience that was similar to Denton's—a system of concentric circles with scientists at the center, ranging out to university graduates and the general lay public with reading skills to handle Johnson's moderately advanced vocabulary.[30]

Johnson's overall rhetorical strategy was not dependent solely on his writing. Rather, he placed a high priority on his public speaking as a key to legitimizing his critique. Immediately after the publication of *Darwin on Trial* and in the fifteen years since, he took well over one hundred opportunities to speak in college or university venues.[31] One of the goals of his public speaking was causing students and professors to think, *There is more to this issue than I realized,*[32] and thereby generate sufficient curiosity for them to read his critique or at least engage the issues he was raising. As Johnson crafted *Darwin on Trial* to "arouse from dogmatic slumber," he realized that two formidable obstacles stood in his way, and he had to devise a plan to overcome them.

Johnson's Obstacles: Creationism and Competence

Phillip Johnson's critique did not sail into a rhetorical vacuum; rather, it had to lurch over smoke-filled battlefields pockmarked with craters left by attacks and counterattacks between creation scientists and the defenders of evolution. As a result, the atmosphere around *Darwin on Trial* was thick with negative emotions—many of them deep and bitter.[33] Johnson himself acknowledged to a colleague, "The great problem is in the sciences. The evolutionary biologists are bitterly hostile, as might be expected, and other scientists are reluctant to go out on a limb on something outside their field."[34]

By 1991 there had built up an acrid and enduring cluster of feelings, attitudes, and emotions that surged within evolutionary biologists as they encountered the Genesis-based creationism that dominated the earlier debate. Evolutionists' hostility ranged on the mild end from amusement, derision, disapproval, frustration and nervousness to the more severe reactions on the other end such as disgust, anger, deep resentment and loathing.[35]

Undoubtedly this emotional sensitivity to creationism produced a major rhetorical obstacle, especially among readers with strong evolutionary convictions. Johnson had to forestall an early rejection or even an automatic dismissal before he could gain a hearing. Scientists who harbored such negative attitudes often perceived in Johnson's sharp attack on Darwinism that his critique was cut from the same cloth as "creation science." This would quickly shut down a hearing.

The situation was complicated by the fact that Johnson never tried to hide his own metaphysical assumptions as a Christian, including his belief in a basic (but nonfundamentalist) idea of creation—that the biosphere owes its existence ultimately to the will and action of the Creator. Because of this acknowledged bias, Johnson had to brace himself for the charge that he was religiously driven, thus incapable of dealing objectively with the evidence.[36] Johnson used a simple plan to surmount this "creationist obstacle." Primarily, he made a number of direct statements (as well as verbal clues) that distanced him from Genesis literalism and pictured the speed and timing of creation as less important issues that distract from the main questions.

In addition to the possible perception of creationism, a second obstacle loomed—questions about the competence of a legal scholar in criticizing an area of science outside his own field of criminal law. A common question in media interviews with Johnson was, What is a law professor doing, writing about problems in evolutionary biology?[37]

Johnson confronted this "competence obstacle" both with explicit remarks addressing this concern and with scholarly qualities within the book that project competence. The most persuasive strategy for establishing competence of the author lay within the substance of the book itself. Thus perceptions of Johnson's excellence as a critic are projected primarily by his apparent breadth and depth of research and his accuracy of understanding, especially when dealing with the scientific evidence. In addition to conveying the sense of "having done one's homework,"[38] the author who is criticizing a field other than his own is helped enormously if he displays the qualities of a nimble, robust intellect and a communicative vibrancy. By nearly unanimous consent, one of Johnson's strongest qualities as a rhetorician is his blend of logic, insight, use of metaphors and analogies, and wit, all of which project the wisdom, intellectual excitement, and verve in the author's mind and personality. Many reviewers, including several who were generally negative, commented positively on his vigorous writing style and clever turns of phrasing.[39]

Surmounting the two rhetorical obstacles is part of the strategy of *Darwin on Trial,* yet it is only preparatory for the meat of Johnson's argument. His critique was crafted not only as a case of persuasion by evidence but, more astonishingly, as a matrix of compelling stories.

Johnson's System of Narrative Arguments

Johnson's most compelling dimension of rhetoric is his intricate system of narratives, which entails the destruction of old stories and construction of new ones. Each narrative Johnson tells has a purpose that supports one or more of the theses outlined above (T-1 through T-4).[40] It is helpful here to note that there are three basic types of factual narrative in the scientific discourses about origins: (1) genesis or cosmological stories (what happened in the distant past), (2) history-of-science stories (how, over decades, have our insights and understanding of these matters been developed), and (3) personal or individual stories (what are the specifics of each person or event or finding that fed into the scientific saga of enlightenment). Working with these three basic types as a "story toolkit," I see within *Darwin on Trial* three veins of narrative argument.

One vein of Johnson's narrative argument works with the first type of story—the evolutionary cosmology. Johnson seeks to question, degrade or destroy that story. In other words, his argument is a shredding of the Darwinian genesis narrative. Let's call this the "antigenesis" genre of argument. Such a line of argument involves many coordinated acts, such as storytelling, the exposure of bad reasoning or the setting forth of evidence that plagues the orthodox story line of evolutionary descent. These credibility problems gradually overwhelm as they accumulate.[41]

In the place of the dissolved genesis story, Johnson puts almost *nothing*—that is, no detailed or specified cosmology at all. The reader is given only a vague or minimal replacement story line—a generalized act (creation)[42] and an agent (the Creator). In chapter one, when he identifies his own bias that a creating agent—God—exists, he states that in his investigation he has not prejudged whether that agent has created using evolutionary mechanisms or more direct, interventional methods. "I do not exempt myself from the general rule that bias must be acknowledged and examined. I am a philosophical theist and a Christian. I believe that a God exists who could create out of nothing if He wanted to do so, but who might have chosen to work through a natural evolutionary process instead."[43]

Eight chapters later he refers to his theistic cosmology in the most general of terms:

> Why not consider the possibility that life is what it so evidently seems to be, the product of *creative intelligence?* Science would not come to an end, because the task would remain of deciphering the languages in which genetic information is communicated, and in general finding out how the whole system *works.* What scientists would lose is not an inspiring research program, but the illusion of total mastery of nature. They would have to face the possibility that beyond the natural world there is further reality which transcends science.[44]

Johnson implies a tentative agnosticism over how creation happened. This view is portrayed as a proper, honorable and intellectually honest position given the present fragmentary state of knowledge. In Johnson's prototype of ID strategy, the timing of creation is not so much unknown as it is deliberately not discussed.[45]

A second vein of narrative argument is destruction of the triumphant history-of-science story in the particular arena he is targeting—evolutionary biology. I will refer to these narrative arguments, which invert and retell the history of evolutionary biology (HEB), as the anti-HEB type. (The prefix *anti* seems doubly appropriate since it can imply both "opposition" and "replacement.") To build these arguments, Johnson writes into his script two kinds of guest appearances: first, Darwinists whose statements and acts reveal the philosophical or religious foundations of their thought or who exhibit shoddy arguments (supporting theses T-2, T-3 and T-4), and second, respected scientists, such as Darwin himself or even Gould, who expressed doubts about some aspect of the orthodox theory or who alluded to serious problems in the scientific evidence (supporting T-1).

As some narrative arguments undermine the old HEB narrative, others build in its place a new history-of-science (HOS) narrative, which carves out a place for design theory. I will call this third type of argument new-HOS. In this revision or replacement of the prevailing history, normal roles are reversed. Darwinism becomes a dogma-driven, deteriorating paradigm. Within biology, the heroes of the new story are the scientists, like Michael Denton, who had the intellectual courage to raise radical questions and propose heretical ideas. Johnson inserts numerous other stories of skeptics of the modern synthesis in biology, such as French zoologist-critic Pierre Grassé, MIT mathematical-skeptic Murray Eden and Eden's fellow mathematicians at the gathering at the Wistar Institute in 1965. In the new-HOS narratives there are interwoven new goals and a projected vision: ID's proposed paradigm will be taken seriously and eventually triumph. As a result, this sort of "historical argument" has an intimate and dynamic connection with the more imaginative type of story—the projection themes, which mix fact with faith and picture ultimate paradigm triumph.[46]

We should note in passing that all three of Johnson's narrative arguments incorporate and are powered by numerous personal narratives—the third key story type in the toolkit of stories.[47] There are two kinds of personal stories. One revolves around a single individual whose actions or words (and reactions to them) either undermine old master narratives[48] or contribute to new ones. Another kind is an "incident narrative" that revolves not around a person but a single incident—a dramatic scene that has some effect on a master narrative but involves several persons whose interaction constructs the meaning. A good example of an incident narrative in *Darwin on Trial* is the 1981 British Museum brouhaha over unsettling captions placed in a new display about evolution. This one story with commentary takes up sixty percent of the text of chapter eleven![49]

In my writing on intelligent design, I have pictured the relationship between specific stories and a larger narrative that links them together as a tree with limbs jutting out. Another use of the tree metaphor is to see the tree as the scheme that organizes the whole "macronarrative" of Johnson's presentation of his case in *Darwin on Trial*. I am suggesting that the spectacle of a leading intellectual figure on the Berkeley campus, clothed and in his right mind, presenting publicly his case against the factual status of Darwinism, was widely perceived as a significant story in and of itself.[50] It is the rare spectacle of a scientific manifesto with the potential to do serious damage to the reigning orthodoxy. Johnson's inclusion of an epilogue in 1993 summarizing the response of leading Darwinists underscores the book as a notable "event."

Let me clarify what I mean about the book as a "story" itself. In *Darwin on Trial,* Johnson does not tell in any detail the story of how he was led to doubt Darwin. Here and there we pick up hints of that story, of course, but the actual biographical story of Johnson from 1987 to 1991 (which I told in chapter four of *Doubts About Darwin*) was one that lived in the rhetorical consciousness of the Ad Hoc Origins Committee, not in the pages of *Darwin on Trial.* Rather, Johnson's manifesto is a "trial story"—somewhat like the summarization of evidence and analysis coming from the lips of a prosecuting attorney before a vast grand jury. Johnson's book—with eight chapters summarizing the state of the scientific evidence and four chapters on the philosophical, religious and educational dimensions of the debate—is a running commentary on *why* and *how* he has come to doubt virtually everything the Darwinists are telling America about macroevolution. As the product of a nonfundamentalist intellectual, his book's body of argumentation counts as a provocative narrative event in the context of the long-running global debate over creation and evolution.

Darwin on Trial, then, is an exotic "tree" among the forest of other commentaries in the creation-evolution debate. In my analysis of *Darwin on Trial,* I occasionally apply the tree metaphor to the book itself as a massive *communication event.* Johnson's barrage of narratives, quotes and analyses constitute the dozens of tree limbs.

Science writers do not hesitate to tell stories; they are among the most important tools of their trade.[51] So it is not unusual for Johnson to build his case on narrative bedrock. In fact, in all his books he uses storytelling frequently with persuasive effect. His five other books after *Darwin on Trial* are heavily populated with fascinating and instructive stories from scientific, legal and educational controversies.[52]

The point, of course, is not the *quantity* of narrative discourse used by Johnson but *how his narrative arguments work*—how they advance his four main theses. Perhaps the central narrative question in *Darwin on Trial* is this: If the empirical evidence for undirected macroevolution is so unimpressive, then how, historically, did we arrive at this "confident conclusion" of Darwinian evolution, and what can be done to restore intellectual integrity?

By deploying eight chapters on evidentiary issues, two chapters on the philosophy of science (chaps. 9 and 12) and two on religious and educational matters (chaps. 10 and 11), Johnson provides a balanced and challenging *narrative answer* to that central question. The epilogue chapter, which discusses the range of academic criticism of the book (added in the 1993 rev. ed.) shows that Johnson's "new HOS" is indeed a story in the making. This new story, energized by a lone Berkeley questioner—a vigorous and persistent "seeker after truth"—has propelled the field of origins into a season of testing from which there is no turning back.

PART II

THE WEDGE AND ITS DESPISERS

5

DEALING WITH THE BACKLASH
AGAINST INTELLIGENT DESIGN

WILLIAM A. DEMBSKI

Ten years ago, the *Quarterly Review of Biology* (December 1995) gave the following plug to the book *Darwinism: Science or Philosophy?*

> The editors deserve credit for a very fair book. Without editorializing or bias, the book lets everyone have their say. . . . In fact, it has a nice tone of "give and take," mostly polite, but in places amusingly peppery. . . . Moreover, the book is a readable primer on scientific philosophy, and provides a relatively sophisticated and invigorating philosophical challenge.

It is a measure of the success of our movement that no biology journal would give our books such respectful treatment any longer.

Why is that? The stakes are now considerably higher. *Darwinism: Science or Philosophy?* is the proceedings of a symposium that took place at Southern Methodist University in the spring of 1992. The focus of that symposium was Phillip Johnson's then recently published book *Darwin on Trial.* At the time, Johnson was a novelty—a respected professor of criminal law at the University of California, Berkeley, who was raising doubts about evolution. All harmless, good fun, no doubt. And Berkeley has an illustrious history of harboring eccentrics, kooks and oddballs.

Ten years later, any amusement about Johnson's critique of Darwinism has long since vanished. All sides now realize that Johnson was, from the start, deadly earnest, not content merely to tweak Darwin's nose but intent, rather, on knocking him down for the ten count. Johnson is, after all, a lawyer, and lawyers think contests are not simply to be enjoyed but also to be won.

This has not set well with the academic community, which thrives on ir-resolution. I once discussed with some philosophers the difference between mathematics and philosophy. One philosopher remarked that whereas in mathematics one finds a problem and solves it, in philosophy one finds an itch and scratches it. It would have been one thing if Johnson had raised doubts about Darwinism and then gestured at some ways of supplementing or reinterpreting evolutionary theory to take the materialist edge off. But Johnson was convinced that Darwinism had become a corrupt ideology that was being enforced by a dogmatic and authoritarian scientific elite, and that the proper course of treatment for Darwinism was not refurbishment or ref-ormation but removal and replacement.

Thanks to Johnson, we now have a cultural, intellectual and scientific movement that gives voice in the academic world to multiple millions of people who find it plausible or even self-evident that the world and its living forms were brought about by a designing intelligence. That movement is now so effective that evolutionists have to spend a lot of time writing articles and even whole books attacking intelligent design (ID), and, in some cases, like Robert Pennock, they even make an academic career attacking it.

In contrast to the respectful review of *Darwinism: Science or Philoso-phy?* a decade ago, we now face an academic and scientific world that is increasingly hostile to intelligent design and that seeks to crush it rather than engage it as a serious intellectual project. This may seem unfair and mean-spirited, but let's admit that our aim, as proponents of intelligent de-sign, is to beat back naturalistic evolution and the scientific materialism that undergirds it to the Stone Age. Our opponents, therefore, are merely returning the favor.

We have this going for us, however, which the evolutionary naturalists don't, namely, the evidence and arguments are on our side. It's therefore to our advantage to discuss intelligent design and naturalistic evolution on their merits. Conversely, the other side needs to delegitimate the debate between intelligent design and naturalistic evolution, casting intelligent design as a pseudoscience and characterizing its significance purely in political and re-ligious terms. As a consequence, critics of intelligent design engage in all forms of character assassination, ad hominem attacks, guilt by association and demonization.

Indeed, evolutionists are increasingly outraged over intelligent design, es-pecially over its cultural and political inroads. To appreciate the extent of the outrage, check out the anti-ID blog at <www.pandasthumb.org>. Paul Myers, a contributor to that blog, captures the spirit of the day:

> The verdict is in. Intelligent Design creationism is a load of horseshit. What has happened is that the movement has made some inroads *solely* in the political and legal arenas, where the absence of a scientific basis for the belief is little handicap, and now scientists are rousing themselves to point out its glaring deficiencies. This is not a sign of its growing importance. It's a sign of growing corruption that demands a response. Read the books. Scientists are not coming out and saying that there is something to this intelligent design idea; they are announcing, with near unanimity, that it is worthless crap, junk that has no place in the lab or the schoolroom.

Myers also had some words for the Minnesota House lawmakers who recently amended the Minnesota Department of Education's social studies and science standards. The lawmakers amended the document to reflect the views of critics of evolution. Myers expressed his disapproval succinctly: "Bastards. Craven, ignorant, despicable bastards."

Granted, Myers will strike most outsiders as unbalanced. But, increasingly, respectable people and organizations are weighing in against intelligent design. The American Association for the Advancement of Science and the Society for Neuroscience have issued formal statements denouncing intelligent design. The president of the Institute for Religion in an Age of Science, Michael Cavanaugh, has now issued a formal warning about intelligent design, the wedge, and Seattle's Discovery Institute, urging that people take seriously the threat to education and democracy that these pose.

My favorite is Marshal Berman's December 2003 piece in the *American Biology Teacher* titled "Intelligent Design Creationism: A Threat to Society— Not Just Biology." The epigraph to that article is the well-known quote by Edmund Burke: "The only thing necessary for the triumph of evil is for good men to do nothing." Increasingly, design theorists and their program are regarded not merely as misguided and pseudoscientific but also as perverse and evil. In a quote widely attributed to Arthur Schopenhauer, "All truth passes through three stages: First it is ridiculed. Second, it is violently opposed. Third, it is accepted as being self-evident." There's no question that we've now entered Schopenhauer's second stage.

Scoring as the Key

Faced with increasing attacks by evolutionists, our natural tendency is to become defensive and to try to justify ourselves. We might even worry that perhaps there's something we've overlooked and that the evolutionists might have a point. After all, there's no question that these people are seriously worked up over intelligent design. Many of them even experience a deep moral revulsion at us and our program. What are we to make of this revulsion? I submit that we should not take it seriously.

Peter Medawar once remarked, "The intensity of a conviction that a hypothesis is true has no bearing over whether it is true or not." Humans are intensely moral creatures, and their moral sensibilities will find expression, especially in the areas that are most important to them. Consequently, we can expect materialistic scientists to react viscerally to intelligent design. But consider, on Darwinian grounds, moral sensibilities are not to be trusted, certainly not as a guide to truth. Nor are they to be trusted on Judeo-Christian grounds, according to which human corruption is pervasive and warps even our moral sensibilities. Far from becoming upset or defensive, let's take it as a recommendation of our program that people like Paul Myers are against us. Indeed, with enemies like these, we must be doing something right.

How then do we effectively handle the attacks and abuse that increasingly are being sent our way? A sports analogy, for me, captures the essential insight. Consider an athletic contest between two teams. For definiteness, let's say soccer. The other team is abusing your team, especially your star players. They're constantly talking trash, constantly trying to trip you up. When the referee isn't looking, expect a knee in the groin or an elbow in the eye. In response, you've got three options: (1) respond in kind, (2) complain to the referee, or (3) score. The first two options are dead ends. The third is supremely satisfying and moves the ID program forward. Some recent notable "scores" for the ID movement have been the PBS broadcast of *Unlocking the Mystery of Life,* the decision by the Ohio board of education to permit weaknesses and criticisms of evolutionary theory to be taught, and the publication of *The Privileged Planet* by Guillermo Gonzalez and Jay Richards.

We score whenever we do something decisive that develops our intellectual and scientific program or that increases our sphere of influence in the wider culture. The two are mutually reinforcing: as we develop our intellectual and scientific program, we make our program more winsome to the wider culture; alternatively, as our program gains ground in the wider culture, we attract talent that helps develop that program intellectually and scientifically.

There are temptations to get sidetracked. The biggest temptation by far is to get bogged down in a war of words with people who are sold out to the old way of thinking. As Thomas Kuhn clearly taught us, the old guard is not going to change its mind. By being wedded to a failing paradigm, they suffer from the misconceptions, blind spots and prejudices that invariably accrue to a dying system of thought. This, in turn, limits their usefulness as conversation partners. What's more, insofar as they regard intelligent design as evil, and therefore as something to be destroyed, they adopt a purely adversarial stance that short-circuits fruitful interchange. As Edward Sisson points out in a book I edited *(Uncommon Dissent),* "A psychology I see everyday in litigation is that opposing lawyers are primed to reject every statement by the other side because there is no advantage to considering that the statements might be true. I also see that psychology again and again within institutional science in the debate over the origin and subsequent diversification of life."

I've witnessed this psychology in the attacks on my own work and that of my colleagues. By any objective standards, the principal players in the ID movement are reasonably intelligent people. Phillip Johnson, for instance, graduated first in his law school class at the University of Chicago and clerked for Chief Justice Earl Warren. Jonathan Wells got double 800s on his SATs and was awarded a full, merit-based undergraduate scholarship at Princeton in the 1960s. Guillermo Gonzalez, though a young assistant professor, has over sixty articles in refereed astronomy and astrophysical journals. These are just a few examples off the top of my head. And yet, when critics attack our work on intelligent design, we seem to get nothing right. You'd think that somewhere, somehow we might make a valid point critical of evolutionary theory. You'd think that no scientific theory can be as good as evolution's defenders make it out to be. Alas, no, the design community is entirely misguided and confused in finding fault with evolution.

The evolutionists' reaction to Jonathan Wells's book *Icons of Evolution* exemplifies this psychology. In that book Wells analyzes ten "icons of evolution." The reason for calling them "icons of evolution" is that they are presented in high school and college biology textbooks as slam-dunk evidence for evolutionary theory. Included here are the Haeckel embryo drawings, the Miller-Urey experiment and changes in coloration of the peppered moth. In every instance, when these icons are carefully examined, they do not confirm evolutionary theory. Haeckel's embryo drawings, for instance, are now known to have been faked, a fact that has been widely admitted in the peer-reviewed biological literature. And yet these icons continue to appear in the textbooks. What has been the response of the evolutionary community? To issue an apology for misleading our young people? To fix the mistakes in the textbooks? No, but to cast Jonathan Wells as a lunatic. Go to the National Center of Science Education website, and you'll find that there is no admission that any of these icons represent a problem for evolutionary theory or should be corrected in the textbooks. Rather, the fault lies with Wells for inventing problems where there are no problems.

Two Policies

Our critics have, in effect, adopted a *zero-concession policy* toward intelligent design. According to this policy, absolutely nothing is to be conceded to intelligent design and its proponents. It is therefore futile to hope for concessions from critics. This is especially difficult for novices to accept. A bright young novice to this debate comes along, makes an otherwise persuasive argument and finds it immediately shot down. Substantive objections are bypassed. Irrelevancies are stressed. Tables are turned. Misrepresentations abound. His or her competence and expertise are belittled. The novice comes back, reframes the argument, clarifies key points, attempts to answer objections and encounters the same treatment. The problem is not with the argument but with the context of discourse in which the argument is made. The solution, therefore, is to change the context of discourse.

Hard-core critics who've adopted a zero-concession policy toward intelligent design are still worth engaging, but we need to control the terms of engagement. Whenever I engage them, the farthest thing from my mind is to convert them, to win them over, to appeal to their good will, to make my cause seem reasonable in their eyes. We need to set wishful thinking firmly to one side. The point is not to induce a cognitive shift in our critics, but instead to clarify our arguments, to address weaknesses in our own position, to identify areas requiring further work and study, and, perhaps most significantly, to appeal to the undecided middle that is watching this debate and trying to sort through the issues.

For now, evolutionists are sitting pretty. They hold the reigns of power in the academy, they control federal research funds, and they have unlimited access to the media. But, like English colonialists trying to keep a colony in check, they are in a distinct minority. A feature of colonialism is that colonists are always vastly outnumbered by the people they are controlling, and maintaining control depends on keeping the requisite power structures in place. The reason intelligent design has become such a threat is that it is giving the majority of Americans, who don't buy the atheistic picture of evolution peddled in all the textbooks, the tools with which to effectively challenge the evolutionists' power structures.

A critic reading the last paragraph will immediately retort that we are flattering ourselves, and that except for the burger-eating, fries-munching, coke-swilling moronic masses, who refuse to accept evolutionary theory for purely religious reasons and, to their further embarrassment, take the public policy recommendations of design theorists like me to heart (recommendations that, for instance, challenge the place of naturalistic evolution in the high school science curriculum), there is absolutely nothing of intellectual substance to intelligent design. Yes, it is a pernicious threat, but for all the wrong reasons.

Let's talk about this. Obviously, this criticism flows directly out of the zero-concession policy. Indeed, it is merely a restatement of the zero-concession policy. How shall we respond to it? The temptation here is to engage in a war of words, justify intelligent design, recapitulate its program, lay out its research agenda or perhaps even complain that the critic is being unfair. *Stop and think.* The critic will be satisfied at no point, deny every claim that supports intelligent design, ask for endless detail, throw in countless red herrings and, whenever possible, turn the tables and accuse you and your program of the very faults that you are raising against evolution. So our job is not to try to justify to such critics why intelligent design has a right to exist, but rather to justify to the outsiders listening in on our debate why intelligent design has more going for it than the hard-core critics are willing to concede.

The proper answer to the critics' zero-concession policy is therefore a "there might be something to it after all" policy. In other words, it is enough to indicate to nonpartisans listening to the debate that there's more going on here than meets the eye. Often it suffices to plant in the minds of nonpartisans a reasonable doubt suggesting that the critic's blanket dismissal of intelligent design is less than credible. Of course, this is not to sidestep the hard work of developing intelligent design as a rigorous intellectual and scientific program. That work must proceed. Rather, I'm talking about how we can make best use of our hard-core critics.

Putting Critics to Use

Critics and enemies are useful. The point is to use them effectively. In our case, this is remarkably easy to do. The reason is that our critics are so assured of themselves and of the rightness of their cause. As a result, they rush into print their latest pronouncements against intelligent design when more careful thought, or perhaps even silence, is called for. The Internet, especially now with its blogs (web logs), provides our critics with numerous opportunities for intemperate, indiscreet and ill-conceived attacks on intelligent design. These can be turned to our advantage, and I've done so on numerous occasions. I'm not going to give away all of my secrets, but one thing I sometimes do is post on the Web a chapter or section from a forthcoming book, let the critics descend, and then revise it so that what appears in book form preempts the critics' objections. An additional advantage with this approach is that I can cite the website on which the objections appear, which typically gives me the last word in the exchange. And even if the critics choose to revise the objections on their website, books are far more permanent and influential than webpages.

An illustration might be helpful here. As I was working on my book *No Free Lunch,* I wrote a section critical of Thomas Schneider's article "Evolution of Biological Information," which appeared in *Nucleic Acids Research.* I would have liked to get from Schneider a well-considered response to my criticisms. But with Darwin fish crawling over his website, I frankly doubted that he could serve as a fair-minded respondent. I therefore posted the relevant section on the science-religion listserv META, framing the discussion around some remarks on design by the German theologian Wolfhart Pannenberg. I posted my three-thousand-word critique one day. Wesley Elsberry immediately alerted Schneider. Schneider posted his rebuttal the following day. I love the Internet!

Schneider failed to address the substance of my critique but provided some useful details about his work that I was able to incorporate into my section. He also engaged in some hair-splitting that could only look ridiculous to outsider observers: no, he was not claiming evolution brought about biological complexity as a "free lunch," but, yes, he was claiming that it brought about biological complexity "from scratch"; no, he did not generally find fitness a useful concept, but, yes, his research did require error-counting functions that scored errors monotonically with respect to survivability (that sure sounds like a fitness function to me). This hair-splitting made it into my book and made for amusing reading, though not at my expense.

As far as possible, I try to steer the attacks of my critics by the judicious dissemination of information. This has the advantage that I know what to expect. Often, however, the attacks by our critics blindside us. A common feature is that they are vicious and personal. Sometimes they are written to employers to discredit us or even to destroy our reputations and careers. (I write from personal experience.) Our natural tendency in response to such attacks is to get upset and react in one of three ways, none of which is advised.

One reaction is to placate: *My, you really are angry. I must have done something wrong. Help me make things right with you.* Another reaction is to flee: *What did I get myself into? This kitchen is a lot hotter than I can stand. I don't need to be dealing with this level of conflict. Let me out of here.* Still another reaction is to fight: *You no good so-and-so. I'll show you. You want to play hardball? You came to the right place.* I submit that none of these reactions is helpful in advancing our cause. The hard-core critics with whom I regularly deal are intellectual bullies, and they don't deserve to be placated. What's more, they are not very frightening, especially when you get past their initial defenses, so there's no reason to flee.

Fighting, however, is not advised either. The problem with fighting is that it consumes valuable energies and is motivated by anger, which always distorts mental clarity and distracts from the real issues. As John Cassian noted over 1,500 years ago (in his discussion of the eight vices, as recorded in *The Philokalia*):

> No matter what provokes it, anger blinds the soul's eyes, preventing it from seeing the Sun of righteousness. Leaves, whether of gold or lead, placed over the eyes, obstruct the sight equally, for the value of the gold does not affect the blindness it produces. Similarly, anger, whether reasonable or unreasonable, obstructs our spiritual vision. Our incensive power can be used in a way that is according to nature only when turned against our own impassioned or self-indulgent thoughts.

So let's put anger aside. Let the other side fume with indignation. Indeed, many of them have turned indignation into a full-time occupation. From our vantage, however, we need to take their vitriol as par for the course. These valiant defenders of evolution are just that—valiant defenders. It would be unworthy of them not to use every means at their disposal to try to stop us. They are as committed to their program as we are to ours (sometimes I wonder if they are not more so).

Think of their no-concession policy in pure business terms. When, for instance, gas prices go down, we don't congratulate oil companies for their generosity. So, when gas prices go up, we would be out of line to accuse them of greed. Oil companies and the prices they charge are constrained by market forces. So too the market of ideas is constrained by ideational forces (especially the inertia of entrenched ideas), and our opponents are simply playing their part. I find this perspective freeing. Far from wanting to curl up in a corner when attacked, I'm grateful for my critics. Truth be known, their attacks are my idea of a good time. Indifference is a far worse form of violence.

The appropriate response to attacks by critics is to see the attacks as opportunities to advance our cause. Think of them as gifts. As a student of the Old Testament, I've always been fascinated with the Israelite conquest of the Promised Land. The pattern that kept repeating itself was this. The Israelites would approach a fortified city. Instead of entrenching themselves in their city and allowing their countryside to be ravaged, the inhabitants of the city would come out for battle. Once outside their positions of safety, however, they were fair game, and the Israelites were able to make short work of them. That's the pattern I see in this debate. The proponents of evolution would very much prefer to stay in their fortified positions. They don't want to dignify us by devoting time and energy to refute us. They would prefer to ignore us. They wish we would just go away. But the challenge to evolution in the schools and public square is real and threatens their monopoly. The unwashed masses are not with them. The evolutionists cannot leave these crazy design theorists unanswered. So, out they come from their positions of safety to challenge us. But, in the very challenge, they open evolutionary theory to a scrutiny it cannot withstand.

Richard Dawkins is a case in point. Dawkins refuses, as a matter of principle, to debate me and my colleagues because it would, in his view, dignify our position. Yet he cannot resist criticizing us in print. Notwithstanding, whenever he does so, he makes himself vulnerable. This was brought home to me in a foreword that Dawkins wrote for a recent book attacking intelligent design—Niall Shanks's *God, the Devil, and Darwin: A Critique of Intelligent Design Theory*. In that foreword, Dawkins asks who owns the argument from improbability. His answer: Not those crazy design theorists but evolutionists like himself. Thus, he writes, "Darwinian natural selection, which, contrary to a deplorably widespread misconception, is the very antithesis of a chance process, is the only known mechanism that is ultimately capable of generating improbable complexity out of simplicity." At the risk of immodesty, I'm the guy who wrote the book on the argument from improbability—it's called *The Design Inference: Eliminating Chance Through Small Probabilities*. Dawkins is a great popular science writer, and he is expert in certain aspects of biology, but he is a duffer when it comes to the argument from improbability. He's now on my turf, and I'm only too happy to instruct him.

Attacks Relating to Logos

Although the attacks against us by evolutionists can be turned to advantage, our success here is not assured and depends on us knowing what we're doing. Recall the soccer analogy. Our side is not the only one that's capable of scoring. The other side is quite capable of scoring as well. Our game plan, therefore, requires not just an effective offense (one that enables us to score points for our team) but also an effective defense (one that keeps the other side in check and hinders them from scoring points). To understand how to defend ourselves in this debate, we need first to understand the forms that the attacks take. They take three forms, corresponding to the three traditional aspects of rhetoric: logos, ethos and pathos.

Logos refers to the reasoned case that is being advanced. Think of it as the formal argument that can be written out on a sheet of paper. The identification of presuppositions, the marshaling of evidence and the drawing of inferences all fall under logos. The evolutionist threatens to score here by making an argument that is not adequately answered by our side. Whether the argument is sound or fallacious does not matter. The important thing is that the argument is allowed to stand and that nonpartisan bystanders think the argument raises a substantive difficulty for intelligent design. (Note that evolutionists are just as intent on winning the undecided middle as we are.) Next comes ethos. Ethos refers to the perceived character, integrity and accomplishment of the rhetor. Ethos is what inspires confidence and establishes credibility in the eyes of the audience. Conversely, it is what destroys confidence and erodes credibility. And finally there is pathos. Pathos refers to the emotion or passion that the rhetor is able to elicit from the audience. The rhetor may be able to play on the audience's heartstrings, thereby eliciting sympathy. Alternatively, the rhetor may inspire anger or fear in the audience. Pathos is especially important if the rhetor is attempting to get the audience to take action.

Evolutionists attack intelligent design by appealing to each of these three aspects of rhetoric. Accordingly, they attack intelligent design with respect to logos by claiming that science utterly fails to support it, whether on evidential or theoretical grounds. What's more, they attack intelligent design with respect to ethos by charging its proponents with being morally and intellectually deficient. Finally, they attack intelligent design with respect to pathos by instilling the fear that intelligent design means not just the end of science but also the end of rational discourse in a free and open society. Let's look at these attacks more closely, especially how to counter them.

Usually, in keeping with the no-concession policy, an attack relating to logos starts with some blanket dismissal such as "intelligent design offers no testable hypotheses" or "intelligent design is just an argument from ignorance" or "intelligent design is incoherent because of the poor design evident in biological systems." The first thing to do when confronted with such an attack is to ask for elaboration of the objection so that it's clear what exactly is under dispute. This can be quite illuminating. Take the objection that intelligent design offers no testable hypotheses. Implicit in this objection is that biological systems are inherently incapable of exhibiting any feature that could more adequately be explained as the result of intelligent design than material mechanisms. But if so, Darwinian and materialistic accounts of evolution become themselves untestable because, in that case, they trump design regardless of evidence. Once such objections are fully laid out, they often reveal a double standard and succumb to internal contradiction.

Not all objections to intelligent design fall this easily. Some require a more in-depth analysis. Take the trio of objections that constitutes Kenneth Miller's standard attack on intelligent design. Miller's main interest is in unseating Michael Behe and his notion of irreducible complexity. Behe argues that certain types of functionally integrated systems, those exhibiting irreducible complexity, resist Darwinian explanations. Miller argues that they don't. To make his case, Miller focuses on three points:

1. Because irreducibly complex systems invariably contain subsystems that are functional in their own right and therefore subject to natural selection, the Darwinian mechanism faces no obstacle in bringing about irreducible complexity.

2. Genetic knock-out experiments that disable a key component of an ir-

reducibly complex system and then successfully (re)evolve a substitute component that restores function support the evolution of irreducibly complex systems.

3. Evolution is a proven instrument for bringing about biological complexity, a fact that can be seen from biological structures that serve the same basic function but that exist at various levels of complexity (e.g., the eye in its many incarnations).

None of Miller's three points holds up under scrutiny. With regard to the first point, just because a functionally integrated system includes a subsystem that can be functional in its own right does not mean that the system evolved from the subsystem. To confirm the evolution of the subsystem into the system requires that a continuous sequence of functional intermediates be exhibited and that a nontelic process be specified that could plausibly connect the intermediates. Miller offers neither. With regard to the second point, whenever Miller cites such experiments, he fails to underscore that gene(s) coding for the substitute component were either already present or introduced by the experimenter. Far from showing how irreducibly complex systems might have evolved in the first place, these experiments at best show how sensitive such systems are to perturbation. With regard to the third point, Miller is presupposing precisely the point in question, namely, whether evolution, a materialistic form of it, can bring about biological complexity. Sample enough organisms, and you'll find structures in different states of complexity that perform the same basic function. But arranging such structures according to some similarity metric and then drawing arrows marking supposed evolutionary relationships does nothing to show whether these systems in fact evolved by material mechanisms. Similarity may suggest evolutionary relationships, but evolution is a process, and the evolutionary process connecting similar structures needs to be made explicit before the similarity can legitimately be ascribed to evolution. Miller's analysis never gets that far. He gestures at similarities but never demonstrates how evolution accounts for them.

When defending intelligent design with respect to logos, I can't overstress the importance of staying on topic. This is a nasty debate. For instance, one of my colleagues, who previously was involved with the abortion controversy and now works on public policy aspects of intelligent design, finds the level of hostility here even greater than with abortion. It's therefore tempting to respond in kind. Our work is not interpreted charitably, so let's not interpret our opponents' work charitably. They nitpick, so let's nitpick in turn. They capitalize on insignificant mistakes and oversights, so let's return the favor. Responding this way hurts us. We come across as churlish and catty. Precisely when the other side throws civility and courtesy to the wind is when we need to bend over backward to address any legitimate concerns that our opponents might be raising. This keeps us on topic and maintains our composure. This is important because maintaining composure under pressure is especially effective for establishing one's credibility.

What does it mean to stay on topic in this debate? The central question that must always be kept front and center in addressing intelligent design's critics is this: *Why might material mechanisms (such as Darwinian natural selection and random variation) lack the creative capacity to bring about the full complexity and diversity of living forms?* The materialist scientist resists this question (and I include here the scientist who is a religious believer but who thinks that science must understand the natural world entirely in terms of material processes that give no evidence of design). Indeed, from a materialist vantage point, what else could be responsible for life's complexity and diversity except material mechanisms? The materialist sees designing engineers as appearing only after evolution—a materialistic form of it—has run its course. That's why Daniel Hillis remarks, "There are only two ways we know of to make extremely complicated things, one is by engineering, and the other is evolution. And of the two, evolution will make the more complex." But whether a purely materialistic form of evolution is able to perform what otherwise would require superengineers to perform amazing feats of design is precisely the point at issue. What's more, unless we are able to press this point, the evolutionist will win by default, having defined science as the study of material processes that by logical necessity disqualify design and that, again by logical necessity, ensure that some materialistic account of evolution must be true.

Attacks Relating to Ethos

Let's turn to attacks against intelligent design with respect to ethos. Attacks here tend to focus on peripheral issues, such as whether design theorists have published their ideas in the right places, whether the scientific community is accepting intelligent design in sufficient numbers to render it credible, whether intelligent design is being unduly politicized, whether design theorists are religiously motivated and so on. Such questions are, to be sure, interesting, but they don't touch on the validity of intelligent design as an intellectual and scientific project, nor do they go to its truth. Nonetheless, such questions are important to people on the sidelines. We therefore need to make certain that we are not misrepresented here.

Take the question of peer review. It's certainly the case that intelligent design is a minority position and that it is only now beginning to gain a hearing in the mainstream, peer-reviewed literature. But our critics contend that intelligent design has no presence in the peer-reviewed literature whatsoever. For instance, Eugenie Scott at the National Center for Science Education claims that my book *The Design Inference* is not peer-reviewed. Nevertheless, that book appeared as part of a Cambridge University Press monograph series: Cambridge Studies in Probability, Induction, and Decision Theory. That series has an academic editorial board (which includes members of the National Academy of Sciences as well as one Nobel laureate), and my manuscript had to be passed on by three anonymous referees before Cambridge University Press could agree to publish the book in that monograph series. In a similar vein, Scott has disparaged my work by claiming that it is not favorably cited in the peer-reviewed literature. But mainstream mathematics and biology journals have cited my work favorably.

At the Design and Its Critics conference (Concordia University, Mequon, summer 2000), Kenneth Miller claimed that Michael Behe's notion of "irreducible complexity" was nowhere to be found in the mainstream peer-reviewed biological literature. Yet, in fact, Richard Thornhill and David Ussery had several months earlier published an article in the *Journal of Theoretical Biology* on that very topic. Currently, the most popular strategy for discrediting intelligent design with regard to peer review is to admit that it is represented in the peer-reviewed literature, but not in any literature that matters. Thus, in particular, it is claimed that design theorists are not publishing work that supports intelligent design in the peer-reviewed *biological* literature. But this claim too is false (the articles by Stephen Meyer, and Scott Minnich and Stephen Meyer in this volume being cases in point). Nevertheless, in keeping with their zero-concession policy, our critics won't concede that this claim is false. They can accept that the papers in question are by design theorists and that they appear in respectable, peer-reviewed biology journals. What they can't accept is that the papers *support* intelligent design.

This raises an important point. We are accustomed to think that what it means for data to count as evidence supporting a hypothesis is uncontroversial. But, in fact, it can be highly controversial. What it means for something to count as evidence is not itself decided by evidence. Rather, it depends on certain cognitive predispositions, and these are heavily influenced by our views on the ultimate nature of reality (metaphysics) and the scope of human knowing (epistemology). In particular, for the materialist, no facts of biology can count as evidence for intelligent design. Thus, when it is claimed that there are no articles supporting intelligent design in the peer-reviewed biology journals, it is appropriate to ask whether any data from biology could even in principle provide such evidence and, if so, what these data might look like. If the answer is that no data could even in principle provide support for intelligent design, then that's a good indication that our conversation has moved from biology and the natural sciences to epistemology and metaphysics.

In science there are no raw data. Data are always collected in light of background knowledge and assumptions. These condition the aspects of nature to which we attend and from which we collect our data. Once collected, we interpret these data. At one level of interpretation, we see facts. At a higher level of interpretation, we see patterns connecting these facts. At still higher levels of interpretation, we formulate hypotheses and theories to make sense of these patterns. It follows that as an inherently hermeneutical enterprise science can never guarantee consensus, especially at the higher levels of interpretation. More and more, critics of intelligent design are outraged by what they call "quote-mining." Accordingly, they fault design theorists for going to the biological literature to pull out quotes and ideas that support intelligent design. The critics are outraged because they see the design theorists as shamelessly exploiting the hard scientific work of others and interpreting it in ways that the scientists who originally did the work would reject. We have nothing to be ashamed of here. As Nobel laureate William Lawrence Bragg remarked, "The important thing in science is not so much to obtain new facts as to discover new ways of thinking about them." Intelligent design is doing just that—discovering new ways of thinking about and interpreting the well-established facts of science that pertain to biological complexity and diversity.

I've stressed that we need to clear up misrepresentations of our work by critics, and so we do. At the same time, we also need to clear up misrepresentations of evolutionary theory. It is, for instance, completely unacceptable that in edition after edition of high school biology textbooks, not only do the Haeckel embryo drawings get recycled but so does the misconception responsible for those drawings, namely, the mistaken idea that similar structures in the adult result from similar developmental pathways in the embryo. It is a fact of embryology that similar adult structures can arise via vastly different developmental pathways. With such misrepresentations—especially when they appear in textbooks, mislead our young people and are supported by our tax dollars—we need to hold the evolutionists' feet to the fire.

Nevertheless, in clearing up misrepresentations let's not become obsessive or pedantic. We don't need to respond to every misrepresentation that the other side makes. It's typically enough to respond to those that are troublesome to the undecided middle. And even here, let's be careful not to become defensive. In line with our "there might be something to it after all" policy, it's usually enough to indicate that there's more to the story than the other side lets on. John Angus Campbell puts it this way: *A draw is a win!* The other side wants to obliterate intelligent design. Yet to persuade the undecided middle, we just have to show that intelligent design has something going for it. As much as possible, therefore, let's always return to the main point at issue, which is that material mechanisms lack the creative capacity to bring about the complexity and diversity of living forms and that intelligent design is helping to elucidate this central issue in biology.

Attacks Relating to Pathos
Finally, let's turn to attacks against intelligent design that appeal to pathos. The strategy of the other side here is clear: induce fear and loathing of intelligent design in the undecided middle—fear that science and society will be subverted and loathing that intelligent design is just a tool for advancing religious and political extremism. By contrast, to promote intelligent design with regard to pathos, the most effective approach is to appeal to the undecided middle's sense of fairness and justice, especially its tendency to root for the underdog and its predilection for freedom of expression.

In practice, to induce fear and loathing of intelligent design, the other side invokes pejorative labels that are rich in negative associations. *Creationism* is by far the preferred pejorative, though *antievolution, antiscience, fundamentalism, right-wing extremism* and *pseudoscience* are great favorites as well. My advice is that as far as possible we resist being labeled. To do this effectively, however, it's not enough simply to deny a label. In fact, being too vocal and adamant about denying a label can be a good way of attaching it more firmly ("thou dost protest too much"). Denial works best if we are explicitly asked to comment on a label and then can explain why the label is inappropriate. For instance, most reporters who interview me ask how intelligent design differs from creationism. That gives me a perfect opening, and I can explain how intelligent design is not a religious doctrine about where everything came from but rather a scientific investigation into how patterns exhibited by finite arrangements of matter can signify intelligence.

Denial can also be effective when it's clear that a pejorative label has been attached maliciously or unfairly. Take Robert Pennock, for instance. Pennock prefers the term *intelligent design creationism* over *intelligent design*. His preference is his thing, and it doesn't do us any good to argue about it with him. Nonetheless, it's a different matter when he tries to force that label into our mouths. A few years ago he published a collection of essays titled *Intelligent Design Creationism and Its Critics*. When the book appeared, I was surprised to learn that I had two essays in it. Pennock, without my knowledge, had approached the publishers of those two essays and gotten their permission to reprint them. Yet, the fact that he went around my back to procure these essays was nowhere evident in the collection. Thus, when the book appeared, it would have seemed to the average reader that I had given my permission to have the essays appear in it. In fact, there is no way I would have given my permission with that title. Imagine if someone critical of Darwinian evolutionary theory decided to publish a book titled *Dogmatic Darwinian Fundamentalists and Their Critics,* managed to obtain copyright permissions for pieces by prominent Darwinists without their knowledge, and then situated their pieces within a collection of critical replies designed to make them look foolish. Substitute intelligent design for Darwinism, and that's what Pennock did. I pointed this out in a press release. Because the issue was one of fairness, public opinion went in my favor rather than Pennock's. Nonetheless, if I had simply complained about Pennock inaccurately labeling me a creationist, I would only have reinforced the label.

The best way to resist being labeled, however, is not by denying the labels but by developing our own vocabulary and ideas that set the agenda for the debate over biological origins. In this way, the other side is increasingly forced to engage us on our terms. Consider the following terms: (1)

irreducible complexity, (2) *specified complexity,* (3) *design inference,* (4) *explanatory filter* and (5) *empirical detectability of design.* The other side now spends an enormous amount of time discussing these terms and the ideas underlying them. Insofar as the other side engages us on our terms, it is in no position to label us. Of course, the other side sees this and therefore self-consciously makes a point of labeling us and our program. Labeling is therefore inevitable. Still, we do ourselves good service by, as much as possible, steering the discussion to matters of substance and away from labels. I've found that clarity and consistency in how we express our ideas is the best antidote to labeling by the other side. Increasingly, the media are grasping our ideas and expressing them not with tendentious labels but in our own words. For instance, the media now consistently refer to "intelligent design" and not to "intelligent design creationism."

Thus far, I've characterized labeling by the other side as a purely negative activity that needs to be resisted. Not all labels, however, have the intended negative effect. To be sure, some do. There's no way, for instance, to give the labels "antiscience" or "pseudoscience" a positive spin. But what about "antievolution" or "wedge"? The evolutionists who are our main critics think evolution is the greatest concept ever conceived. Daniel Dennett, for instance, in *Darwin's Dangerous Idea,* writes: "If I were to give an award for the single best idea anyone has ever had, I'd give it to Darwin, ahead of Newton and Einstein and everyone else. In a single stroke, the idea of evolution by natural selection unifies the realm of life, meaning, and purpose with the realm of space and time, cause and effect, mechanism and physical law." For most of the population, however, the term *evolution* holds no such positive associations. For most people, evolution is an implausible and controversy-riven theory of biological origins, one that gives comfort to atheists and undermines religious faith. To characterize intelligent design as a form of antievolution is therefore a positive advertisement in some circles.

Even so, there's an important clarification to keep in mind here. Intelligent design is antievolution not in the sense of rejecting all evolutionary change. Indeed, some design theorists, like Michael Behe, accept the universal common ancestry of all organisms. Rather, intelligent design is antievolution only in the limited sense that it regards blind material forces as inadequate for explaining all evolutionary change. Michael Behe and I debated Kenneth Miller and Robert Pennock at the American Museum of Natural History in the spring of 2002. The debate was initially titled "Blind Evolution or Intelligent Design?" Yet when the debate actually took place on April 23, the program bulletin that was distributed at the event quietly dropped the word "blind" and titled the debate simply "Evolution or Intelligent Design?" The original title was more accurate. Intelligent design is opposed to blind evolution, not to evolution simpliciter. But since by "evolution" our critics mean a blind form of it, that is, a form of evolution underwritten entirely by undirected material mechanisms, they label us as antievolutionists. It would be more accurate if they labeled us as being against blind evolution. But calling attention to the blindness, or absence of teleology, in the evolutionary process is clearly not in their interest.

The term *wedge* is also a mixed curse. Normally, when this term is cast our way, our opponents want to stress the political and religious dimensions of intelligent design—that its proponents are on a crusade to stamp out materialism and that evolution is the first item on their agenda. The consistent reaction by materialists to the wedge is therefore outrage. Here they are, these proponents of intelligent design, claiming to advance science, but in fact they are just pursuing an ideological agenda to which they themselves have attached the moniker *wedge*. Adopt a nonmaterialist perspective, however, and the situation looks quite different. From our vantage, materialism is not a neutral, value-free, minimalist position from which to pursue inquiry. Rather, it is itself an ideology with an agenda. What's more, it requires an evolutionary creation story to keep it afloat. On scientific grounds we regard that creation story to be false. Moreover, we regard the ideological agenda that has flowed from it to be destructive to rational discourse. Our concerns are therefore entirely parallel to the evolutionists'. Indeed, all the evolutionists' worst fears about what the world would be like if we succeed have, in our view, already been realized through the success of materialism and evolution. Hence, as a strategy for unseating materialism and evolution, the term *wedge* has come to denote an intellectual and cultural movement that many find congenial.

I want to make one last point about attacks against intelligent design that appeal to pathos. Normally, when intelligent design is attacked, the attackers are in positions of power and authority, and proponents of intelligent design are the underdogs. That's not always the case, however. The world of evangelical Christianity, for instance, seems to prefer intelligent design over theistic evolution. Theistic evolutionists, therefore, feel increasingly beleaguered among evangelical Christians. Thus, at a meeting of evangelical scientists a few years back (the American Scientific Affiliation meeting at Westmont College in the summer of 1997), an interesting reversal occurred. Phillip Johnson had been speaking, and Keith Miller, a theistic evolutionist (who recently edited *Perspectives on an Evolving Creation*), challenged him during the question-and-answer period. For several minutes, Miller read from notes and, in bullet-point fashion, listed the faults that he found in Johnson's program. There was no way, in the allotted time, for Johnson to respond adequately to Miller's many objections. Thus, after Miller finished, Johnson simply remarked that he and Miller saw things differently. At this, Miller burst into tears and ran out of the auditorium.

To Johnson's supporters, Miller's tears amounted to a histrionic display not worthy of reasoned discourse in an academic setting. Yet that misses the point—the appeal of tears is not to logos but to pathos. Moved by his tears, several members in the audience rallied around Miller to console him. Further, they cast Johnson as a villain. The lesson for us here is that when appealing to the undecided middle, don't allow our opponents to cast themselves as underdogs or intelligent design proponents as villains. I see a dynamic increasingly at work among theistic evolutionists, whose science, let us always bear in mind, is no different from that of a Richard Dawkins or a Stephen Jay Gould. Accordingly, they cast themselves as the kind face of religion, and they characterize intelligent design as theologically naive and misguided. Theistic evolutionists have now become marvelously adept at rationalizing not only how their religious faith makes sense in light of evolution but also how evolution enhances their religious faith. Let's not play this game. The issue for us is not how evolution relates to religious faith but whether evolution, as currently understood by science, is true. If, as we argue, it is not true, then exploring its religious ramifications constitutes a vain exercise.

The Principle of Radiating Confidence

In closing this essay, I want to speak to a topic I've already addressed in passing but now wish to focus on in earnest, namely, our composure under pressure. Years ago, as a teenager in the 1970s, I read a book titled *Winning Through Intimidation*. The title was slightly misleading because the point of the book was not to instruct readers on how to win by intimidating others. Rather, the point was to instruct readers on how to win by not allowing others to intimidate them. This was made somewhat clearer when the book was recently reprinted under the title *To Be or Not to Be Intimidated? That Is the Question.* (By the way, its author, John Ringer is himself a critic of Darwinism.) To this day, I love the book. In fact, I regard it as required reading for anyone who challenges evolution and has to deal with the inevitable backlash. Although the book offers many useful principles and insights (the "leapfrog principle" is especially important in our debate—i.e., don't let the other side define the rules by which you and your ideas are permitted to advance but instead leap right over such artificial impediments), its primary lesson is this: once your opponent has intimidated you and knows it, you've lost. In closing, therefore, I want reflect on how we can avoid being intimidated and maintain our composure in the face of evolutionist opposition.

Two extremes need to be avoided. On the one hand, we must refuse to allow evolutionists to send us cowering into a corner. Instead, we need to be mentally and emotionally tough enough to withstand their attacks and to avoid being cowed. This depends on doing our homework so that we know what we're talking about. It also depends on going out and mixing it up with enough evolutionists so that we know what we're up against. The other extreme to be avoided is this, namely, we must refuse to allow evolutionists to get under our skin, make us angry and thereby upset our equanimity. Once that happens, we lose self-control. This in turn typically leads us to denounce our opponents, issuing harsh and bitter words, and these never help our cause. Aggressiveness and argumentativeness are almost always interpreted as defensiveness, and rightly so. As Seneca noted two millennia ago, harshness is always a sign of weakness, not of strength. Victor Hugo put it this way: "Strong and bitter words indicate a weak cause." The reason is plain: we act harshly because people are not spontaneously behaving the way we would like and that, in turn, is because our strength of personality is not enough to fetch compliance.

I do a fair amount of public speaking and know from experience the feeling of a questioner getting under my skin and the urge to "let him have it." Unless I'm overtired or otherwise not firing on all cylinders, however, I now resist that urge. The simplest way I've found to do that is simply to stay on topic, answering the questioner's actual questions rather than being distracted by his animus or rudeness. Staying on topic, being courteous throughout and, as much as possible, attributing to the questioner sincere motives has several advantages: (1) it prevents you from seeming defensive (as noted, aggressiveness and argumentativeness indicate defensiveness);

(2) it wins the respect of the audience (and they're the ones we're trying to reach); and (3) it usually is the best way to slap some sense into a recalcitrant questioner, whose aim is to distract you from your message—by refusing to be distracted, you reinforce your message.

The most effective people on the other side always maintain their composure. They come across as neither servile nor aggressive. Instead, they always exude confidence. Even when they are speaking far outside their area of expertise, the confidence is still there. Even when their confidence has no basis in reality, it's still there. There's a principle at work here that John Maynard Keynes saw with special clarity in the circle surrounding the Cambridge moral philosopher G. E. Moore. As Toby Young describes it in *How to Lose Friends and Alienate People* (in many ways a quite serious book despite its flippant title):

> According to Keynes, Moore succeeded in dominating his disciples not because he could outmaneuver them in argument but simply because he had the loudest voice. Moore carried the day because he appeared to be so much more confident than anyone else. In Keynes's view, the key to persuading someone of the rightness of your moral point of view lay in asserting it as *emphatically* as possible. There was nothing more to it than that. Once your opponents got a whiff of just how *unambivalent* you were, they'd come round to your way of thinking.

I see this principle of "radiating confidence" at work every day among evolutionists, whether it be in print, on the Internet or in person.

But what about us? In closing, I want to suggest that we on the design side are in an even better position than the evolutionists to radiate confidence. Here's why. The evolutionists are essentially in a defensive posture. If they could demonstrate the power of material mechanisms to generate biological complexity and diversity, we wouldn't be having this discussion—*Darwin on Trial* would never have been written and the intelligent design movement would not exist. But they have nothing here, and that despite possessing, as far as they are concerned, the greatest scientific theory ever put forward—the theory of evolution. Furthermore, given Richard Dawkins's claim in the *Blind Watchmaker* that "biology is the study of complicated things that give the appearance of having been designed for a purpose," we have no burden to show that *every* feature of biology is actually designed. To demonstrate that there is more to biology than merely apparent design, it is enough for us to show that certain key biological systems (such as the bacterial flagellum) are actually designed. As evolutionists continue to fail to explain the origin of these systems in materialist terms and as the design characteristics of these systems become increasingly evident, the evolutionists will themselves face severe pressures to maintain their composure.

But there's more. Evolution has become totally status quo. Its supporters, therefore, tend to be stodgy and humorless. (See the video *Icons of Evolution* and decide for yourself.) They continually need to instruct the benighted masses on why criticisms of evolution should be disregarded, especially criticisms by those crazy design theorists. We, on the other hand, can afford to keep our sense of humor. We don't have anything to lose. We don't have positions of authority to preserve. We don't have public moneys to administer. We don't have a professional guild that we need to keep happy for the sake of our careers. We can be free spirits. This sits especially well with young people, who thrive on rebelling against the status quo and don't like it when an authoritarian elite tells them what they must think and believe. And these young people are the scientists of tomorrow.

Perhaps most significantly, however, we can admit our mistakes and receive instruction. The evolutionists cannot. Indeed, the moment they admit that we might have a point, they let the genie out of the bottle. To open evolutionary theory to critical scrutiny would destroy their monopoly over the study of biological origins. It is simply not an option as far as they are concerned. All the same, their no-concession policy is a loser as well. The problem with this policy is its heavy-handedness in stifling inquiry. I quoted Edward Sisson earlier about the psychology he sees at play as a litigator in which "opposing lawyers are primed to reject every statement by the other side because there is no advantage to considering that the statements might be true." Although Sisson sees this psychology at play among evolutionists, he does not find it in our ranks. As he writes in my book *Uncommon Dissent,* "I do *not* see that psychology in the work of intelligent design proponents. The fact that this psychology is missing from their work is one reason why I have come to trust them more than their opponents in the debate." Our ability to inspire such trust is the key to victory in this debate.

IT'S THE EPISTEMOLOGY, STUPID!
SCIENCE, PUBLIC SCHOOLS
AND WHAT COUNTS AS KNOWLEDGE

FRANCIS J. BECKWITH

——

In the opening pages of his masterful work, *Natural Rights and the Right to Choose*,[1] political philosopher Hadley Arkes recounts the thoughts of the Great Emancipator, Abraham Lincoln, soon after the Union army, under the command of General Meade, had defeated the Confederate forces at the Battle of Gettysburg. Unlike his generals, "Lincoln understood . . . that the tactical objective was not to 'take Richmond' but to destroy [General] Lee's army, the military force that alone sustained that 'pretended government' known as the Confederate States of America." Because "Lee and his forces could not yet cross the Potomac River while the tide was high," the president "urged Meade to strike at Lee before the general could" cross the river to return to Virginia. "Meade, however," writes Arkes, "held back, and in holding back, lost the moment." The general soon afterward telegraphed Lincoln, proclaiming "that they could take consolation at least in this: that the army had been successful in 'driving the invader from our soil.'" It was for the president, "just like McClellan all over again—the same spirit that led the general to proclaim a great victory because 'Pennsylvania and Maryland were safe.'" "Lincoln," Arkes writes, "wondered how he could covey the point to his officers: 'Will our generals never get that idea out of their heads? The whole country is our soil.' "[2]

The generals, as Lincoln saw the problem, had assimilated "the premises of the other side."[3] This incident from military history has within it an important lesson for opponents of philosophical naturalism (which includes intelligent design [ID] advocates) that is sometimes lost on people who, with the noblest of motives, believe they are advancing the cause of fairness or "balanced treatment" by offering policies that absorb the premises of scientific materialism without knowing it.[4]

To illustrate what I mean, we will look at two court cases and the contemporaneous and subsequent legal and policy arguments offered by advocates in both of them.

Segraves v. California Board of Education (1981)

Segraves[5] concerned the children of Kelly Segraves, cofounder of the Creation Science Resource Center. Mr. Segraves, whose children attended public schools, "argued that the State of California had violated the religious freedom of his children by teaching evolution as fact."[6] According to the court, the issue "is whether or not the free exercise of religion by Mr. Segraves and his children was thwarted by the instruction in science that his children had received in school, and if so, has there been sufficient accommodation of their views?"[7] The court held that the state's Antidogmatism Policy, adopted by the California Board of Education, if incorporated into the Science Framework and practiced by teachers in the science classroom, is an adequate compromise that protects the Segraves's free exercise rights without violating the constitutional prohibition of religious establishment. The state's guideline called for "(1) Dogmatism to be changed to conditional statements where speculation is offered as explanation for origins, and (2) Science should emphasize 'how' and not 'ultimate cause' for origins."[8] That is, the question of origins should be taught conditionally. Although it is not clear how this would be accomplished, at least one philosopher, Alvin Plantinga, has seen the wisdom of this holding and has employed its logic in his own proposal to resolve the debate over origins in public education.[9]

Plantinga argues that the key condition doing the intellectual work is epistemological, that is, what one believes counts as "knowledge" will shape what one thinks about the origin and nature of the universe. For example, if one embraces a naturalist epistemology, believing that only natural, nonagent-directed or caused explanations count as knowledge in the hard sciences, then naturalistic evolution is more likely true than not. However, if one embraces a different epistemology, one that allows for nonnatural, agent-directed or caused explanations in the hard sciences, then some form of antinaturalism (including creationism or intelligent design) is more likely true than not. Plantinga, in his other writings, does not maintain that one's epistemology is merely a matter of stipulation. For he offers arguments as to why there should not be a presumption in favor naturalism[10] as a worldview and that naturalist epistemology, which contends that our cognitive faculties have not been designed for the proper function of acquiring knowledge, is self-stultifying under the aegis of a Darwinist account of how our minds came to be.[11] That is, a mind resulting from the nonrational forces of natural selection and random mutation is no more trustworthy in delivering to one the truth—including the truth of materialism—than a random collection of Scrabble tiles that appears to instruct one to "accept Darwinism."

Citing the testimony of Dr. Mayer (no first name given) in support of its holding, the *Segraves* court seems to accept, with Plantinga, that epistemology is doing all the intellectual work. Unfortunately, Dr. Mayer's presentation, though superficially consistent with Plantinga's view, seems to imply that "knowledge" not derived from the hard sciences cannot count against the apparent deliverances of the hard sciences. For example, Mayer asserts, "I would not like to see my theology and my science to get mixed. I have never dealt with a scientific process where somebody says, 'I believe.' I have dealt with theological processes where one believes. In short, I think at that point you begin to mix epistemologies."[12] But what happens when "science" conflicts with "theology" when each is describing the exact same phenomenon, event or entity? For example, the claim that there is an immaterial ground to the human being, such as a soul, is *inconsistent with* the claim of materialist philosophers of mind who argue that an exhaustive materialist accounting of the human person is in principle possible. Suppose I have good reasons to believe in the existence of the soul, a conclusion inconsistent with the deliverances of materialist science. Who wins? I suspect that Dr. Mayer would say that they are not in conflict but are two "different ways" of "knowing." He may *say* that, just as one may *say* that Fred is both married and a bachelor. But it is not clear how contrary accountings of precisely the same entity can *both* be accurately described as knowledge. Granted, one may offer categorically different descriptions of the same entity—for example, an aesthetic one ("Mt. Rushmore contains the sculptures of four U.S. presidents") and a geological one ("Mt. Rushmore is composed of rock, sand, granite and some vegetation")—but that is different than offering metaphysically contrary descriptions of the same entity—for instance, a materialist one ("Abraham Lincoln's person consisted entirely of matter") and an immaterial one ("Abraham Lincoln's person consisted of both soul and matter"). The conflicts arising in the debate over origins are more often than not like the latter rather than the former. Moreover, Dr. Mayer's distinction between theological and scientific knowledge is something he claims to know that is neither theology nor science, for it is a distinction arrived at through philosophical reflection.

Stephen Jay Gould commits the same error when he suggests what he calls the NOMA principle, "nonoverlapping magisteria."

> Each subject [science and religion] has a legitimate magisterium, or domain of teaching authority—and these magisteria do not overlap. . . . The net of science covers the empirical universe; what it is made of (fact) and why does it work this way (theory). The net of religion extends over questions of moral meaning and value.[13]

But to what magisterium does NOMA belong? It seems to be a philosophical principle by which Gould assesses the nature of science and religion, and thus Gould is implying that philosophy is logically prior to science and thus the appropriate discipline by which to assess questions of the nature of science. If that is what he is implying, then it is not clear on what grounds he could object to, or not seriously consider, intelligent-design arguments against scientific materialism, for they typically include philosophical challenges to the prevailing view of the nature of science.

The *Segraves* court ordered that a copy of the state guidelines "shall be disseminated to all the publishers, institutions, school districts, schools, and persons regularly receiving the science framework."[14] According to the court, this order applies to those who have received the framework in the past as well as those who will receive it in the future.[15] In 1989, however, the California Board of Education adopted the State Board of Education Policy on the Teaching of Natural Sciences,[16] which does not explicitly mention evolution,[17] but nevertheless presents a framework that, according to some commentators, effectively removes the underlying principle of the Antidogmatism Policy and is thus contrary to the order of the *Segraves* court.[18] For even though the 1989 framework decries dogmatism, it does not require that the science of origins be taught conditionally as did the 1972 policy. Rather, the 1989 framework asserts that only *natural* explanations, even ones that attempt to explain the order and nature of being itself, count as "science" as well as real knowledge:

The domain of the natural sciences is the natural world. Science is limited by its tools—observable facts and testable hypotheses.

Discussions of any scientific fact, hypothesis, or theory related to the origins of the universe, the earth, and life (the how) are appropriate to the science curriculum. . . . A scientific fact is an understanding based on confirmable observations and is subject to test and rejection. . . . Scientific theories are constantly subject to testing, modification, and refutation as new evidence and new ideas emerge.[19]

On the other hand, nonnatural explanations, even ones that attempt to explain the order and nature of being itself, *precisely the same* phenomena the board states natural explanations are employed to account for, *do not* count as "science," and thus the board implies that they cannot count as real knowledge that could serve as a defeater to naturalistic explanations:

Discussions of divine creation, ultimate purposes, or ultimate causes (the why) are appropriate to the history-social science and English-language arts curricula. . . . Philosophical and religious beliefs are based, at least in part, on faith and are not subject to scientific test and refutation. . . . If a student should raise a question in a natural science class that the teacher determines is outside the domain of science, the teacher should treat the question with respect. The teacher should explain why the question is outside the domain of natural science and encourage the student to discuss the question further with his or her family and clergy.[20]

It is difficult to take seriously such educational pronouncements from a document whose authors cannot even present their views without relying on self-refutation as their ground of principle: "Nothing in science or in any other field of knowledge shall be taught dogmatically. A dogma is a system of beliefs that is not subject to scientific test and refutation. Compelling belief is inconsistent with the goal of education; the goal is to encourage understanding."[21] So, the California Board of Education, a government body, employs the coercive power of the state to compel its educators to adhere to a belief—"nothing in science or in any other field of knowledge shall be taught dogmatically"—that is itself not subject to scientific test and refutation and is thus affirmed dogmatically, in order to instruct its teachers to teach only "science" and not engage in compelling others to hold beliefs that are dogmatic and not subject to scientific test and refutation. Consequently, if school districts are to obey their state board's framework and incorporate it into their science curricula, each district must, ironically, *reject* the board's definition of what counts as science or knowledge, since it is a claim that is either self-refuting (i.e., it is a claim *of* science that is inconsistent with itself) or it is a philosophical claim (i.e., it is a claim *about* science and thus cannot be part of the science curriculum because it is not a claim *of* science), a "belief based, at least in part, on faith and" is "not subject to scientific test and refutation." Of course, each board member is free "to discuss the question further with his or her family and clergy."[22]

Freiler v. *Tangipahoa Parish Board of Education* (1999)

This 1999 case of *Freiler* v. *Tangipahoa Parish Board of Education*[23] was heard by the Fifth Circuit Court of Appeals and denied certiario by the U.S. Supreme Court in 2000 after the Fifth Circuit in 2000 denied a rehearing *en banc*. In the 1999 case, the Fifth Circuit affirmed the District Court opinion against the defendant,[24] Tangipahoa Parish Board of Education (Louisiana), which had in 1994 passed a resolution that required the following:

> Whenever, in classes of elementary or high school, the scientific theory of evolution is to be presented, whether from textbook, workbook, pamphlet, other written material, or oral presentation, the following statement shall be quoted immediately before the unit of study begins as a disclaimer from endorsement of such theory.
>
> It is hereby recognized by the Tangipahoa Board of Education, that the lesson to be presented, regarding the origin of life and matter, is known as the Scientific Theory of Evolution and should be presented to inform students of the scientific concept and not intended to influence or dissuade the Biblical version of Creation or any other concept.
>
> It is further recognized by the Board of Education that it is the basic right and privilege of each student to form his/her own opinion and maintain beliefs taught by parents on this very important matter of the origin of life and matter.[25] Students are urged to exercise critical thinking and gather all information possible and closely examine each alternative toward forming an opinion.[26]

In assessing the resolution the Fifth Circuit applied the first two prongs of the Lemon test, a standard formulated by the Supreme Court by which it assesses whether a law violates the constitutional prohibition of religious establishment.[27] According to the first prong, the resolution must have a secular purpose. The court maintained that the school board asserted three purposes for the disclaimer: "(1) to encourage informed freedom of belief, (2) to disclaim any orthodoxy of belief that could be inferred from the exclusive placement of evolution in curriculum, and (3) to reduce offense to the sensibilities and sensitivities of any student or parent caused by the teaching of evolution."[28] If the disclaimer furthers only one of these purposes and that purpose is secular, the resolution passes the first prong of the Lemon test.[29]

According to the court, the resolution passes Lemon's first prong because the second and third purposes are secular and are furthered by the disclaimer.[30] Disclaiming orthodoxy and reducing offense are legitimate secular purposes.[31] The first purpose, however, though secular, is not furthered by the disclaimer because the resolution does not really encourage freedom of belief or critical thinking, but rather "the disclaimer as a whole furthers a contrary purpose, namely the protection and maintenance of a particular religious viewpoint."[32] Because the disclaimer, according to the court, instructs school children "that evolution as taught in the classroom need not affect what they already know,"[33] and "[s]uch a message is contrary to an intent to encourage critical thinking, which requires that students approach new concepts with an open mind and a willingness to alter and shift existing viewpoints,"[34] freedom of thought is not furthered by the disclaimer.

This is a curious argument. For it is not clear how advocating freedom of thought and critical thinking is inconsistent with reminding school children that they need not believe the state-endorsed metaphysics taught to them by their science teachers, and that these children in fact have a right to believe otherwise, whether they are beliefs held by their parents or ones arrived at through extended and critical reflection. It seems to me that the authors of the resolution were dismayed by the poverty of their science curriculum's metaphysical offerings. For it offered only one point of view, materialism, a view that by its very nature excludes certain points of view from the privileged status of "knowledge." And in reply to this metaphysical monopoly— this ontological imperialism—the school board was more than gracious and tolerant, for it did not coerce its teachers to offer another point of view as part of its curriculum. Rather, the board took an overly modest strategy: it required its teachers to merely mention to their students, for a few moments, that the students have a right to disagree with the only permissible viewpoint that will be presented without rebuttal, reply or counterargument for the remainder of the course. Yet this unpretentious prerequisite, according to the court, was inconsistent with freedom of thought and critical thinking, for *it* and not its removal "furthers a contrary purpose, namely the protection and maintenance of a particular religious viewpoint."[35]

Nevertheless, the court may still have a defensible point to its reasoning in light of both the line of cases that preceded this one[36] as well as the language employed in the resolution to convey the school board's advocating of critical thinking. In order to see this point, we must examine why the court concluded that the resolution failed Lemon's second prong, its sole basis for rejecting the disclaimer as unconstitutional. It failed, according to the court, for *precisely the same reason* why the first purpose failed Lemon's first prong: "the primary effect of the disclaimer is to protect and maintain a particular religious viewpoint, namely belief in the Biblical version of creation."[37] The court provided three reasons for this: "(1) the juxtaposition of the disavowal of endorsement of evolution with an urging that students contemplate alternative theories of life; (2) the reminder that students have the right to maintain beliefs taught by their parents regarding the origin of life; and (3) the 'Biblical version of Creation' as the only alternative theory is referenced in the disclaimer."[38] According to the court, "The disclaimer, taken as a whole, encourages students to read and meditate upon religion in general and the 'Biblical version of Creation' in particular,"[39] and this "impermissibly advances religion."[40] So, it is not that the court opposes a disclaimer in principle, but rather that this particular disclaimer mentions by name only one option to evolution, even if only illustratively,[41] creation-science, the view rejected in *Epperson* v. *Arkansas* (1968), *McLean* v. *Arkansas* (1982) and *Edwards* v. *Aguillard* (1987). (For summaries of these cases, see note 36.)

Edward McGlynn Gaffney Jr. replied to the court's reasoning by arguing that the "chief difficulty with this move is that it conveniently ignored the next four words in the disclaimer: 'or any other concept.' "[42] I disagree. I don't think it is accurate to say that the court ignored this phrase in its analysis, for it does address the phrase in three places.[43] In addition, the resolution's claim that what is taught in the classroom need not alter the beliefs one learns at home could reasonably be seen as *inconsistent* with critical thinking. Moreover, the school board did nothing to further its case when it chose to include in the resolution the phrase "Biblical version of Creation," especially in light of the judicial history of the concept. This is why Gaffney is mistaken when he claims that the court is saying that "promoting critical thinking about the origins of the universe must be equated with advancing religion."[44] Rather, the court is saying that it is inconsistent to call for critical thinking and at the same time say that what is learned in the classroom need not alter one's beliefs, *including* religious beliefs. Granted, there are deep philosophical questions about belief and rationality that are percolating be-neath the court's reasoning in this case.[45] Nevertheless, as I argue below, the school board did not help matters much by employing language that re-inforces the widely held though defeasible view that creation and evolution are not two perspectives on the same subject but two different subjects, re-ligion and science respectively. Although an argument could be made that the court's decision was flawed and that it exaggerates the extent to which such a mild and modest disclaimer impermissibly advances religion,[46] it seems to me that the court, given the genre of case it was given, had little wiggle room to rule otherwise. Moreover, the school board did not seem to have an overabundance of insight or cleverness among its well-meaning members. And here, I'd like to make two points that are germane for the case for the defeat of the materialist hegemony in the public schools.

First, in drafting its resolution the school board seemed oblivious to the four issues that federal courts have focused on when assessing creation-science statutes, as I have argued elsewhere:[47]

- the statute's historical continuity with *Scopes* as well as the creation-evolution battles throughout the twentieth century
- how closely the curricular content required by the statute paralleled the creation story in Genesis or whether the curricular content prohibited by the statute is proscribed because it is inconsistent with the creation story in Genesis
- the motives of those who supported the statute in either the legislature or the public square
- whether the statute was a legitimate means to achieve appropriate state ends

By placing in its resolution "Biblical version of Creation" it raised a red flag that could not be argued away. Such a resolution should have focused on the underlying metaphysical dispute in the debate over origins—materi-alism versus design—rather than on viewpoints that the courts have already declared, whether rightly or wrongly,[48] religious (creation science) and sci-entific (evolution). Consequently, given the jurisprudential hand dealt to the school board, they should have opted for a more circumspect strategy, one that addressed the deeper philosophical issues for which the concepts of critical thinking and freedom of thought were likely designed.

Second, though the resolution is no doubt the result of the deliberations of well-meaning religious people, it seems to presuppose that religion or theology is not a branch of real knowledge, and that evolution is in fact real knowledge, for the latter is labeled as "scientific": "[T]he lesson to be presented, regarding the origin of life and matter, is known as the Scientific Theory of Evolution and should be presented to inform students of the scientific concept and not intended to influence or dissuade the Biblical version of Creation or any other concept."[49] But to imply to students that something that is scientific ought not to interfere with their religious beliefs seems to denigrate the epistemological status of those beliefs. For the message being conveyed is that science, which is said to presuppose materialism, is the highest (and perhaps the only) form of knowledge, and because religious beliefs cannot touch or be touched by that knowledge, religious beliefs are not really knowledge. This is why, for example, in the preface to a 1984 pamphlet published by the National Academy of Sciences, its then-president Dr. Frank Press can write without fear of affirming an inconsistency:

> It is false . . . to think that the theory of evolution represents an irreconcilable conflict between religion and science. A great many religious leaders accept evolution on scientific grounds without relinquishing their belief in religious principles. As stated in a resolution by the Council of the National Academy of Sciences in 1981, however, "Religion and science are separate and mutually exclusive realms of human thought whose presentation in the same context leads to misunderstanding of both scientific theory and religious belief."[50]

When courts accept this epistemology of religious belief—as apparently the *McLean*[51] and *Peloza*[52] courts (as well as the Tangipahoa school board) have—they are in fact proposing a metatheory on how to understand the nature of religious beliefs, which comes perilously close to violating the Supreme Court's rule against assessing the truth of a religion[53] or the interpretation of doctrines and creeds.[54] Ironically, the *Freiler* court may have rejected the view of religion and science found in *McLean* and *Peloza*. For the *Freiler* court said that *it was wrong* for the school board to say that evolution, a scientific theory, should not be considered by the students to count against creationism and other concepts they may have learned from their parents. This implies that evolution and other views on origins, including antinaturalist views (including ID), are contrary perspectives on the same subject rather than separate spheres as the *McLean* and *Peloza* courts seem to think.

Conclusion

I began this chapter by telling the story of President Lincoln's frustrations with his generals, who had absorbed the premises of the Confederacy without knowing it. Antinaturalism's confederacy consists of a collection of beliefs and assumptions about knowledge that have served as cognitive gatekeepers that allow materialists the luxury to dismiss any adversaries (such as ID) and the arguments for them without serious inspection or consideration. In the debates over *Seagraves* and *Freiler,* apparent opponents of dogmatic materialism—both pretended and actual—had assimilated the premises of the epistemological confederacy, and for that reason offered suggestions and arguments that provided sustenance to beliefs and assumptions that in principle will never allow criticisms of materialism, including ID, to flourish.

Bibliography

Beckwith, Francis J. *Law, Darwinism & Public Education: The Establishment Clause and the Challenge of Intelligent Design.* Lanham, Md.: Rowman & Littlefield, 2003.

———. "Science and Religion 20 Years After *McLean v. Arkansas:* Evolution, Public Education, and the Challenge of Intelligent Design." *Harvard Journal of Law & Public Policy* 26, no. 2 (2003): 456-99.

———. "Rawls' Dangerous Idea? Liberalism, Evolution, and the Legal Requirement of Religious Neutrality in Public Schools." *Journal of Law and Religion* 20, no. 2 (2004-2005): 101-37.

Campbell, John Angus, and Stephen C. Meyer, eds. *Darwin, Design, and Public Education.* East Lansing: Michigan State University Press, 2003.

Greenawalt, Kent. "Establishing Religious Ideas: Evolution, Creationism, and Intelligent Design." *Notre Dame Journal of Law, Ethics & Public Policy* 17, no. 2 (2003): 321-97.

"Not Your Daddy's Fundamentalism: Intelligent Design in the Classroom." *Harvard Law Review* 117, no. 3 (2004): 364-71.

Pennock, Robert T. "Reply to Plantinga's 'Modest Proposal.' " In *Intelligent Design Creationism and Its Critics: Philosophical, Theological, and Scientific Perspectives,* edited by Robert T. Pennock. Cambridge, Mass.: MIT Press, 2001.

Plantinga, Alvin. "Creation and Evolution: A Modest Proposal." In *Intelligent Design Creationism and Its Critics: Philosophical, Theological, and Scien-*

tific Perspectives, edited by Robert T. Pennock. Cambridge, Mass.: MIT Press, 2001.

Reppert, Victor. *C. S. Lewis's Dangerous Idea: In Defense of the Argument from Reason*. Downers Grove, Ill.: InterVarsity Press, 2003.

Wexler, Jay D. "Of Pandas, People, and the First Amendment: The Constitutionality of Teaching Intelligent Design in the Public Schools." *Stanford Law Review* 49 (1997).

CUTTING BOTH WAYS

The Challenge Posed by Intelligent Design
to Traditional Christian Education

TIMOTHY G. STANDISH

—

Phillip Johnson, as a founder and leader of the intelligent design (ID) movement, has persuasively advocated thoughtful reexamination of the philosophical presuppositions inherent in modern science; particularly the philosophy of naturalism from which modern neo-Darwinian evolutionary theory springs. Johnson's advocacy of change within the scientific academy has profound potential for ripple effects in other areas, among the most obvious of which is science education. The impact of Johnson's critique of Darwinism, along with those of other ID advocates, is currently a hotly debated topic within the public education systems of several states in the United States[1] as well as some other countries.[2]

On occasion Johnson's call for openness in teaching about the neo-Darwinian synthesis has been conflated with his advocacy of the intelligent design alternative to Darwinism. In most recent high profile cases, it is a fair, honest and reasonable approach to teaching about evolution that has been advocated, not the teaching of an alternative theory. For example, when the Ohio State Board of Education voted to liberalize the way evolution is taught in Ohio public schools, Bruce Chapman, president of the Discovery Institute (an ID-friendly think tank) is quoted as calling the vote, "a victory for common sense against the scientific dogmatism of those who think evolution should be protected from any critical examination."[3] At no time did anyone affiliated with the Discovery Institute advocate replacement of evolution with ID. When Discovery Institute's Stephen C. Meyer and Jonathan Wells were invited by the Ohio School Board to debate the issue in 2002, both of them explicitly stated that ID should not be required alongside evolution.[4] Instead, they both advocated a "teach the controversy" approach to evolution, informing students fully about the evidence and arguments for the theory as well as evidence and arguments against it.[5] While most sensible people can clearly see the difference between critical examination and teaching a new theory, this has not stopped some from trying to conflate the two as they attempt to protect Darwinism from rational evaluation.[6]

The debate about how evolution should be taught in public schools raises profound questions about state sponsorship of education, the relationship between education and students' religious beliefs and the moral upbringing that parents provide their children. In the United States thorny questions about the meaning of the Establishment clause of the First Amendment[7] to the Constitution are also raised. Because of these philosophical and legal factors, along with the deeply held religious beliefs of many citizens, how and what public school students are taught about the history of life is likely to remain a hotly debated issue for some time.

Hidden in the dust kicked up by the debate over public education has been the challenge presented by Johnson's thinking to private Christian[8] parochial school systems. As it turns out, ID is a two-edged sword with potential to cut both ways, eliciting positive but painful changes in both secular and Christian schools. ID is clearly a different creature from the Christian doctrine of creation or the creationism embraced by many Christians. Despite this, Christian educators may be as eager to incorporate the sophisticated ID design argument into their biology lesson plans as pressure groups like the National Center for Science Education (NCSE) are to ensure that no such thing happens in public schools. Before incorporating ID, Christian educators should be careful to ensure that they are willing to swallow the entire ID pill. Because, if they don't, they will be left with an interesting but incoherent package. Johnson dedicates an entire chapter to this issue in his most recent book, *The Right Questions*.[9]

Intelligent design unavoidably cuts to a profound issue about the role and purpose of education: indoctrination versus freedom of choice. Are students being molded for a purpose (indoctrination) or prepared to discover their purpose? The thesis of this chapter is that ID, as formulated by Johnson, requires the latter, and that this is consistent with the Christian understanding of the nature of humanity and the place of humankind in the creation. Whether Christian educators will embrace the kind of open liberal education that logically follows from ID may take as long to resolve as those questions it raises about public education.

All education inescapably springs from a foundation within the worldview or philosophy of those providing it. For purposes of comparison, secular humanism and Christianity represent two contrasting views. Both reason from profoundly different starting points to arrive at their array of beliefs. Secular humanism rejects a priori the supernatural, embracing instead ontological materialism: belief that the material world is all that has ever existed. Naturalism, the belief that all phenomena can be accounted for by natural causes, is a corollary of materialist assumptions: because there is nothing outside of the material world to cause what we observe, all that we see must result from natural laws and chance.

Christianity is not burdened with the requirement that everything result from natural processes. The Christian worldview allows for supernatural intervention in the material world, and thus, depending on which is most logically consistent with the data, either natural or supernatural explanations of nature are allowed. In the study of biology this means that, like other theists, Christians have a broader palette of explanations to draw on than do materialists. Not surprisingly, this may mean that Christians are willing to entertain explanations that are anathema within materialist dogma.

Unfortunately, the willingness of Christians to entertain supernatural explanations for phenomena observed in nature has been construed as a "science stopper." As Jeffrey Jordan put it, "Theological beliefs can act as a kind of 'science stopper' by making it seem that no naturalistic explanation is needed."[10]

A historical analogy illustrates the fallacy of this position. After heroic efforts, the combined work of several European explorers placed the source of the Nile river at Lake Victoria, bordered by modern-day Uganda, Tanzania and Kenya. The discovery of the ultimate source of the Nile did not bring the science of geography to an end or stop any other science that deals with rivers, how they operate, where they are located, how they can be best managed or why they exist. Discovery of Lake Victoria provided a valuable insight into the nature of the Nile. In the same sense, discovery that God is the ultimate source and cause of life on earth does not bring the study of nature to a halt. In fact, discovering the ultimate cause of phenomena is the Holy Grail of science, not because it causes the work of science to cease but because it provides profound insights suggesting new potentially productive lines of research. The idea that allowing intelligent causes as potential explanations for phenomena observed in nature somehow stops science has been thoroughly refuted by many scholars, prominent among them the mathematician, philosopher and intelligent design advocate William Dembski.[11]

When the long list of sincere Christian believers who have left an indelible mark on science is taken into consideration, the falsehood of the claim that belief in God somehow removes all motivation to study nature is demonstrably false. From Newton to Linnaeus, from Pasteur to von Braun, the history of scientific progress is rich with characters who openly professed faith in the Creator God.[12] In the words of Johannes Kepler:

> To God there are, in the whole material world, material laws, figures and relations of special excellency and of the most appropriate order. . . . Those laws are within the grasp of the human mind; God wanted us to recognize them by creating us after his own image so that we could share his own thoughts.[13]

In other words, at least one motivation for Christian believers to study the creation is that this allows them to share God's thoughts.

The fundamentally different approaches to understanding nature adopted by Christians as opposed to materialists naturally result in different approaches to teaching in the sciences. Within the materialist paradigm, only natural causes are allowed; so only natural causes, no matter how improbable they may seem, are presented in the classroom.[14] This "materialist rule," commonly referred to as "methodological naturalism," may be reflected in the very way science is defined. For example, during 2002 a controversy arose in the state of Ohio over how biology should be taught. The state sets science standards that are to be followed by all state-sponsored schools. In an early draft of these standards, tenth-grade students were to "Recognize that scientific knowledge is limited to natural explanations for natural phenomena based on evidence from our senses or technological extensions."[15]

Because only "natural explanations" are allowed in this definition of science, it exemplifies a materialist bias and presents a number of philosophical dilemmas. Due to circular reasoning inherent in this definition, the problem of differentiating natural from unnatural phenomena becomes complicated. If something is "natural," it must have a natural explanation. But what happens when whether something is natural or not is unknown?

Determining whether life is "natural"—the product of natural laws and chance alone—is not possible within this definition of science. Instead, life must first be assumed to be natural, and if that is done, it must be explained via natural causes. When problems inherent in materialist definitions of science were recognized by those writing the Ohio science standards, they wisely modified this definition to read: "Recognize that science is a systematic method of continuing investigation, based on observation, hypothesis testing, measurement, experimentation, and theory building, which leads to more adequate explanations of natural phenomena."[16]

This definition of science is not burdened with materialist dogma, emphasizes the process of science and allows for a more realistic understanding of how tentative knowledge is gained by studying the empirical world. While this is clearly a definition of science superior to the materialist one, Christian educators may find it unsatisfying if they believe that the empirical world reveals the power and wisdom of the Creator God they worship. To address this objection, we must step back and address the profound, purpose-of-education question.

Phillip Johnson presents two questions about education in his book *The Right Questions:*

> [1] Should a college education prepare students to understand the ultimate purpose or meaning for which life should be lived and to choose rightly from among the available possibilities? [2] Alternatively, should this subject be left out of the curriculum on the ground that the choice among ultimate purposes involves only subjective preferences and not knowledge?[17]

There is nothing restricting these questions to only college education; in fact, they are inherent in all educational endeavors. The historical association of education with both religion and government suggests that, at least in the past, education has been a tool for both investigation of and indoctrination into specific worldviews. The line between investigation and indoctrination has not always been clearly drawn. In modern Western thought, the very idea of indoctrination seems coercive—a denial of an individual's right to make sovereign and informed decisions about the nature of reality and humanity's place in it. This view of individuals' freedom to make informed decisions about the most fundamental aspects of life's meaning has a firm foundation in Christian thinking. The Bible begins with a story of this freedom and how it was exercised: a tree in the garden of Eden was provided with a warning that "in the day that thou eatest thereof thou shalt surely die" (Gen 2:17 KJV). Not only was the opportunity to disobey God offered to humans, they could make an informed decision to do it because they knew the consequences.

Answering the second question posed by Johnson is easy in the context of Christian education: No, the ultimate purpose of life is not based on subjective preferences but is founded in objective knowledge—knowledge of nature and of its Creator, both of which point toward knowledge of the Savior of humanity. In the Christian worldview, meaning is inherent in understanding that humankind was created in the image of God (Gen 1:27; 9:6). Answering Johnson's first question may warrant more consideration.

If Christian education is to ensure students come to the "right answers," but those "right answers" are decided before the education begins, then this "education" may well turn out to be the very kind of coercion that should be anathema to Christian thinking. Should Christian educators sacrifice the Christian principle of freedom to ensure students only choose to live lives consistent with other Christian principles? The story of the first sin and fall of humans tells us that the God Christians worship is not willing to make this kind of compromise. In any case, most experienced educators can recount examples of how this approach may fail, especially with those students possessing the greatest academic potential.

In teaching biology, coercion may take the form of simply providing only evidence pointing to a single conclusion instead of laying out all the evidence and explaining how those starting from different viewpoints might interpret it. Unfortunately, restricting or miscasting evidence to support a single view is an approach commonly used in biology textbooks, particularly when they cover controversial topics like ecological issues or evolution.[18] Before attempting to lay out all the information on a controversial topic so that students can develop an informed opinion, Christian educators can and should honestly state their predispositions. Because hidden philosophical agendas make all claims subject to them suspect, this is what ID demands of both materialists and theists. Possibly most frightening about unrecognized, hidden or ignored philosophical presuppositions is their tendency to blind adherents of a position to the true foundation of their beliefs. This should be as disturbing to Christian educators as it should be to those teaching from a materialist perspective.

Christians start with a particular view of nature and humanity's place in it; pretending otherwise would be dishonest. Christians are not unbiased, and neither is anyone else. Reasoning from the ninth commandment—Thou shalt not bear false witness (Ex 12:16 KJV)—Christians value honesty, and that means admitting bias and taking it into account when explaining data. This does not mean that the biases of others should be ignored, and it does not mean that biases should be used as an ad hominem argument against the interpretations of others. Being open about one's own partiality is a major advantage when attempting objective analysis, and objective analysis seems to be a reasonable objective when doing science. Pretending no bias exists is a dangerous self deception.

Significant risk is inherent in presenting information in a way that allows students to make their own judgment; they may not make the judgment that Christian teachers believe to be correct. For example, when presented with information about the fossil record, a student may conclude that the biblical account of creation is false. This is a real risk. However, the alternative is even more problematic; students who have never learned about the fossil record, only that the beauty of the flowers testify to God's love, may cease to believe in God when their faith in Scripture is challenged by the fossil record—or at least certain aspects of it. When data from nature that are consistent with biblical Christian understanding of history are openly discussed along with that which at first inspection appears inconsistent, the risk is real that students will either never develop faith or lose it altogether. On the other hand, attempting to reduce the risk by keeping students ignorant denies them their God-given right to a free and informed choice, and risks precipitating a crisis of faith when they discover that their faith is based on only a partial picture of reality.

Christian educators are not alone in believing that a certain understanding of history is an objective of biology education. The eminent evolutionist Richard Dawkins wrote, "It is absolutely safe to say that if you meet somebody who claims not to believe in evolution, that person is ignorant, stupid or insane (or wicked, but I'd rather not consider that)."[19] While an educated person may be wicked, stupidity, insanity and ignorance are not hallmarks of education. The product of a successful education will, according to Dawkins, be people who "believe in evolution." Ernst Mayr echoes this sentiment: "No educated person any longer questions the validity of the so-called theory of evolution, which we now know to be a simple fact."[20]

Because bias is impossible to eliminate and good teachers endeavor to provide what they consider to be the best for their students, the most that can be achieved is a clear statement of bias so that students can evaluate data knowing that they have been presented by one who wishes to convince them of a certain position. This position may be that God created life, or it may be that natural laws and chance created life. In either case, if students are to make an informed decision about the meaning of what they are learning, there must be full disclosure. This means abandoning the illusion of objectivity suggested in Johnson's second question while embracing the honesty inherent in the first.

If the purpose of education is, as Johnson puts it, to "understand the ultimate purpose or meaning for which life should be lived and to choose rightly from among the available possibilities," this suggests that education in the sciences, and specifically in biology, should not be divorced from the humanities. Science is not only about learning the facts about nature as they are currently understood; these facts provide a foundation for addressing bigger questions dealing with how life should be lived. Rote memorization of facts and figures or mastery of techniques can be achieved independently of understanding the principles involved and implications of what is being learned. Thus science teaching can be divorced from the truly big questions faced by all students, but the result is a hollow, uninspiring and brittle understanding of science. Because the humanities help address those big questions that transcend technology and facts, they should serve as vital tools in the arsenal of Christian science teachers. It would not be overstating the point to say that without the humanities, Christian biology teachers lack essential tools to teach their material. With the humanities, biology education transcends memorization of facts and figures to become a dynamic informed investigation of life and its meaning.

A concrete example may illustrate this point. In college, a student may learn about human embryos, how they develop, how they can be manipulated, their dynamic interaction with women's bodies and their basic chemical composition. All of this knowledge is important, but a far greater lesson will be lost if this information is not applied to questions about the basic nature of humanity and the ethical implications of what is being learned. If they only know the what and how of embryology, students are unprepared to think about the implications of producing artificial embryos by inserting the DNA of a human into the egg of a cow,[21] combining human and mouse cells into a single embryo,[22] or combining male and female cells to produce "she-male" embryos.[23] In other words, students may know how things can be done, but this does not guarantee they are equipped to ethically apply their knowledge. There is a very tight connection between the knowledge gained in biology and questions of good and evil or right and wrong.

If students are to wrestle with questions of purpose and meaning in biology, the false fact-value or science-humanities dichotomy must not prevail in Christian science classrooms. This view is not restricted to a Christian approach to science and science education. As Paul Ehrlich put it, "The idea that science should (or can) be value-free is wrong. . . . Being steeped in values is part of being human."[24]

Dealing with value type questions may be a powerful motivator for some students who would not otherwise be attracted to the sciences. Teaching understanding of the empirical world within a value framework suggests some techniques for learning and evaluation may be more effective than others. For example, limitations of so-called objective testing in encouraging and evaluating student's analysis of information and its meaning in the context of larger questions are self evident. Multiple-choice and true-false testing may be a quick and dirty way of evaluating whether certain facts or opinions have been memorized, but they do not measure students' integration of this information into a global understanding of life and its meaning. Because essay writing has been developed as a means of exploring and evaluating the value and meaning of information, authoring essays in the sciences may be a more effective tool for both learning and evaluation than multiple-choice tests. The problem is that evaluation of essays is a time-consuming process and when first introduced to information, students may not be ready to form an opinion about its meaning. They must first know what the information is before they can evaluate it. Thus, at a practical level, objective testing may be used to encourage and measure mastery of the facts, while essay writing may be reserved for evaluation of the student's ability to apply, assess and use the information.

So far in this chapter, five principles have been established. (1) The Christian worldview affects the way science is defined. (2) Fundamental questions about the nature of education affect how education will be done, and answers to these questions influence answers to the narrower question of integrating Christian beliefs in the science classroom. (3) Informed freedom of choice is a Christian principle that should strongly influence the Christian approach to education. (4) No approach to education is without bias and the best way to address this bias is to be open about it rather than pretending that it does not exist. (5) Asking students to wrestle with questions of purpose and meaning in science classes encourages treatment of the various academic disciplines as part of a coherent whole rather than as distinct fields of knowledge, and also suggests certain methods of evaluation may have greater utility than others. How might these principles be practically applied in a Christian biology classroom? Many examples could be chosen, but the most problematic of these has to be the question of biological evolution. Should evolution even be taught in Christian schools?

Many Christians view arguments to and from design to be a superior framework on which to structure understanding of nature. If ID is better than Darwinism, why not teach only what is best, ID, and ignore Darwinism? Practical and ethical considerations strongly argue against this position. On a practical level, students ignorant of alternative ideas may be ill equipped to deal with them when they do encounter something different from their previous learning. In addition, the strength of any position is best judged relative to the strongest alternative. Understanding of the robust nature of ID is greatly enhanced when compared with the strengths and weaknesses of the Darwinian alternative. Ethically, ignoring Darwinism is an untenable option because denying students knowledge of alternative views of nature denies them the opportunity to evaluate and choose between options.

The Christian principle of informed freedom of choice dictates that even "bad" ideas must be taught in Christian schools. Phillip Johnson puts it this way: "The way to deal with timidity and self-deception in Christian education is not to try to prevent bad ideas from being taught but rather to ensure that the bad ideas are effectively countered by better ideas in an atmosphere of open deliberation."[25] Students must be given a choice; it is the job of educators to so clearly lay out the information, logic and issues involved so that students see the clear advantages of better ideas over those with less merit. In doing this, the impact of the teacher's life as a testament to the power and beauty of Christian living cannot be underestimated. If Christianity truly offers something better than the alternatives, all the theoretical and practical advantages offered to students cannot outweigh their empirical observation of the work of Christ in the life of the teacher. But this does not mean that logic plays no role in the decisions students make. In fact, to provide the information necessary for students to make informed choices, the personal testimony of a Christian life should include a clear and logical understanding of what evolution is, the philosophical presuppositions it springs from and its epistemological limitations. In short, Christian teachers who want to integrate their faith with the teaching of evolution must first allow the Holy Spirit to work in their lives and then ensure that they understand the subject at a level that exceeds that given in typical high school and college biology textbooks. Educators must be educated before they can educate!

Assuming a teacher is already adequately equipped to logically and honestly teach the neo-Darwinian model, how might he or she go about it? Johnson has consistently argued that the philosophical underpinnings of Darwinism should be exposed so that people can evaluate the quality of the arguments made in favor of the theory in the full knowledge of whence it springs.[26] Thus it makes no sense to dive into data consistent and inconsistent with a scientific theory before adequate prior preparation. Students first need to understand what science is, its tentative nature, the philosophical presuppositions behind various definitions of science and their implications. A thorough understanding of the scientific method is necessary for students if they are to understand the confidence they can put in scientific conclusions and how Darwinism, intelligent design or any other scientific topic fits into the general model of how science is done. Thus the foundation for understanding theories about the nature and origin of life is laid down long before the topic is introduced. When the origin of life is first discussed, the teacher's bias should be clearly stated. This is not a negative admission, it is an opportunity to evaluate the relative worth of one philosophical approach to science over another

Along with both a theoretical framework for understanding science, a theological foundation for evaluation should also be integrated into the class before dealing with the difficult question of evolution. This is essential if students are to understand the actual position that Christian educators and materialists are reasoning from. This should help prevent students from becoming easy prey to the straw-man type arguments so frequently employed when debating for materialistic over theistic views of origins.

Another methodological foundation may also be laid; when other subjects are discussed before getting to evolution, questions should be asked about the broader meaning of the information being learned. Students may thus learn to view questions of meaning and purpose as a natural part of science, and specifically biology classes. In addition, students should be encouraged to consider how the information they are currently learning integrates with information learned earlier. For example, students typically learn about osmosis, the movement of water across membranes, early in biology classes. Later these principles may be applied to understanding how kidneys operate. This approach encourages students to see the knowledge they are learning as part of a much larger picture instead of isolated facts to memorize.

When evolution is discussed, the following points should be made: (1) Questioning Darwin's ideas does not imply a lack of respect for him or that his thinking is irrelevant. Questioning others' theories is a normal part of science. (2) The theological implications that make Darwinism unpalatable should be explained. (3) The word *evolution* may be used in a number of distinctly different ways. Some of these meanings are not objectionable. The disagreement is not over evolution per se but over the specific materialist theory of Darwinism. For example, when *evolution* is used to mean only change over time, generally Christians have no problem with this concept. Christians typically do not believe that the earth or the life forms on it are the same today as when God created them. (4) Disagreement with some meanings of the word *evolution* do not spring exclusively from theology but from science. (5) Those aspects of evolution that are most troubling arise not from science but from the philosophy of materialism.

Most textbooks present a very similar collection of information supporting the theory of evolution, typically with little, if any, critique. If students are to make an informed choice about the value of evolutionary theory, they must engage in critical thinking. This is problematic because sometimes what textbooks present is factually incorrect.[27] When errors of fact appear in text books, for example the still commonly used fraudulent embryo drawings by Ernst Haeckel,[28] this provides an opportunity to teach students that their textbooks should not be expected to always have all the facts straight. For some students this may be a disturbing revelation that causes them to question the validity of all they are learning.

Awareness of different uses of the term *evolution* helps students evaluate the logic of how various data are presented in support of both ID and Darwinism. Population genetics, changes in allele frequencies within populations, is frequently presented as directly measurable evolution in action. For this reason textbooks commonly place population genetics squarely in the middle of the discussion of evolution instead of in chapters covering genetics.[29] This provides an excellent opportunity to ask questions about the nature of the evolution being discussed. Is it reasonable to expect that different environments will favor members of a species that have one genetic makeup while others are selected against? How might this process be related to production of new types of organisms? Does it address the question of where genetic variability that selection may act on came from? Encouraging students to wrestle with and ask questions about the meaning of population genetics promotes something more than uncritical memorization of the assumptions made when calculating Hardy-Weinberg equilibria.

Understanding that changes in allele frequencies may be caused by multiple factors, one of which is natural selection, but that natural selection does not account for the origin of the alleles that selection may act on helps students to see that a logical gap exists between empirical investigation of population genetics and the theoretical production of new kinds of organisms. Understanding this "empirical evolution" in populations points out the gap between what is empirical and what is theoretical in science. It also shows that what is empirical does not necessarily conflict with the claim of Scripture that God created the various kinds of organisms or the purely scientific notion of intelligent design.

Allowing students to evaluate the challenge evolution poses to the Christian understanding of origins does not mean simply dismissing the evidence presented as either false, as in the case of Haeckel's drawings, or tangential to the central question, as in the case of population genetics. Some evidence is well explained within an evolutionary paradigm. For example, order in the fossil record is clearly something logically consistent with evolution of life from organisms less like those living today at the bottom of the geological column to those more like the living things we know today near the top. It is tempting to try minimizing the significance of order in the record or to argue that it is not real.[30] Either tactic would be unfortunate because it would be dishonest. Order in the fossil record does not disprove history as told in Scripture, although students should be able to understand that when reasoning from a materialist starting point it may be better explained, as currently understood, by a process of change over time in which organisms start out different from those living today and evolve into the organisms now extant.

Students need to be aware that evolution, when it means common descent, does explain evidence, and in some cases it may be a more reasonable appearing explanation than a history involving creation, the fall and a global flood. Having said that, they should not be left to believe that all the evidence is either inconclusive or well explained within the evolutionary paradigm. Because it seeks to provide students with the freedom to make informed choices, Christian education cannot ignore evidence within the fossil record that is consistent with the story of creation as told in Scripture. This may mean broadening the content of courses beyond the information contained in textbooks, and this is certainly the case when discussing evidence relating to evolution. In the case of the fossil record, while it appears to be true that order exists, other evidence appears inconsistent with the concept of common descent. For example, sudden appearance of fossils is also a generally agreed-on characteristic of the fossil record. The sudden appearance of many profoundly different organisms in Cambrian strata is inconsistent with Darwinian predictions but consistent with intelligent intervention.[31] The same could be said for the complexity evident in the first fossil animals, like trilobites, as well as for systematic gaps between both living and fossil groups of organisms.[32]

Presented with the best and most comprehensive understanding of what the fossil record is, students can judge for themselves what explanations make the most sense. They should always be encouraged to base their decisions on more than one narrow data set, the fossil record in this case, but also on the much greater set of knowledge gained in previous learning. For example, if they have already learned about the ways organisms and the cells from which they are made work, this evidence can also be brought to bear when evaluating evolutionary theory. Is the neo-Darwinian mechanism really adequate to explain not just the complexity but the specific kind of complexity evident in living things? Going beyond the realm of science and incorporating the humanities, students may be encouraged to ask what theological implications the data suggest: where might tensions exist that should stimulate further research?

Conclusions

Materialism and the Christian worldview of a God intimately involved with the material world logically lead to profoundly different views of science. Inherent within each worldview are ideas about how people can most productively lead their lives. Education is widely understood to be a process in which students are taught not just facts and skills but also about the meaning and purpose of life. Because of this, it is tempting to restrict education to a process of indoctrination into a worldview, but the Christian principle of informed freedom of choice should preclude yielding to this temptation. Instead, Christian educators who seek to integrate their faith into their teaching must provide students with the best possible personal example of Christian living, the best information and thinking skills available, and encourage them to apply the information they learn to the larger question of life's meaning and purpose. This means that Christian science teachers are called to genuine mastery of their fields and understanding of where their specialty fits with other fields of knowledge, especially those in the humanities.

Within the context of human knowledge, biology presents special challenges to Christian faith. These are best faced by providing a more comprehensive approach to the subject than by avoiding discussion of biological evolution and other challenging areas. This provides an opportunity to better educate students about the value of scientific understanding and a more detailed knowledge of information that bears on questions like the origin and history of life. Thus Christian biology education should be both broader and deeper in its scope than some other approaches. The linking of knowledge with questions of meaning and values may also serve as a powerful motivator for students to acquire biological knowledge. Given these advantages, incorporation of the Christian faith with the teaching of biology provides greater opportunities for both teachers and students than teaching biology under the misconceived notion that it is independent of faith, values and meaning.

Discussing practical ways in which teaching about the origin and nature of life may be influenced by ID illustrates the importance and implications of the seminal observation that philosophical presuppositions are highly relevant to scientific understanding of nature. On the one hand, ID may be the simple observation that certain natural phenomena are best explained in terms of intelligent rather than natural causes. On the other hand, rationally discussing the possibility of ID requires setting aside cherished presuppositions and replacing them with an ideal that does not answer the question of the origin of life before it is even asked. This is as true for Christian presuppositions as it is for the materialist presuppositions of secular humanists. From this recognition, a cascade of effects logically follows: an interdisciplinary approach to teaching the sciences that honors and incorporates the rich tradition of the humanities, a different testing pedagogy, courage to explore alternative views, a broader understanding of the three dimensional nature of science rather than the test-tube-clutching, lab-coat-clad, bespectacled caricature so frequently presented in classrooms and in the popular media.

The educational impact of ID thinking as formulated by Phillip Johnson is broad and deep. In public schools dominated by materialist thought, these changes have the potential to be significant. The changes may be equally significant if ID is embraced in Christian schools. Embracing the full ID package means abandoning well-meaning attempts to shield students from dangerous or wrong ideas, replacing it with the aim of equipping young minds to discriminate between good and bad ideas. It also means ensuring that teachers are genuinely prepared to teach science subjects from an informed position and to argue their case for a Christian understanding of nature rather than simply teaching the received wisdom. This also means that science teachers must take the time and make the effort to understand the humanities and their contribution to our understanding of science and its place in culture. While these are potentially positive things, implementation is another matter; one that might at a very practical level test any claims of divine blessing on Christian teachers.

PART III

TWO FRIENDLY CRITICS

Two Fables by Jorge Luis Borges

David Berlinski

On the Derivation of Ulysses from Don Quixote

This story was told to me by Jorge Luis Borges one evening in a Buenos Aires café.

> His voice dry and infinitely ironic, Borges remarked that "the *Ulysses*, mistakenly attributed to the Irishman James Joyce, was, in fact, derived from the *Quixote*."
>
> I must have raised my eyebrows.
>
> Borges paused to sip discreetly at the bitter black coffee our waiter had placed in front of him, guiding his hands to the saucer.
>
> "The details of this remarkable series of events," he said, "may be found at the University of Leyden. They were conveyed to me by the Freemason Alejandro Ferri in Montevideo." Borges wiped his thin lips with a white linen handkerchief that he had withdrawn from his breast pocket.
>
> "As you know," he continued, "the original handwritten text of the *Quixote* was given to an order of French Cistercians in the autumn of 1576."
>
> I held up my hand to signify to our waiter that no further service was needed.
>
> "Curiously enough, for none of the brothers could read Spanish, the order was charged by the Papal Nucio Hoyo dos Monterrey, a man of great refinement and implacable will, with the responsibility for copying the *Quixote,* the printing press having then gained no currency in the wilderness of what is now known as the department of Auvergne. Unable to read or even to speak Spanish, a language that they not unreasonably detested, the brothers copied the *Quixote* over and over again, re-creating the text but, of course, compromising it as well, and so inadvertently discovering the true nature of author-

ship. They thus created Fernando Lor's *Los Hombres d'Estado* in 1598 by means of a singular series of copying errors, and then in 1654, Juan Luis Samorza's remarkable epistolary novel, *Por Favor* by the same means, and then in 1685, the errors having accumulated sufficiently to change Spanish into French, Molière's *Le Bourgeois gentilhomme,* their copying continuous and indefatigable, the work handed down from generation to generation as a sacred but secret trust, so that in time, the brothers of the monastery, known only to members of the Bourbon house and, rumor has it, the Englishman and psychic Conan Doyle, copied into creation Stendhal's *Le Rouge et le noir* and Flaubert's *Madame Bovary,* and then, as the result of a particularly significant series of errors, in which French changed into Russian, Tolstoy's *The Death of Ivan Ilyich* and *Anna Karenina.* Late in the last decade of the nineteenth century there suddenly emerged, in English, Oscar Wilde's *The Importance of Being Earnest* and then the brothers, their numbers reduced by an infectious disease of mysterious origin, finally copied the *Ulysses* into creation in 1902, the manuscript lying neglected for almost thirteen years and then making its way to Paris in 1915, just months before the British attack on the Somme, a circumstance whose significance remains to be determined."

I sat there amazed by what Borges had recounted. "Is it your understanding, then," I asked, "that *every* novel in the West was created in this way?"

"Of course," Borges replied imperturbably. "Although every novel is derived directly from another novel, there is really only one novel, the *Quixote.*"

On the Intelligent Design of the Spanish Language

This story was told to me by Jorge Luis Borges on another evening in a Buenos Aires café.

"It is a pleasure," Borges remarked serenely, "to express oneself again in Spanish."

He had that afternoon delivered a lecture in fluent but heavily accented German devoted to the influence of Goethe's later love lyrics on certain nineteenth-century papal encyclicals.

"As the American psychologist and South Slav sympathizer William James was heard to remark," Borges observed with satisfaction, "German is a language that lacks all of the modern conveniences."

Borges was sipping his third bitter espresso while puffing sedately at a thin cheroot.

"There is, in fact," he affirmed, savoring his consonants elegantly in the south American style, "only one intelligently designed instrument of thought and communication, and that is, of course, the Spanish language."

"Intelligently designed?" I must have murmured as a small orchestra began to play tango music.

"Contemporary research undertaken at the Byelorussian Academy of Science," Borges said, "has assigned the design and creation of the Spanish language to the last decades of the thirteenth century."

I said, "My dear Borges, Spanish is a romance language, and like the others, it evolved from Latin. Of this there is surely no doubt?"

"The thesis that the Spanish language emerged as the result of intelligent design," Borges said calmly, "as Ptolemy remarked in defending the *Almagest* against criticisms that were later to prove insubstantial, is such that 'absolutely all phenomena are in contradiction to any of the alternate notions that have been propounded.'"

"If, in fact," he continued, "there were a connection between the Spanish language and the Latin tongue, one would expect to see records of thousands of intermediate languages between the disappearance of the Latin vernacular in the West, which as the French medievalist and gold speculator Henri Pirenne established took place well before the Battle of Poitiers in 754, and the quite sudden emergence of the fully formed Spanish language among members of the Castilian aristocracy late in the thirteenth century. There are none, the gaps in the bibliographic record sufficiently striking to prompt the archbishop of Toledo, in a circular letter otherwise devoted to predictions made by the Moslem astrologer Abu Ma Shar, to refer to it as a "trade secret among scholars of antiquity."

"What of the hundreds and hundreds of homologous words in Spanish and Latin," I asked, perhaps in some frustration, "the obvious structural similarities between Latin and Spanish?"

"It is true," Borges answered, puffing serenely on his cheroot, "that certain words in Spanish may suggest certain words in Latin—a matter discussed in the second volume of Gertz's *Scriptores minores historiae danicae*. But as Gertz himself admits, perhaps with the asperity prompted by his unfortunate participation in certain fiscal irregularities pertaining to the Stavisky affair, the thesis that there are *homologues* between the Latin and the Spanish lexicon depends, of course, on the assumption that Spanish evolved from Latin, and that," Borges observed with satisfaction, "is precisely the thesis at issue."

"There is, in addition, the fact, mentioned in footnotes by the French lexicographer Claude Duneton, and credited to letters exchanged between Gottfried Leibniz and the Count Ernst von Hessen-Rheinfels in 1686, that Spanish contains structures of such irreducible complexity as to defeat any attempt to create them by incremental means. There is the example of the future subjunctive, a mood not found in Latin, and one reflecting an organization of experience completely beyond the power of the Latin language to express."

The café was by now more or less empty, but the orchestra continued to play tangos, Borges nodding his head to keep time with the music.

"Is it your view," I finally asked, "that *all* of the Romance languages emerged as the result of some sort of deliberate design?"

"Certainly," Borges said. "As the result of work undertaken by Alfred Delvau, it is possible authoritatively to date the creation of the French language to the years between 1276 and 1292, results that you may find recorded in his monograph 'Un vieux verbe français encore employé en Normandie' (1867)."

"Quite incredible," I murmured skeptically, "and just who do you imagine undertook the design of the Romance languages?"

"The identity of the designer," Borges said imperturbably, "is a matter upon which current research has shed little light, but the fact that the Romance languages *were* designed is surely of greater importance than the designer's identity."

Borges paused to sip at his espresso and then wipe his lips with a white linen handkerchief. He smiled bleakly. "While it is true that all of the Romance languages were designed," he said, "it is a fact worth noting, that only one of the Romance languages was *intelligently* designed, and that is, of course, the Spanish language."

DARWINISM AND THE PROBLEM OF EVIL

MICHAEL RUSE

—

Many of today's Darwinians are convinced that if you accept the theory of evolution through natural selection, then it is impossible to be a sincerely believing Christian. To take only the most extreme example, England's most ardent and vocal atheist, Oxford biologist Richard Dawkins, has said: "I'm a Darwinist because I believe the only alternatives are Lamarckism [the inheritance of acquired characteristics] or God, neither of which does the job as an explanatory principle. Life in the universe is either Darwinian or something else not yet thought of."[1] Elsewhere, he has said: "The universe we observe has precisely the properties we should expect if there is, at bottom, no design, no purpose, no evil and no good, nothing but blind, pitiless indifference."[2]

Notoriously, Dawkins has spoken of the "cowardly flabbiness of the intellect [that] afflicts otherwise rational people confronted with long-established religions."[3] He says: "The kinds of views of the universe which religious people have traditionally embraced have been puny, pathetic, and measly in comparison to the way the universe actually is. The universe presented by organized religions is a poky little medieval universe, and extremely limited."[4] Dawkins takes pride in the harshness of his judgments. "I am considered by some to be a zealot. This comes partly from a passionate revulsion against fatuous religious prejudices, which I think lead to evil."[5]

Life's Tape Replayed

As it happens, I probably have no more positive religious belief than does Dawkins. But that is not my topic now. Rather, I want to consider the relationship between Darwinism and religion, more specifically between Darwinism and Christianity. I believe that there are a number of issues here, and I shall not pretend to treat of all of them. One of the most pressing, it seems to me, is the problem of contingency.[6] If Christianity be true, then the existence of humans (or of something very humanlike) cannot be mere chance but must be inevitable or necessary in some way. The late Stephen Jay Gould has put things clearly, as always. Referring to the comet that hit the earth about sixty-five million years ago and wiped out the dinosaurs, he wrote:

> Since dinosaurs were not moving toward markedly larger brains, and since such a prospect may lie outside the capabilities of reptilian design, . . . we must assume that consciousness would not have evolved on our planet if a cosmic catastrophe had not claimed the dinosaurs as victims. In an entirely literal sense, we owe our existence, as large and reasoning mammals, to our lucky stars."[7]

In a new book, *Life's Solutions: Inevitable Humans in a Lonely Universe,* the Cambridge paleontologist Simon Conway Morris offers an answer to this objection, suggesting that the arrival of humans on earth was a lot more determined, even in the light of Darwinian evolution, than one might expect. Pointing to the fact that evolving life seems to seek out niches, and that often different life forms find (converge on) the same niches, Conway Morris writes:

> If brains can get big independently and provide a neural machine capable of handling a highly complex environment, then perhaps there are other parallels, other convergences that drive some groups towards complexity. Could the story of sensory perception be one clue that, given time, evolution will inevitably lead not only to the emergence of such properties as intelligence, but also to other complexities, such as, say, agriculture and culture, that we tend to regard as the prerogative of the human? We may be unique, but paradoxically those properties that define our uniqueness can still be inherent in the evolutionary process. In other words, if we humans had not evolved then something more-or-less identical would have emerged sooner or later.[8]

I am not sure that this will be the last word, but I do think that Conway Morris has offered at least a starting point to argue back against those like Gould who think that life is all chance, and that if you run the tape again you will never ever get anything remotely humanlike.

Is Atheism Possible?

Another matter that is important but I will not really discuss—mainly because I have given it extended discussion in my book *Darwin and Design: Does Evolution Have a Purpose?*—is the question of the possibility of nonbelief in Christianity and the relevance of Darwinism. The Greeks set the problem for the nonbeliever. There is something striking about the world, especially the world of organisms. They are not just thrown together higgledy-piggledy but seem rather to be complex and ordered. They work, they function, they show organization, they show what Aristotle called "final causes." The parts of the body—the hand, for instance, or the eye—do not just exist but seem framed for specific purposes or ends—for grasping or for seeing.

Of course, in trying to understand the hand or the eye causally, we need to consider such things as embryological growth, but this in itself is not enough. In the *Phaedo,* Plato's great dialogue about the death of Socrates, the condemned philosopher considers the question of why it is that a man grows.

> I had formerly thought that it was clear to everyone that he grew through eating and drinking; that when, through food, new flesh and bones came into being to supplement the old, and thus in the same way each kind of thing was supplemented by new substances proper to it, only then did the mass which was small become large, and in the same way the small man big.[9]

But then Plato (using Socrates as his mouthpiece) continues that this kind of explanation unaided will not do. It is not wrong, but it is incomplete. One must address the question of why someone would grow. How is it that those necessary things like hands and eyes come about? Blind forces, without direction, will not do. One needs special kinds of forces to make such features. One has to invoke a thinking mind or intelligence to get such an effect. The hand and the eye work because someone (or Someone) thought about the needs of humans and made them—designed them—expressly for this end or purpose. As Socrates says: "the ordering Mind ordered everything and placed each thing severally as it was best that it should be; so that if anyone wanted to discover the cause of anything, how it came into being or perished or existed, he simply needed to discover what kind of existence was *best* for it, or what it was best that it should do or have done to it."[10]

Richard Dawkins argues—and I agree with him—that this organic organization demands an explanation.[11] Before Darwin the only adequate explanation was God. Famously, David Hume ran a skeptical truck right over the argument for God's existence from the designlike nature of the world. He pointed out that, at best, by analogy with human artifacts, we should expect a squad of designers at work on the world, and this world would be but one of a long series from very crude models to worlds far more sophisticated than ours. At worst, with the evil and pain in the world, who could think that the designer is the Christian God of love: "what racking pains . . . arise from gouts, gravels, megrims, tooth-aches, rheumatisms; where the injury to the animal-machinery is either small or incurable?"[12] Yet for all the skepticism, even Hume realized that final causes demand some special kind of explanation. "If the proposition before us is that *the cause or causes of order in the universe probably bear some remote analogy to human intelligence,*" then "what can the most inquisitive, contemplative, and religious man do more than give a plain, philosophical assent to the proposition, as often as it occurs; and believe that the arguments, on which it is established, exceed the objections, which lie against it?"[13]

Darwin shattered all of this. In his great work, *On the Origin of Species,* he argued for the evolutionary origins of all organisms, including us humans. Animals and plants, living and dead, are the end process of a long, slow, natural (that is, law bound) process of development from a few, original, primitive forms. More than this, Darwin argued for a mechanism, the natural selection of the fitter over the less fit. He first introduced and proved that between organisms there is an ongoing struggle for existence.

A struggle for existence inevitably follows from the high rate at which all organic beings tend to increase. Every being, which during its natural lifetime produces several eggs or seeds, must suffer destruction during some period of its life, and during some season or occasional year; otherwise, on the principle of geometrical increase, its numbers would quickly become so inordinately great that no country could support the product. Hence, as more individuals are produced than can possibly survive, there must in every case be a struggle for existence, either one individual with another of the same species or with the individuals of distinct species or with the physical conditions of life. It is the doctrine of Malthus applied with manifold force to the whole animal and vegetable kingdoms; for in this case there can be no artificial increase of food, and no prudential restraint from marriage.[14]

And then on to natural selection:

> Let it be borne in mind in what an endless number of strange peculiarities our domestic productions, and, in a lesser degree, those under nature, vary; and how strong the hereditary tendency is. Under domestication, it may be truly

said that the whole organization becomes in some degree plastic. Let it be borne in mind how infinitely complex and close-fitting are the mutual relations of all organic beings to each other and to their physical conditions of life. Can it, then, be thought improbable, seeing that variations useful to man have undoubtedly occurred, that other variations useful in some way to each being in the great and complex battle of life, should sometimes occur in the course of thousands of generations? If such do occur, can we doubt (remembering that many more individuals are born than can possibly survive) that individuals having any advantage, however slight, over others, would have the best chance of surviving and of procreating their kind? On the other hand we may feel sure that any variation in the least degree injurious would be rigidly destroyed. This preservation of favourable variations and the rejection of injurious variations, I call Natural Selection.[15]

I agree with Dawkins that natural selection is the first mechanism that today's professional evolutionists call upon to explain some interesting or important aspect of the living world. Moreover, I agree that selection speaks, as Darwin himself certainly intended, not just to evolution as such but also to the nature of organisms. In particular, selection explains final causes, or what Darwin himself more commonly called "adaptations." The reason why some organisms are more successful than others is because they have features that the losers do not have. The eye and the hand, sight and grasping, are good things to have in the struggle for existence. More than this, within the Darwinian picture, the features themselves do not come about in any way through intention or direct design. Darwin had not much idea about the nature of organic variation and its origins—those raw building blocks that are needed to make ongoing evolutionary change. But his hunch was that they were entirely natural and unguided, and this—in the age of DNA—is just what is agreed by modern evolutionists. Selection working on undirected variation makes for adaptation. Final causes have been given a natural explanation.

Is Atheism Necessary?

So I agree with Dawkins when he says that after Darwin—and only after Darwin—is it possible to be an intellectually fulfilled atheist. But does one have to be an atheist? Should an honest Darwinian be an atheist? Dawkins and many others think this. Although people may have been sincere in embracing both Darwinism and Christianity, sincerity is not enough. As Darwinians, they should have rejected Christianity. University of Chicago evolutionist Jerry Coyne speaks for many: "If one applies the same empirical standards to Christianity as scientists do to Darwinism, religion suffers: we have far more evidence for the existence of dinosaurs than for the divinity of Christ."[16] Reviewing *Can a Darwinian Be a Christian?*—a book that answered positively the question posed in the title—Coyne wittily quoted George Orwell. "One has to belong to the intelligentsia to believe things like that. No ordinary man could be such a fool."[17]

Obviously, if one equates Christianity with a crude literalistic reading of the Bible, including the early chapters of Genesis, there is no more to be said. But what if one has a more nuanced understanding of Scripture, including—especially including—the Old Testament? What then? For all of his emotional loathing of religion, Dawkins does not just get to his conclusion with an unspoken and unproven assumption. He has a strong argument to back his position. This is the venerable argument from evil, which Dawkins (following Darwin himself) thinks is much strengthened by Darwinism. In a letter written just after the publication of the *Origin,* to his American friend and supporter Asa Gray, Charles Darwin wrote:

> With respect to the theological view of the question; this is always painful to me.—I am bewildered.—I had no intention to write atheistically. But I own that I cannot see, as plainly as others do, & as I sh^d. wish to do, evidence of design & beneficence on all sides of us. There seems to me too much misery in the world. I cannot persuade myself that a beneficent & omnipotent God would have designedly created the Ichneumonidae with the express intention of their feeding within the living bodies of caterpillars, or that a cat should play with mice. Not believing this, I see no necessity in the belief that the eye was expressly designed.[18]

Dawkins concurs. Even if God does exist, he is certainly nothing like the Christian God: He is unkind, unfair, totally indifferent.

If nature were kind, it would at least make the minor concession of anesthetizing caterpillars before they are eaten alive from within. But nature is neither kind nor unkind. It is neither against suffering nor for it. Nature is not interested one way or the other in suffering, unless it affects the survival of DNA. It is easy to imagine a gene that, say, tranquilizes gazelles when they are about to suffer a killing bite. Would such a gene be favored by natural selection? Not unless the act of tranquilizing a gazelle improved that gene's chances of being propagated into future generations. It is hard to see why this should be so, and we may therefore guess that gazelles suffer horrible pain and fear when they are pursued to the death—as most of them eventually are. The total amount of suffering per year in the natural world is beyond all decent contemplation. During the minute it takes me to compose this sentence, thousands of animals are being eaten alive; others are running for their lives, whimpering with fear; others are being slowly devoured from within by rasping parasites; thousands of all kinds are dying of starvation, thirst and disease. It must be so. If there is ever a time of plenty, this very fact will automatically lead to an increase in population until the natural state of starvation and misery is restored.

Dawkins concludes:

> As that unhappy poet A. E. Houseman put it:
>
> For Nature, heartless, witless Nature
>
> Will neither know nor care.
>
> DNA neither knows nor cares. DNA just is. And we dance to its music.[19]

Moral Evil

Despite Dawkins's savage rhetoric, he has never been overburdened with an undue knowledge of philosophy or Christianity. Perhaps things are not quite as cut and dried as he thinks. Darwin did not think so, for, to continue his letter to Asa Gray just at the point where Dawkins finishes it:

> On the other hand I cannot anyhow be contented to view this wonderful universe & especially the nature of man, & to conclude that everything is the result of brute force. I am inclined to look at everything as resulting from designed laws, with the details, whether good or bad, left to the working out of what we may call chance. Not that this notion *at all* satisfies me. I feel most deeply that the whole subject is too profound for the human intellect. A dog might as well speculate on the mind of Newton.—Let each man hope & believe what he can.[20]

In the spirit of Darwin rather than Dawkins, giving the Christian at least a chance to respond, let us make the traditional distinction between moral evil—the evil brought about by Hitler—and natural or physical evil—the Lisbon earthquake. Let us note that Christians have traditional answers to both of these issues, and the matter for us must be whether Darwinism—the mechanism of natural selection—destroys one or both of these counterarguments.

The Christian response to moral evil is that of Saint Augustine.[21] God gave us free will, that is a great gift for good, and it is better that we have it, even though we will then do evil, than that we do not have free will and do nothing of our own accord. There are two issues here. First, does science as such—and Darwinism is part of science—make free will impossible? Second, is there anything in Darwinism itself that makes free will impossible? The answer to the first question is that science and free will can go together, and indeed there are reasons to think that they must go together. David Hume is the authority here.[22] If there are no laws governing human behavior, then we are not free—we are crazy and do things without cause, without rhyme or reason. The compatibilist argues that the true distinction is not between freedom and law but between freedom and constraint. The person in chains is not free, nor is the person under hypnosis. It is true that they are subject to law, but so also is the free person not in chains or under hypnosis.

To take up the second question, doesn't Darwinian science specifically have something about it that puts us all under hypnosis—genetic hypnosis? Or as critics like Harvard biologist Richard Lewontin have put it, doesn't Darwinism deny freedom by making us "genetically determined"?[23] We have no freedom, good or ill, because our genes made us do it. Hitler is not to blame. He just had a lousy genotype, and it is natural selection that put that in place. Blame the process, not us. Likewise, of course, Mother Teresa is not to be praised. She drew a good genotype (set of genes).

This argument does not stand up. Some things are surely genetically determined. Ants for instance. They are preprogrammed by the genes, as produced by selection. Philosopher Daniel Dennett gives a beautiful case of genetic determinism, that (because of its chief player) he call "sphexishness." A wasp (Sphex) digs a hole, finds and brings in a cricket that she stings to paralyze but not to kill, lays her eggs next to this food store, and finally closes off the hole, never to return. A wonderful case of thoughtfulness and intention, until something goes wrong and the mechanical nature of the whole process is revealed.

> The wasp's routine is to bring the paralyzed cricket to the burrow, leave it on the threshold, go inside to see that all is well, emerge, and then drag the cricket in. If the cricket is moved a few inches away while the wasp is inside making her preliminary inspection, the wasp, on emerging from the burrow, will bring the cricket back to the threshold, but not inside, and will then repeat the preparatory procedure of entering the burrow to see that everything is all right.

This can go on and on indefinitely. "The wasp never thinks of pulling the cricket straight in. On one occasion this procedure was repeated forty times, always with the same result."[24]

But as Dennett (as much a Darwin booster and critic of Christianity as Dawkins) stresses, we humans are not wasps—we are not genetically determined in this way. Our evolution has been such as to give us the power to make decisions when faced with choices and to revise and rework when things go wrong. In the language of evolutionists, wasps and ants were produced by "r-selection." They produce lots of offspring, and when something goes wrong they can afford to lose them because there are more. We humans are "K-selected." We produce just a few offspring, and we cannot afford to lose them when things go wrong. Hence, we have the abilities to make decisions, in order to avoid obstacles. That is why we have big brains.

Ants and wasps are like cheap rockets—many are produced and they cannot change course once fired. We humans, by contrast, are like expensive rockets—just a few are produced but we can change course even in mid-flight if the target changes direction or speed or whatever. The expensive rocket has a flexibility—a dimension of freedom—not possessed by the cheaper rocket. Both kinds of rockets are covered by laws, and so are ants, wasps and humans. We have freedom over and above genetic determinism, and this freedom was put in place by—not despite—natural selection. Hence the argument from evil against free will fails—at least, it fails if you are making the case based on Darwinism.

Natural or Physical Evil

This is Dawkins's big argument, and one would be insensitive were one not to agree that he does have a point. Darwinism does highlight pain and suffering. But is this a counter to Christianity? The traditional saving argument is one that is usually associated with the great German philosopher Leibniz. He pointed out that being all powerful has never implied the ability to do the impossible. God cannot make $2 + 2 = 5$. No more can God, having decided to create through law (and there may be good theological reasons for this), make physical evil disappear. Indeed, it may well be that physical evil simply comes as part of a package deal.

> For example, what would it entail to alter the natural laws regarding digestion, so that arsenic or other poisons would not negatively affect my constitution? Would not either arsenic or my own physiological composition or both have

to be altered such that they would, in effect, be different from the present objects which we now call arsenic or human digestive organs?"[25]

Paradoxically and somewhat amusingly, Dawkins himself rather aids this line of argument. He has long maintained that the only way in which complex adaptation could be produced by law is through natural selection. He argues that alternative mechanisms (notably Lamarckism) which produce adaptation are false, and alternative mechanisms (notably evolution by jumps, or saltationism) which do not produce adaptation are inadequate: "If a life-form displays adaptive complexity, it must possess an evolutionary mechanism capable of generating adaptive complexity. However diverse evolutionary mechanisms may be, if there is no other generalization that can be made about life all around the Universe, I am betting that it will always be recognizable as Darwinian life."[26] In short, if God created through law, then it had to be through Darwinian law. There was no other choice. (This of course is not to say that, knowing the subsequent pain, God was right to create at all, but that is another matter and none of Darwinism's business.)

In other words, just as the thinking of the arch-Darwinian atheist Dan Dennett can be turned to good account to save Christianity in the face of moral evil, so now the thinking of the arch-Darwinian atheist Richard Dawkins can be turned to good account to save Christianity in the face of physical evil!

Conclusion

I have said that I probably have as little positive religious faith as Richard Dawkins, and that is true. My aim has not been to defend Christianity, but to defend the integrity of the Darwinian who wants to be a Christian. But I do differ from Dawkins in where I stand, overall, on matters of faith. He is convinced that there is no God, no realm of existence beyond science. He is an out-and-out atheist. I am probably an atheist with respect to Christianity—I cannot bring myself to believe that Jesus died on the cross for my sins—but in the larger dimension I prefer to think of myself as an agnostic or skeptic. Perhaps there is something more. Perhaps there is not. I do not know. What I do know is that Darwinism tells me that I am a mid-range primate, with adaptations to let me (or rather my ancestors) get out of the trees and live on the plains and find food (probably scavenging), and to defend myself from enemies (probably fellow hominids). I have no reason to think that I have adaptations necessary to peer into the mysteries of the universe—although I am sure that I have the adaptations necessary to make me cockily confident that I do have such mystery-peering adaptations. Selection does not favor intellectual humility.

What I am saying is that Darwinism bids me to be modest about what I know and can know.[27] Ultimate reality may be very peculiar indeed. Better to cover one's bets and be agnostic. In support, I can do no better than quote another Darwinian who apparently agrees with me: "Modern physics teaches us that there is more to truth than meets the eye; or than meets the all too limited human mind, evolved as it was to cope with medium-sized objects moving at medium speeds through medium distances in Africa."[28] Amen!

Bibliography

Augustine. *The City of God Against the Pagans*, edited and translated by R. W. Dyson. Cambridge: Cambridge University Press, 1998.

Brockman, J. *The Third Culture: Beyond the Scientific Revolution*. New York: Simon & Schuster, 1995.

Conway Morris, Simon. *Life's Solution: Inevitable Humans in a Lonely Universe*. Cambridge: Cambridge University Press, 2003.

Coyne, Jerry. *Intergalactic Jesus. London Review of Books*. 2002. Accessible at <www.lrb.co.uk/v24/n09/coyn01_.html>.

Darwin, Charles. *On the Origin of Species*. London: John Murray, 1859.

Dawkins, Richard. *The Blind Watchmaker*. New York: Norton, 1986.

———. *A Devil's Chaplain: Reflections on Hope, Lies, Science and Love*. Boston and New York: Houghton Mifflin, 2003.

———. "Obscurantism to the Rescue." *Quarterly Review of Biology* 72 (1997).

———. *A River Out of Eden*. New York: Basic Books. 1995.

———. "Universal Darwinism." *Molecules to Men*, edited by D. S. Bendall. Cambridge: Cambridge University Press, 1983.

Dennett, Daniel C. *Elbow Room*. Cambridge, Mass.: MIT Press, 1984.

Gould, Stephen Jay. *Wonderful Life: The Burgess Shale and the Nature of History*. New York: W. W. Norton, 1989.

Hume, David. *Dialogues Concerning Natural Religion*, edited by N. K. Smith. 1779. Reprint, Indianapolis: Bobbs-Merrill, 1947.

———. *A Treatise of Human Nature*. Oxford: Oxford University Press, 1978.

Lewontin, Richard C. *Biology as Ideology: The Doctrine of DNA*. Toronto: Anansi, 1991.

Plato: Complete Works, edited by J. M. Cooper. Indianapolis: Hackett, 1997.

Reichenbach, Bruce R. "Natural Evils and Natural Laws: A Theodicy for Natural evil." *International Philosophical Quarterly* 16 (1976).

Ruse, Michael. *Can a Darwinian be a Christian? The Relationship Between Science and Religion.* Cambridge: Cambridge University Press, 2001.

———. *Mystery of Mysteries: Is Evolution a Social Construction?* Cambridge, Mass.: Harvard University Press, 1999.

———. *On a Darkling Plain: The Evolution-Creation Struggle.* Cambridge, Mass.: Harvard University Press, 2005.

———. *Taking Darwin Seriously: A Naturalistic Approach to Philosophy.* 2nd ed. Buffalo, N.Y.: Prometheus, 1998.

Wooldridge, D. *The Machinery of the Brain.* New York: McGraw-Hill, 1963.

PART IV

JOHNSON'S REVOLUTION IN BIOLOGY

10

THE WEDGE OF TRUTH VISITS
THE LABORATORY

DAVID KELLER

—

Imagine if you will that Phillip Johnson visited one of the many molecular biology labs down the hill from his Boalt Hall office at Berkeley. In our mind's eye, we can envision what most likely took place . . .

It was five o'clock and the evening cleaning crew was just getting started. Professor Johnson poked his nose into a high-ceilinged room filled with various spinning, sucking and electrifying gadgets. He walked up to the lab bench in the center of the room and began thumbing through a grad student's lab book, carefully noting the unintelligible scratchings and the taped-in computer printouts. Setting down the lab book, Prof. Johnson strolled past an enclave of computer screens and book-laden desks and spied an opportunity to insert his famous "wedge of truth" into the mind of an unsuspecting postdoc.

"Whacha reading there, doc?"

The startled postdoc reached for his glasses to better size up this wizened inquirer.

"Just a copy of Maniatis,"[1] he sighed, then added (to his obviously non-biology-minded guest), "It's a book about lab techniques."

"Looks like Maniatis has the upper hand today."

"Yeah, I've been trying to subclone this guinea pig GULO gene for weeks! Anyway, were you looking for my adviser?"

"No, just dropped by to see what's new in the world of molecular biology. Sounds like it's all GULOs and guinea pigs these days?" Johnson said with a twinkle in his eye.

The postdoc smiled, closing his big blue book. Then returning the twinkle, exclaimed, "Ah, you must be the infamous Phil Johnson from up the hill! I've seen your picture on the dartboard in our lounge."

"Dartboard? I thought I would at least be voodoo doll material by now." The postdoc laughed and offered Johnson a chair.

"So, Professor Johnson, what mischief can I help you with today?"

"Well, that depends. Tell me a little more about this GULO clone that you are unleashing on the civilized world."

Using the back of a Fat Slice pizza box, the postdoc sketched his research plan. Over the next ten minutes he explained how GULO stands for L-Gulono-gamma-lactone oxidase and how, in humans, some primates and guinea pigs, it's become a pseudogene, a gene that has lost its function, thereby blocking the synthesis of vitamin C. His strategy, he told Johnson, was to clone (i.e., copy) a guinea pig's gene for GULO into bacteria (hence the term subclone) so that the bacteria could make oodles of this gene—which would, among other things, be sent to the local DNA sequencing facility.

According to the postdoc, some Japanese scientists created a hubbub with regard to the phylogeny of the GULO gene.[2] They showed that guinea pig and human GULO both shared the same set of mistakes relative to a rat's functional GULO gene. This was troubling since guinea pig ancestors and human ancestors were supposed to have had their GULOs incapacitated independently, and any mutational errors accumulating after that point should have been completely random because natural selection does not operate on nonfunctional genes.

"We all know that pseudogenes provide unequivocal support for evolution, since they prove how humans and primates have a shared set of mistakes in the useless DNA that they inherited from a common ancestor," the postdoc concluded, "but these results from Japan are messing everything up. That's why my adviser recommended I try to reproduce their experiment and see if they might have interpreted things incorrectly."

"Lemme get this straight," Johnson probed in lawyerly fashion. "You are slaving away with ol' Maniatis just to redo someone else's published work? Doesn't that seem like a waste of time?"

"Yeah," the postdoc sighed, scratching his head. "Sometimes I wonder if my adviser hates me."

"The other thing I don't get," Johnson said, pointing at the pizza box, "is that this family tree you've drawn shows rats and guinea pigs over here, and primates and humans over there. And the unequivocal support for this tree is pseudogenes, like the kind you're studying."

"Right. And your point is?"

"Well, if the fact that humans and chimps share the same DNA mistakes can only be explained by positing a shared human-chimp ancestor that carried those mistakes, what about the DNA mistakes that guinea pigs share with us? Do we share a nonvitamin-C-producing ancestor with them too?"

"Of course not!"

"So there must be some other way that the shared mistakes can get there besides having a common ancestor, right?"

"A gene hot spot, according to the Japanese researchers."

"A hot spot?"

"Yes, certain sites in the GULO sequence that preferentially accumulate mutations."

"OK, a hot spot or something." Johnson rose to his feet and moved in for the kill. "But that means that pseudogenes aren't the unequivocal support for evolution everyone knows them to be, since you could just as easily explain the shared mistakes by invoking a hot spot instead of a common ancestor. Isn't that true?"

With the law professor leaning over him, the postdoc suddenly felt a bit unsure of his years of training. Was evolution really unequivocally supported by all the studies he had read? This simple country lawyer was starting to get under his skin.

Once again the wedge of truth splits assumptions about science from science itself.

The Right Questions About Science

What are the right questions to ask about science? The story of Phil Johnson visiting a Berkeley lab illustrates how scientists, on their own, often don't ask themselves the right questions. As long as they are spinning their test tubes, sucking off their supernatants and electrifying their DNA gels, the scientist's bag of questions is perfectly sufficient. Questions like "Should I use Tris buffer or PBS for this reaction?" or "Has my standard gone bad, or is my DNA really twice as long as it should be?" get the job done for solving problems in day-to-day research.

What about solving problems when the stakes are higher? How would a postdoc know if he or she is barking up the wrong phylogenetic tree? When is it time to give up on the appeal to chance mutations and natural selection in favor of the intelligent design (ID) hypothesis? In these situations, the tried-and-true bag of questions is not enough. Scientists need to get beyond the Tris-buffer-inhibits-biotinylation-so-I'd-better-use-PBS kind of logic and look at the very assumptions that drive them. "Why am I studying this system?" and "Why do I interpret my results like this instead of like that?" might be a better way to start.

Though Phil Johnson, to my knowledge, never had a conversation with a postdoc about guinea pig GULO, he has had a powerful effect on the way scientists think—at least for the scientists who are willing to hear what he has to say. The main contribution Johnson has made to science was in pointing out that the gatekeepers of science hold religiously to their faith in materialism, just like a Muslim holds to faith in Allah or a Jew to faith in Yahweh. Johnson further demonstrated that without blind allegiance to materialism, Darwinism becomes unsupported.

In this essay, I will examine Johnson's contribution to science from three points of view. First, I will explore the personal benefit experienced by scientists who, at Johnson's urging, have reexamined their assumptions. Second, I will look at Johnson's insight from the broad perspective of scientific progress, with careful attention to the structure of scientific revolutions. Finally, I will investigate the way Johnson has affected and should continue to affect the field of evolutionary biology.

Helping Scientists Help Themselves

"Know thyself," the great Socratic imperative, could be rephrased, "Know thy assumptions." Scientists who have come to terms with their assumptions are the wiser for doing so. From my point of view, most scientists do not take time to understand their core assumptions. Scientists who have read and appreciated Johnson's work are the notable exceptions to this rule. By and large, these exceptions have become members of Johnson's "wedge strategy," aligning themselves with the growing ID community.

Most intelligent-design scientists have spent a good deal of effort excavating the layers of assumptions that guide their everyday decisions. These assumptions may have been handed down from parents, pastors or professors over several decades. A good example of how one scientist came to terms with his assumptions is found in Michael Behe's contribution "From

Muttering to Mayhem: How Phillip Johnson Got Me Moving" (see chap. 2). In Behe's case, his assumptions about the veracity of evolution were suddenly upended after reading Michael Denton's *Evolution: A Theory in Crisis* and were further shaped by reading *Darwin on Trial.*[3]

For others, the process of sifting through hidden assumptions takes much longer. One chemical engineering professor recalled how over the course of several years his materialistic assumptions began to unravel, thanks in part to Johnson's wedge strategy. "Suddenly it dawned on me," he recounts,

> that I was starting to agree with my crazy old lady next door who told me when I was a kid how God made ice float to protect the fish in the winter. With my old assumptions, I could see nothing but the hydrogen bonds that made ice less dense than water, but with my new perspective, I could see how ice and everything else may have been fine-tuned by a designer.[4]

These two professors illustrate how old assumptions can be reexamined and thrown out, with new, openly acknowledged assumptions put in their places. But for some younger scientists (I would count myself in this category), our assumptions were not discarded after Johnson motivated us to examine them, but they were definitely subjected to careful scrutiny.

For example, a former postdoc of mine had a feeling that the universe was the product of a designer. However, his arguments to defend this position always seemed to regress to "Well, you might not agree, but that's just what I think is true." However, after having read *Darwin on Trial,* the arguments could be framed at the level of "My assumptions about the universe don't preclude the possibility of a designer the way yours seem to, and with both options on the table I would say that the evidence points to a designer."

In a similar way, I had long understood that the academy's naturalistic worldview and materialist metaphysics have a big impact in the historical sciences but could not see any practical way to make the case. The quasi-religious nature of Darwinism seemed too well disguised by the outward forms of scholarly objectivity and too well protected by the career-killing consequences of being labeled "creationist." Johnson's book *Darwin on Trial* showed that with a bit of care (and a fair amount of space, and a lot of guts) it was really not that difficult to show convincingly that Darwinism is, in Phil Johnson's phrase, "soaked with materialism."

Although Johnson's writings may have only encouraged the open-minded to examine their assumptions, he still has had an influence on the most reluctant scientists. Most often, Johnson's reasoning finds fertile ground with one scholar who can hold his or her colleagues accountable to examine their own core beliefs. Take William Provine, Cornell University professor in the history of biology, as a prime example. He writes in a book review about the religious assumptions inherent in evolutionary biology: "If the primary effect of teaching about creationism is to advance religion, then it appears to me equally true that teaching modern evolutionary biology inhibits religion."[5]

Provine, a close acquaintance of Phil Johnson, openly admits his religious assumptions. Although he does not align himself with the ID community, he still helps Johnson influence the thinking of more close-minded evolutionists. He concludes his book review with a penultimate paragraph that begins: "And I have a suggestion for evolutionists. Include discussion of supernatural origins in your classes, and promote discussion of them in public and other schools. Come off your high horse about having only evolution taught in science classes. The exclusionism you promote is painfully self-serving and smacks of elitism."[6]

If scientists are unwilling to take advice from a Berkeley lawyer, perhaps they might consider learning from one of Johnson's open-minded foes. Either way, Johnson has influenced, and will continue to influence, the way scientists think on a personal level.

The Revolutionary from Aurora, Illinois

In 1997 Tim Stafford published a story on Phillip Johnson titled "The Making of a Revolution."[7] The article recounted Johnson's experience at a conference for Christian university professors whose Christianity "made no direct impact on their ideas." Johnson told Stafford, "What [they] were all doing was taking a naturalistic approach for intellectual purposes." When asked why the Christian professors saw naturalism as the only way to proceed, Johnson replied, "They think that it's been validated by science. At the very heart of that scientific validation is the story of life, the story of our creation."[8]

The unexamined assumptions of these Christian professors launched Johnson on a revolution, a "program of transformation," as Gandhi would call it. Like Behe and others, Johnson was influenced by Denton's book and saw clearly how the battle must be fought over metaphysical frameworks and hidden assumptions. Though not a scientist, he was catalyzing what

Thomas Kuhn called a "paradigm shift," and as Kuhn pointed out, these catalysts are often very young scientists or people who are very new to the scientific community.[9]

Without a doubt, Johnson could easily poke holes in Kuhn's "nearly perfect" analogy between scientific progress and Darwinian evolution—which, according to Kuhn, has no guide and no objective goal. Moreover, Johnson would eschew Kuhn's relativistic worldview. Quite the contrary, Johnson has repeatedly argued for theistic realism.[10] Just as scientific realism aimed to displace Kuhn's relativism as an alternative to logical positivism, Johnson offers theistic realism as a solution to the quagmire of relativism, constructionism and irrationalism. But unlike scientific realism, Johnson's theistic realism would bring back absolute truth as an entity to be confronted, marking a return to the classical realism of Aquinas and other pre-Darwinian thinkers (see Nancy Pearcey's contribution in chap. 14). Johnson sees theistic realism (combined perhaps with its weaker but more palatable scientific counterpart) as the backbone metaphysics in the Darwin-free paradigm that he works hard to promote.

Ironically, Kuhn used Darwinism as an analogy for scientific revolutions, yet his description of entrenched paradigms that resist change can best be seen in the religiously held views of Darwinists. And the strategies that Kuhn says are essential for the overthrow of an entrenched paradigm are precisely the tactics that Johnson has mastered in his conflict with Darwinism. For example, Kuhn points out that scientific revolutions are affected by the impact of logic and the techniques of persuasive argumentation.[11] This, Kuhn says, is not because scientists value argument over data but because the data will inevitably be seen from two conflicting points of view, and logic—such as the impeccable logic of Phil Johnson—must be brought to bear on the evaluation of each point of view.

The revolution from the paradigm of Darwinism to the paradigm of intelligent design will undoubtedly be accompanied by a metaphysical shift from materialism to theistic realism. Since materialism is the prevailing philosophy of the entire academy, the stakes for this paradigm shift are much higher than what Kuhn envisioned. That is not to say that the shift will be accordingly more difficult. After all, the task will be simply to help scientists examine their assumptions and philosophical frameworks and not to convince them to change their metaphysics.[12] Most scientists are not intrinsically wed to the philosophy of materialism. Once Johnson and others can help them see that their cherished Darwinian worldview is propped up by a philosophy they already find unconvincing, the whole entrenched paradigm will crumble. True to his revolutionary nature, Johnson likens this to the collapse of the Soviet Bloc: one day it seemed an unassailable monolith; the next, it was a crumbling wall toppled by dancing East and West Berliners.[13]

Afflicting the Comfortable

But how can a lawyer whose formal biological education ended in the tenth grade make a whit of difference in the world of evolutionary biology? To answer that question, consider Johnson's effect on the most famous evolutionary biologist of our time, the late Stephen Jay Gould. Gould took it upon himself to write a scathing review of *Darwin on Trial* in the July 1992 issue of *Scientific American.*[14]

Five years later, with Johnson still under his skin (perhaps also because of the hug Johnson gave him at a meeting in 1990),[15] Gould's irritation came out in the form of an essay for *Natural History* magazine titled "Nonoverlapping Magisteria,"[16] which later was expanded in the book *Rocks of Ages: Science and Religion in the Fullness of Life.*[17]

Gould's book, essay and review represented his attempt to stop Johnson's kind of thinking from spreading, especially to younger evolutionary biologists. Why does this icon of evolution—indeed, the only biologist ever to appear in person on *The Simpsons*—take the time to shift his great weight against the paltry challenge of Phillip Johnson's ID arguments? Because this evolutionary heavyweight knew that if young scientists, if small-school scientists working under the shadow of Harvard, if mom and pop at the school board meeting or if any of the average Joes who watch *The Simpsons*[18] found out that the authority of evolutionary biology was nothing more than the authority of metaphysical materialism, Gould's goose would be cooked.

Carrying the Battle Forward: Two Scenarios

So where do we go from here? Phil Johnson has shown us a workable approach: point out the weaknesses in the Darwinian creation story; emphasize the role played by metaphysics and how it covers up those weaknesses; insist on critical thinking, tolerance of "heretical" ideas and letting the evidence itself decide; point out the large, basic differences between the empirical and the historical sciences; make the case that saying "We don't know" is an acceptable outcome. But though ID has made great progress, most of both the public and the scientific community remain unaware of our real arguments and positions. Many of those who are aware have a very distorted picture painted by our opponents.

I can envision two plausible scenarios for the way forward. In the first, we succeed in taking our case directly to the scientific community itself. Few scientists are dogmatic Darwinists; most are closer to kids taught their worldview in Darwinian Sunday school. Unfortunately, most scientists are also (like most nonscientists) rather apathetic about ID-related issues. So the people we mainly engage with are the ideologues and true believers of the church of Darwin. A big part of our task is to get around these gatekeepers and take our case to the average guy at the bench.

One of the main questions every ID person gets from even well-disposed fellow scientists is, How would you go about doing ID? The only really good reply is to give examples. If we want our fellow scientists to accept ID as a player in the sandbox, we need to show them some good fruit. And ID's bigger, freer paradigm has some intrinsic advantages: (1) It's metaphysically much more sophisticated. Few scientists, even among the prominent metaphysicians of the scientific community, like Richard Dawkins, E. O. Wilson and the late Stephen Jay Gould, have really thought through what it means to adopt the ultimately pessimistic implications of materialism. The philosophical side of ID provides that analysis and also points out that there is no need to despair; science really hasn't shown that our existence is meaningless, and suicide is not the only really important decision. (2) It's more rigorous. The way the historical sciences proceed today—assuming a naturalistic mechanism for every event in the past, dreaming up stories based on that assumption and then issuing papers and press conferences full of weasel words ("this is how it *may* have happened")—is so obviously unscientific that the historical sciences have a bad reputation among their more rigorous colleagues in the harder sciences. Under the ID paradigm, assumptions need to be stated and defended up front, so even if we usually can't say how some historical event happened for sure, we are better aware of where the story might go wrong. (3) It's new. I can't help believing that a lot of scientists are tired of the endless stream of variants on the same old story. A couple of splashy successes, pushed by IDers with lots of panache, and it may become fashionable (with young, daring scientists) to poke a finger at the old guard by embracing a little heresy.

All of this follows the Kuhnian template: if we want converts to the new paradigm, we need to show people what it would be like to do ID science. We must show how attractive it is. The most important thing is not so much the conclusions reached or the thesis defended but the palpable competence, rigor and general coolness of the ideas presented. We still suffer from the stereotype of old stick-in-the-mud fundamentalists, but a few widely known counterexamples would dispel that forever. Then we become the young rebels taking on the old generalissimos of the establishment. Being on the outside has its advantages—if we know how to use our distinctives.

In the second scenario the scientific community remains closed to us. I think this is unlikely, but it is possible that tradition, self-interest, spin and politics may be able to keep us out, at least as long as we connect ourselves explicitly to ID. This is the Long March scenario, where, to use an analogy from another revolution, ID remains in the caves of Hunan for a while. I believe that even in this case the strategy should be much the same. The main difference will be that our writings (and financial support) will be more weighted toward books and popular writings and less toward official papers in the science journals. Even so, the scientific community must remain a prime target audience, and as much as possible our writings must maintain the same thoroughness and rigor as would be demanded for journal publications.

With the public, ID seems to have already become very mainstream, especially among evangelical Christians (no surprise here; evangelical worldviews, history and politics all favor ID), and to a lesser extent among Christians generally as well as nonconformists of all stripes. If we influence the worldview of the next generation, ID ideas will eventually gain influence in academia, no matter what the Darwinian orthodox do.

By either scenario, our task is to change worldviews. And that brings us to the final element in Philip Johnson's recipe for revolution. Changing worldviews is not easy to do: it takes a fairly extended argument, which takes time, effort and attention on the part of our audience. About the only way to get people to spend their time listening to us is to make the process enjoyable—to make people want to listen because it's fun. It doesn't matter if it's scientists or the public: everybody wants to be entertained; nobody wants a sermon. Let the other guys be the preachers; we want to be the smiling tweakers of rigid authority. Even more important than exposing the impact of naturalism or critiquing the weaknesses of evolutionary stories, even more important than the big tent of ID, is Philip Johnson's rule number one: keep it interesting.

COMMON ANCESTRY ON TRIAL

JONATHAN WELLS

—

The year was 1991. I had just finished my coursework and exams for a Ph.D. in molecular and cell biology at Berkeley when someone told me that a Berkeley law professor was coming out with a book critical of Darwinism. I myself had rejected some aspects of evolutionary theory since the late 1970s, but most of the books I had read that were critical of Darwinism had been second-rate, and I was not eager to read another one.[1] Besides, some of my biology colleagues told me that this latest book was terrible—and I figured that if it came from a lawyer it probably was.

Then Walt Hearn, a retired biochemist and member of the American Scientific Affiliation whom I had known since 1980, told me the book was actually quite well done. Curious, I bought a copy of Phillip E. Johnson's *Darwin on Trial* and started reading. The book was so good that I read it through to the end almost without stopping. When I finished, I wanted to stand up and cheer.

The next day I telephoned Phil, whose law school office was on the other side of campus, and we met for lunch. (Walt, who lived in Berkeley and knew Phil, was kind enough to come along to introduce us.) Thus began a friendship that has been one of the best things that has ever happened to me. The focus of this essay, however, is not my friendship with Phil but the impact his book has had on my thinking.

Theism Versus Darwinism

Years before, as a seminary student at Unification Theological Seminary in the late 1970s, I had become convinced that there is a fundamental conflict between theistic religions and Darwinian evolution.[2] Among the former I include Christianity, Islam, Judaism, Unificationism and Zoroastrianism.[3] By the latter I mean the theory that natural processes such as genetic mutation and survival of the fittest account for the origin of all features of living things.

While a seminary student I read works by liberal theologians who took Darwinism for granted and thus saw no room for God's intervention in nature or history. I also listened to many talks by Reverend Moon, who was critical of Darwinism for promoting the belief that living things originated without God's creative activity.

The conflict between theism and Darwinism bothered me deeply. Before attending seminary I had been an undergraduate science major at Princeton and Berkeley. There I had been steeped in Darwinian evolution, and I accepted it without question. Now I realized that I couldn't be a theist and a Darwinian. I decided to turn to the evidence. If theism and Darwinism were logically incompatible, and the scientific evidence supported the latter, then so much the worse for theism. But did the evidence really support Darwinian theory?

Several days a week I made the two-hour trip from the seminary to New York City to do research at the Columbia University biology library. I focused on the mechanism of evolution—specifically, the neo-Darwinian mechanism of natural selection acting on random genetic mutations. I found that evidence for selection, with a few highly questionable exceptions, was limited to minor changes within existing species. That left mutations, which supposedly provide the raw materials for evolution and also seemed most directly opposed to theism. As Jacques Monod had said in 1970: "With the understanding of the random physical basis of mutation that molecular biology has provided, the mechanism of Darwinism is at last securely founded, and man has to understand that he is a mere accident."[4]

I quickly learned, however, that genetic mutations are almost always harmful. Sure, a few are beneficial to the organisms that carry them in cases of resistance to antibiotics, pesticides or herbicides. Such cases, however, involve only minor biochemical changes. I found absolutely no evidence that genetic mutations can produce beneficial changes in anatomy, of the sort needed by evolutionary theory. Nor did I find evidence that mutations (any more than selection) could produce new species.

So by 1978 I had become convinced that the neo-Darwinian mechanism of evolution was scientifically unsupported. Yet most biologists continued

to promote and defend it, and liberal theologians continued to accommo-
date their views to it. I considered this a misuse of the scientific enterprise,
and I decided that my mission was to criticize and discredit it. Just as many
of my fellow Unificationists had dedicated themselves to destroying the antitheistic ideology of Marxism, I dedicated myself to destroying the antitheistic ideology of neo-Darwinism.

My research at Columbia gave me no reason, however, to doubt the Darwinian picture of the history of life. I still accepted without question the
standard distinction between the "pattern" and "process" of evolution—the
former referring to the branching-tree pattern of descent with modification
from a common ancestor, and the latter referring to the particular mechanisms responsible for modification. Although I knew that the processes of
evolution were controversial on scientific grounds, I did not yet have any
quarrel with the pattern. Since I accepted descent with modification from a
common ancestor—what Darwinists call the "fact of evolution"—I considered myself a "theistic evolutionist."

Yale and Berkeley

In 1978, Reverend Moon chose me and several other seminary graduates to
receive church scholarships for further graduate study in theology. I decided
to enter a Ph.D. program in religious studies at Yale.

At Yale I learned that the root of the conflict between Christian theology
and Darwinian evolution in the nineteenth century was not the age of the
earth or the fall of Adam and Eve or the origin of the soul. The root of the
conflict was *design*.[5] According to the Christian theological tradition, living
things are designed, and human beings (having been planned from the start
and created in the image of God) are the most designed of all. According to
Darwinism, however, living things are accidental byproducts of purposeless
natural causes, and human beings (as the latest in a long line of accidental
events) are the least designed of all.

Since the root of the conflict between theism and Darwinism was design, and the neo-Darwinian mechanism that seemed to exclude design
was empirically unsupported, I thought that theism could be saved by rejecting the mechanism. On the other hand, since the Darwinian pattern of
descent with modification seemed both theologically neutral and empirically well supported, I continued to accept it. In other words, I remained
a theistic evolutionist.

After completing my Yale Ph.D. in 1986, I worked for two years as director of an interreligious conference organization. (It was through this organization that I had met Walt Hearn years earlier.) In the late 1980s, however, the cultural influence of neo-Darwinism continued to grow, and I felt that it was still my mission to put a stop to this misuse of science. In 1988 I resigned my position and made preparations to return to graduate school—this time in biology.

In 1989 I entered the Ph.D. program in molecular and cell biology at the University of California, Berkeley. I planned to do research on nongenetic factors in embryo development, which I believed would undercut the neo-Darwinian emphasis on DNA mutations as raw materials for evolution. I found that several of my professors at Berkeley were as skeptical as I of the neo-Darwinian mechanism, though they (like me) accepted the overall pattern of evolution. As one of them put it, "Darwinism, yes; neo-Darwinism, no."

So during my first two years at Berkeley, I continued to accept the "fact of evolution." Eager to stay abreast of the latest developments in evolutionary biology and the science-religion dialogue, I joined two organizations dedicated to promoting the former (the National Center for Science Education and the Institute for Human Origins), and two organizations dedicated to promoting the latter (the American Scientific Affiliation and the Center for Theology and the Natural Sciences). Secure in my belief that the antitheistic implications of Darwinism were a limited problem that I could solve without leaving mainstream academia, I dozed comfortably in the lap of theistic evolutionism.

Then along came Phillip Johnson, who woke me from my dogmatic slumber.

Darwin on Trial

Phil's book began with criticisms of natural selection and genetic mutation that were quite familiar to me, but I enjoyed reading about them again from such an able writer. By the time I reached chapter four ("The Fossil Problem"), I had forgotten about my reluctance to read yet another book criticizing Darwinism.

Even the problems with the fossil record that Phil pointed out were old hat to me, but this time they made a new and deeper impression. Maybe I was just ready to hear them, or maybe it was the clarity Phil brought to bear on the subject, or maybe it was the combination of the two.

For example, I already knew that Stephen Jay Gould considered the extreme rarity of transitional forms in the fossil record to be "the trade secret of paleontology," and I had read ad nauseam about punctuated equilibria. Yet while Gould, Niles Eldredge and Steven Stanley were reassuring everyone that sudden appearance and absence of change were exactly what Darwin's theory would predict, Phil was spreading the naked truth: "If evolution means the gradual change of one kind of organism into another kind, the outstanding characteristic of the fossil record is the absence of evidence for evolution."[6] The important thing about the "punk eek" (punctuated equilibrium) controversy, Phil wrote, is not the theoretical solution that Gould, Eldredge and Stanley proposed, but "the problem to which they drew attention."[7]

Phil hung onto his point like a bulldog. Even if we grant that mammal-like reptiles were exactly the sorts of transitional forms needed by Darwin's theory (a point I used to, but would no longer, grant), the fact remains that the fossil record as a whole is strikingly devoid of the innumerable transitional forms posited by Darwin. The most glaring example of this is the Cambrian explosion, which alone (as Darwin himself acknowledged) poses a serious problem for evolutionary theory.

As an undergraduate I had been taught not to worry because fossils were only part of the overwhelming evidence for the pattern of evolution. Other areas such as embryology provided plenty of evidence too. In particular, I had been shown drawings of vertebrate embryos showing that they look almost identical in their early stages. The common ancestry of humans and fish was supposedly obvious in the striking resemblance between their embryos. As a graduate student in developmental biology, however, I learned that the actual embryos didn't look like the drawings in the textbooks. What Charles Darwin had considered "by far the strongest class of facts" in favor of his theory turned out not to be facts at all.[8]

I had also been taught that modern molecular data provided a great deal of evidence. But when I looked into the matter, I could see that the molecular evidence, like the fossil and embryological evidence, was fragmentary, contradictory, and open to conflicting interpretations. While a student in the standard upper-division Berkeley course on evolution, I wrote a research paper on the use of homologous gene sequences in reconstructing evolutionary history. I learned, to my surprise, that the very notion of homology is confused. On the one hand, homologous genes are those with similar sequences; on the other hand, homologous genes are those inherited from a common ancestor. How can we tell whether similar gene sequences come from a common ancestor or from another source (such as horizontal gene transfer)? By looking at fossils and embryos, I was told. But the molecular evidence was supposed to shed light on the fossils and embryos, not vice versa. The whole process began to look like a dog chasing its own tail.

Darwin on Trial didn't teach me all these things, but the book confirmed my vague intuitions about them and encouraged me to confront their implications. Darwinian evolution was more deeply flawed than I had ever imagined. What was really going on here?

Putting the Cart Before the Horse

When I had first set my sights on destroying Darwinism, I thought its offense was limited to drawing antitheistic conclusions from an empirically unsupported mechanism. I thought that once the evidence for that mechanism could be shown to be insufficient and the conclusions to be unwarranted, the problem would be solved. The "fact" of evolution, the overall pattern of decent with modification from a common ancestor, would remain intact. After reading Phil's book, however, I realized that I had been completely mistaken.

The whole Darwinian story—pattern as well as process—was driven by an a priori commitment to materialistic, antitheistic philosophy. The pattern of evolution was not something that had been inferred from overwhelming evidence in paleontology, molecular biology and embryology, but something that was assumed to be true from the start. This certainly seemed to be true in paleoanthropology—the study of hominid fossils. According to paleoanthropologist Misia Landau, "themes found in recent paleoanthropological writing . . . far exceed what can be inferred from the study of fossils alone and in fact place a heavy burden of interpretation on the fossil record—a burden which is relieved by placing fossils into preexisting narrative structures."[9]

So it's theory first, evidence later. At Berkeley I could see the same principle operating in my own field of embryology. The fledgling field of evolutionary developmental biology ("evo-devo") was hot stuff while I was doing my Ph.D. research. Evolutionary theory had neglected embryology for decades, preferring to focus on population genetics instead, but some Darwinists now wanted to bring the two together. As I immersed myself in evo-devo, I realized that the distorted drawings of vertebrate embryos were just the tip of an iceberg. It turned out that the stage portrayed in the drawings was not the earliest stage of development but its midpoint. Early vertebrate embryos actually look radically different from each other, and they converge in appearance only after passing through very different developmental pathways. But Darwin had reasoned that it was the earliest stages that show us common ancestry. The same reasoning applied to actual embryos would suggest that the various classes of vertebrates had separate origins!

Of course, such a conclusion would be thoroughly un-Darwinian. My evo-devo colleagues, who were "evo" first and "devo" second, tried to save Darwin's theory by arguing that early development must be much more pliable than anyone had supposed. According to them, the midpoint at which vertebrate embryos looked vaguely similar was somehow evolutionarily "constrained," but the earlier stages must have been free to evolve in different directions.

There was nothing logically inconsistent in this view, but it meant that early embryos could no longer provide evidence for evolution. The early stages of embryo development—like many features in the fossil record—had become something the theory had to explain away. The only way to do this was to assume the theory to be true on other grounds, then use it to interpret seemingly incongruous evidence.

Theory first, evidence later. I had been led to believe that the driving force in science was the evidence, but this is not what I was seeing. In the case of Darwinian evolution, the cart had clearly been put before the horse. And Phil saw it too. "Naturalism and empiricism are often erroneously assumed to be very nearly the same thing," he wrote, "but they are not. In the case of Darwinism, these two foundational principles of science are in conflict." In fact, for Darwinists, "empiricism is *not* the primary value at stake. The more important priority is to maintain the naturalistic worldview."[10] Since that was the highest priority for Darwinian "science," evidence mattered only to the extent that it could appear to provide plausibility for the theory. As Phil put it, the objective was always "to find confirmation for a theory which was conclusively presumed to be true at the start of the investigation."[11]

What then about the pattern of universal common ancestry that I had taken for granted for so many years? What about the "fact" of evolution?

Is Universal Common Ancestry a Fact?

I had been taught that the Darwinian pattern of descent with modification was a simple fact, like gravity. No sane person doubts the fact that unsupported heavy objects fall to the ground, but reasonable people disagree over theories about why it happens. Similarly, I was taught that no sane person doubts the fact that all living things are descended from a common ancestor, though scientists disagree over theories about how living things became modified from the ancestral form.

But is descent with modification a simple fact? Parents directly observe descent with modification within their own families. Everyone familiar with living things knows that descent with modification happens—within existing species. Watching descent with modification within a species is analogous to watching an object fall to the ground. But no one has ever seen descent with modification produce a *new* species. If I want reports about nonhuman babies being born to human parents, I have to read supermarket tabloids, not *Science* or *Nature*.

So, as a description of the origin of new species, descent with modification from a common ancestor is *not* a fact. If anything about descent with modification is a fact, it is that like begets like. As Phil wrote:

> Ancestors give birth to descendants by the same reproductive process that we observe today, extended through millions of years. Like begets like, and so this process can only produce major transformations by accumulating the small differences that distinguish offspring from their parents. Some shaping force must also be involved to build complex organs in small steps, and that force can only be natural selection. There may be arguments about the details, but all the basic elements of Darwinism are implied in the concept of ancestral descent.[12]

In other words, the Darwinists' distinction between the pattern of descent with modification and the process of natural selection, between fact and theory, is bogus. All the distinction really does is protect Darwin's theory from empirical testing. After all, who (except perhaps a beginning physics student) would bother to test the fact that heavy objects fall to the ground? "Recasting the theory as fact serves no purpose other than to protect it from falsification," Phil wrote. "Nobody needs to prove that apples fall down rather than up," but Darwinists are constantly offering "proofs" of universal common ancestry.[13]

Those "proofs" typically include fossils, embryos and molecules, but (as we have seen) none of them provides good evidence for the hypothesis of common ancestry. Indeed, anomalies in all three present serious problems for the hypothesis. By calling the hypothesis a "fact," Darwinists divert attention from those anomalies and reassert the primacy of their theory over the evidence.

After reading Phil's book, I tried to imagine how things might have happened if Darwin's theory were true. I reflected on the few generations in my own family, of which I had direct knowledge—my grandparents, my parents, my childhood and my own children. Then I tried to stretch my mind to think back thousands, even millions of years, to the apelike creature from which my great-ancestors had supposedly been born. I tried to imagine where in this chain the important changes had occurred. Did the apelike creature have a child that was 1 percent human? Did it have a child that was 2 percent human? Given what I knew about the *real* fact of descent with modification—that like begets like—none of this made any sense. The more I tried to imagine something else—something Darwinian—the more wooly-headed I felt. The effect was similar to the feeling I used to get in the middle of a boring afternoon class in a stuffy, overheated lecture hall. *This* was the slumber into which I had fallen under the beguiling spell of Darwinian dogma. And it was this dogmatic slumber from which *Darwin on Trial* had roused me.

Nothing in Biology Makes Sense Except in the Light of Evidence

When I had started questioning Darwinism in the 1970s, my inclination was to give Darwinian biologists the benefit of the doubt. After all, I had been educated in an environment suffused with the view that "nothing in biology makes sense except in the light of evolution."[14]

So, initially I challenged only as much of evolutionary theory as I thought I had to. As a theology student, I disagreed with the antitheistic implications of neo-Darwinism, but I found that those implications were unwarranted extrapolations from a mechanism that lacked empirical support. I became a critic of that mechanism and those implications, but I remained a Darwinian in my acceptance of the pattern of descent with modification.

Then, as a biology student I got a good look at some of the principal evidence for descent with modification. I was surprised by the problems I found, but it wasn't until I was emboldened by Phil's book that I began to confront them fully. When I became a critic of universal common ancestry, it was not initially on theological grounds but on empirical ones. Nevertheless, as Phil pointed out, common ancestry was not as theologically neutral as I had thought. It was a deduction from the same dogmatic exclusion of divine action that had caught my attention in the first place. Now I knew that this deduction, disguised as a scientific fact, was even *less* supported by the evidence than the neo-Darwinian mechanism I had started out criticizing.

I underwent a radical transformation. Rather than being someone who accepted most of the evolutionary story and challenged only its unwarranted implications, I became someone deeply skeptical of *all* claims made by Darwinists. I realized that I had been granting evolutionary theory entirely too much credence, and I began asking to see the evidence for every aspect of it. The more evidence I saw, the less plausible the theory looked. And the "fact" of universal common ancestry was not a fact at all.

Since then I have become convinced that Darwinism's monopolistic domination of biology has actually obstructed scientific progress. Theories are indispensable to science, but when people elevate a particular theory to the status of unquestioned fact, and they claim the right to dismiss or reinterpret any troublesome evidence, then science enters a dream world where dogma rules and objective reality disappears. I predict that biomedical science will enter a new and fruitful era of discovery once it wipes the Darwinian sand from its eyes. Exactly how this will happen, and what fruits will come, I don't know yet. But I know this: Phillip Johnson's *Darwin on Trial* helped me to wipe the sand from my eyes.

The Origin of Biological Information and the Higher Taxonomic Categories

Stephen C. Meyer

—

In a recent volume of the "Vienna Series in Theoretical Biology," Gerd B. Muller and Stuart Newman argue that what they call the "origination of organismal form" remains an unsolved problem. In making this claim, Muller and Newman distinguish two distinct issues, namely, (1) the causes of form generation in the individual organism during embryological development, and (2) the causes responsible for the production of novel organismal forms in the first place during the history of life.[1] To distinguish the latter case (phylogeny) from the former (ontogeny), Muller and Newman use the term origination to designate the causal processes by which biological form first arose during the evolution of life. They insist that "the molecular mechanisms that bring about biological form in modern day embryos should not be confused" with the causes responsible for the origin (or "origination") of novel biological forms during the history of life.[2] They further argue that we know more about the causes of ontogenesis, due to advances in molecular biology, molecular genetics and developmental biology, than we do about the causes of phylogenesis—the ultimate origination of new biological forms during the remote past.

In making this claim Muller and Newman are careful to affirm that evolutionary biology has succeeded in explaining how preexisting forms diversify under the twin influences of natural selection and variation of genetic traits. Sophisticated mathematically based models of population genetics have proven adequate for mapping and understanding quantitative variability and populational changes in organisms. Yet Muller and Newman insist that population genetics, and thus evolutionary biology, has not identified a specifically causal explanation for the origin of true morphological novelty during the history of life. Central to their concern is what they see as the inadequacy of the variation of genetic traits as a source of new form and structure. They note, following Darwin himself, that the sources of new form and structure must precede the action of natural selection—that selection must act on what already exists.[3] Yet, in their view, the "genocentricity" and "incrementalism" of the neo-Darwinian mechanism has meant that an adequate source of new form and structure has yet to be identified by theoretical biologists. Instead, Muller and Newman see the need to identify epigenetic sources of morphological innovation during the evolution of life. In the meantime, however, they insist neo-Darwinism lacks any "theory of the generative."[4]

As it happens, Muller and Newman are not alone in this judgment. In the last decade or so a host of scientific essays and books have questioned the efficacy of selection and mutation as a mechanism for generating morphological novelty, as even a brief literature survey will establish. K. S. Thomson expressed doubt that large-scale morphological changes could accumulate via minor phenotypic changes at the population genetic level.[5] G. L. G. Miklos argued that neo-Darwinism fails to provide a mechanism that can produce large-scale innovations in form and complexity.[6] Gilbert, Opitz and Raff attempted to develop a new theory of evolutionary mechanisms to supplement classical neo-Darwinism, which, they argued, could not adequately explain macroevolution. As they put it in a memorable summary of the situation:

> Starting in the 1970s, many biologists began questioning its (neo-Darwinism's) adequacy in explaining evolution. Genetics might be adequate for explaining microevolution, but microevolutionary changes in gene frequency were not seen as able to turn a reptile into a mammal or to convert a fish into an amphibian. Microevolution looks at adaptations that concern the survival of the fittest, not the arrival of the fittest. As Goodwin (1995) points out, "the origin of species—Darwin's problem—remains unsolved."[7]

Though Gilbert, Opitz and Raff attempted to solve the problem of the origin of form by proposing a greater role for developmental genetics within an otherwise neo-Darwinian framework,[8] numerous recent authors have continued to raise questions about the adequacy of that framework itself or about the problem of the origination of form generally.[9]

What lies behind this skepticism? Is it warranted? Is a new and specifically causal theory needed to explain the origination of biological form?

This review will address these questions. It will do so by analyzing the problem of the origination of organismal form (and the corresponding emergence of higher taxa) from a particular theoretical standpoint. Specifically, it will treat the problem of the origination of the higher taxonomic groups as a manifestation of a deeper problem, namely, the problem of the origin of the information (whether genetic or epigenetic) that, as it will be argued, is necessary to generate morphological novelty.

In order to perform this analysis and to make it relevant and tractable to systematists and paleontologists, this chapter will examine a paradigmatic example of the origin of biological form and information during the history of life: the Cambrian explosion. During the Cambrian explosion, many novel animal forms and body plans (representing new phyla, subphyla and classes) arose in a geologically brief period of time. The following information-based analysis of the Cambrian explosion will support the claim of recent authors such as Muller and Newman that the mechanism of selection and genetic mutation does not constitute an adequate causal explanation of the origination of biological form in the higher taxonomic groups. It will also suggest the need to explore other possible causal factors for the origin of form and information during the evolution of life and will examine some other possibilities that have been proposed.

The Cambrian Explosion

The Cambrian explosion refers to the geologically sudden appearance of many new animal body plans about 530 million years ago. At this time, at least nineteen and perhaps as many as thirty-five phyla of forty total[10] made their first appearance on earth within a narrow five- to ten-million-year window of geologic time.[11] Many new subphyla, between thirty-two and forty-eight of fifty-six total,[12] and classes of animals also arose at this time with representatives of these new higher taxa manifesting significant morphological innovations. The Cambrian explosion thus marked a major episode of morphogenesis in which many new and disparate organismal forms arose in a geologically brief period of time.

To say that the fauna of the Cambrian period appeared in a geologically sudden manner also implies the absence of clear transitional intermediate forms connecting Cambrian animals with simpler pre-Cambrian forms. And indeed, in almost all cases the Cambrian animals have no clear morphological antecedents in earlier Vendian or Precambrian fauna.[13] Further, several recent discoveries and analyses suggest that these morphological gaps may not be merely an artifact of incomplete sampling of the fossil record,[14] suggesting that the fossil record is at least approximately reliable.[15]

As a result, debate now exists about the extent to which this pattern of evidence comports with a strictly monophyletic view of evolution.[16] Further, among those who accept a monophyletic view of the history of life, debate exists about whether to privilege fossil or molecular data and analyses. Those who think the fossil data provide a more reliable picture of the origin of the Metazoan tend to think these animals arose relatively quickly—that the Cambrian explosion had a "short fuse."[17] Some,[18] but not all,[19] who think that molecular phylogenies establish reliable divergence times from pre-Cambrian ancestors think that the Cambrian animals evolved over a very long period of time—that the Cambrian explosion had a "long fuse." This review will not address these questions of historical pattern. Instead, it will analyze whether the neo-Darwinian process of mutation and selection or other processes of evolutionary change can generate the form and information necessary to produce the animals that arise in the Cambrian explosion. This analysis will, for the most part,[20] therefore, not depend on assumptions of either a long or short fuse for the Cambrian explosion, or upon a monophyletic or polyphyletic view of the early history of life.

Defining Biological Form and Information

Form, like life itself, is easy to recognize but often hard to define precisely. Yet a reasonable working definition of *form* will suffice for our present purposes. *Form* can be defined as the four-dimensional topological relations of anatomical parts. This means that we can understand form as a unified arrangement of body parts or material components in a distinct shape or pattern (topology)—one that exists in three spatial dimensions and that arises in time during ontogeny.

Insofar as any particular biological form constitutes something like a distinct arrangement of constituent body parts, form can be seen as arising from constraints that limit the possible arrangements of matter. Specifically, organismal form arises (both in phylogeny and ontogeny) as possible arrangements of material parts are constrained to establish a specific or particular arrangement with an identifiable three-dimensional topography—one that we would recognize as a particular protein, cell type, organ, body plan or organism. A particular form, therefore, represents a highly specific and constrained arrangement of material components (among a much larger set of possible arrangements).

Understanding form in this way suggests a connection to the notion of information in its most theoretically general sense. When Claude Shannon first developed a mathematical theory of information, he equated the amount of information transmitted with the amount of uncertainty reduced or eliminated in a series of symbols or characters.[21] Information, in Shannon's theory, is thus imparted as some options are excluded and others are actualized. The greater the number of options excluded, the greater the amount of information conveyed. Further, constraining a set of possible material arrangements by whatever process or means involves excluding some options and actualizing others. Thus, to constrain a set of possible material states is to generate information in Shannon's sense. It follows that the constraints that produce biological form also imparted in*form*ation. Or conversely, we might say that producing organismal form by definition requires the generation of information.

In classical Shannon information theory, the amount of information in a system is also inversely related to the probability of the arrangement of constituents in a system or the characters along a communication channel.[22] The more improbable (or complex) the arrangement, the more Shannon information, or information-carrying capacity, a string or system possesses.

Since the 1960s, mathematical biologists have realized that Shannon's theory could be applied to the analysis of DNA and proteins to measure the information-carrying capacity of these macromolecules. Since DNA contains the assembly instructions for building proteins, the information-processing system in the cell represents a kind of communication channel.[23] Further, DNA conveys information via specifically arranged sequences of nucleotide bases. Since each of the four bases has a roughly equal chance of occurring at each site along the spine of the DNA molecule, biologists can calculate the probability, and thus the information-carrying capacity, of any particular sequence n bases long.

The ease with which information theory applies to molecular biology has created confusion about the type of information that DNA and proteins possess. Sequences of nucleotide bases in DNA or amino acids in a protein are highly improbable and thus have large information-carrying capacities. But, like meaningful sentences or lines of computer code, genes and proteins are also *specified* with respect to function. Just as the meaning of a sentence depends on the specific arrangement of the letters in a sentence, so too the function of a gene sequence depends on the specific arrangement of the nucleotide bases in a gene. Thus molecular biologists beginning with Francis Crick equated *information* not only with complexity but also with specificity, where "specificity" or "specified" has meant "necessary to function."[24] Molecular biologists such as Jacques Monod and Crick understood biological information—the information stored in DNA and proteins—as something more than mere complexity (or improbability). Their notion of information associated both biochemical contingency and combinatorial complexity with DNA sequences (allowing DNA's carrying capacity to be calculated), but it also affirmed that sequences of nucleotides and amino acids in functioning macromolecules possessed a high degree of *specificity* relative to the maintenance of cellular function.

The ease with which information theory applies to molecular biology has also created confusion about the location of information in organisms. Perhaps because the information-carrying capacity of the gene could be so easily measured, it has been easy to treat DNA, RNA and proteins as the sole repositories of biological information. Neo-Darwinists in particular have assumed that the origination of biological form could be explained by recourse to processes of genetic variation and mutation alone.[25] Yet if we understand organismal form as resulting from constraints on the possible arrangements of matter at many levels in the biological hierarchy—from genes and proteins to cell types and tissues to organs and body plans—then clearly biological organisms exhibit many levels of information-rich structure.

Thus, we can pose a question, not only about the origin of genetic information but also about the origin of the information necessary to generate form and structure at levels higher than that present in individual proteins. We must also ask about the origin of the "specified complexity," as opposed to mere complexity, that characterizes the new genes, proteins, cell types and body plans that arose in the Cambrian explosion. Bill Dembski has used the term "complex specified information" (CSI) as a synonym for "specified complexity" to help distinguish functional biological information from mere Shannon information—that is, specified complexity from mere complexity.[26] This review will use this term as well.

The Cambrian Information Explosion

The Cambrian explosion represents a remarkable jump in the specified complexity or "complex specified information" (CSI) of the biological world. For over three billion years the biological realm included little more than bacteria and algae.[27] Then, beginning about 570-565 million years ago, the first complex multicellular organisms appeared in the rock strata, including sponges, cnidarians and the peculiar Ediacaran biota.[28] Forty million years later the Cambrian explosion occurred.[29] The emergence of the Ediacaran biota (570 million years ago) and then, to a much greater extent, the Cambrian explosion (530 million years ago) represented steep climbs up the biological complexity gradient.

One way to estimate the amount of new CSI that appeared with the Cambrian animals is to count the number of new cell types that emerged with them.[30] Studies of modern animals suggest that the sponges that appeared in the late Precambrian period, for example, would have required five cell types, whereas the more complex animals that appeared in the Cambrian (e.g., arthropods) would have required fifty or more cell types. Functionally more complex animals require more cell types to perform their more diverse functions. New cell types require many new and specialized proteins. New proteins, in turn, require new genetic information. Thus an increase in the number of cell types implies (at a minimum) a considerable increase in the amount of specified genetic information. Molecular biologists have recently estimated that a minimally complex single-celled organism would require between 318 and 562 kilobase pairs of DNA to produce the proteins necessary to maintain life.[31] More complex single cells might require upward of a million base pairs. Yet to build the proteins necessary to sustain a complex arthropod such as a trilobite would require orders of magnitude more coding instructions. The genome size of a modern arthropod, the fruit fly *Drosophila melanogaster,* is approximately 180 million base pairs.[32] Transitions from a single cell to colonies of cells to complex animals represent significant (and, in principle, measurable) increases in CSI.

Building a new animal from a single-celled organism requires a vast amount of new genetic information. It also requires a way of arranging gene products—proteins—into higher levels of organization. New proteins are required to service new cell types. But new proteins must be organized into new systems within the cell; new cell types must be organized into new tissues, organs and body parts. These in turn must be organized to form body plans. New animals therefore embody hierarchically organized systems of lower-level parts within a functional whole. Such hierarchical organization itself represents a type of information, since body plans comprise both highly improbable and functionally specified arrangements of lower-level parts. The specified complexity of new body plans requires explanation in any account of the Cambrian explosion.

Can neo-Darwinism explain the discontinuous increase in CSI that appears in the Cambrian explosion—either in the form of new genetic information or in the form of hierarchically organized systems of parts? We will now examine the two parts of this question.

Novel Genes and Proteins

Many scientists and mathematicians have questioned the ability of mutation and selection to generate information in the form of novel genes and proteins. Such skepticism often derives from consideration of the extreme improbability (and specificity) of functional genes and proteins.

A typical gene contains over one thousand precisely arranged bases. For any specific arrangement of four nucleotide bases of length n, there is a corresponding number of possible arrangements of bases, 4^n. For any protein there are 20^n possible arrangements of protein-forming amino acids. A gene 999 bases in length represents one of 4^{999} possible nucleotide sequences; a protein of 333 amino acids is one of 20^{333} possibilities.

Since the 1960s, some biologists have thought functional proteins to be rare among the set of possible amino acid sequences. Some have used an analogy with human language to illustrate why this should be the case. Michael Denton, for example, has shown that meaningful words and sentences are extremely rare among the set of possible combinations of English letters, especially as sequence length grows.[33] (The ratio of meaningful twelve-letter words to twelve-letter sequences is $1/10^{14}$, the ratio of one-hundred-letter sentences to possible one-hundred-letter strings is $1/10^{100}$.) Further, Denton shows that most meaningful sentences are *highly isolated* from one another in the space of possible combinations, so that random substitutions of letters will, after a very few changes, inevitably degrade meaning. Apart from a few closely clustered sentences accessible by random substitution, the overwhelming majority of meaningful sentences lie, probabilistically speaking, beyond the reach of random search.

Denton and others have argued that similar constraints apply to genes and proteins.[34] They have questioned whether an undirected search via mutation and selection would have a reasonable chance of locating new islands of function—representing fundamentally new genes or proteins—within the time available.[35] Some have also argued that alterations in sequencing would likely result in loss of protein function before a fundamentally new function could arise.[36] Nevertheless, neither the extent to which genes and proteins are sensitive to functional loss as a result of sequence change, nor the extent to which functional proteins are isolated within sequence space, has been fully known.

Recently, experiments in molecular biology have shed light on these questions. A variety of mutagenesis techniques have shown that proteins (and thus the genes that produce them) are indeed highly specified relative to biological function.[37] Mutagenesis research tests the sensitivity of proteins (and by implication, DNA) to functional loss as a result of alterations in sequencing. Studies of proteins have long shown that amino acid residues at many active positions cannot vary without functional loss.[38] More recent protein studies (often using mutagenesis experiments) have shown that functional requirements place significant constraints on sequencing even at nonactive site positions.[39] In particular, D. D. Axe has shown that multiple as opposed to single position amino acid substitutions inevitably result in loss of protein function, even when these changes occur at sites that allow variation when altered in isolation.[40] Cumulatively, these constraints imply that proteins are highly sensitive to functional loss as a result of alterations in sequencing and that functional proteins represent highly isolated and improbable arrangements of amino acids—arrangements that are far more improbable, in fact, than would be likely to arise by chance alone in the time available.[41] (See below the discussion of the neutral theory of evolution for a precise quantitative assessment.)

Of course, neo-Darwinists do not envision a completely random search through the set of all possible nucleotide sequences—so-called sequence space. They envision natural selection acting to preserve small advantageous variations in genetic sequences and their corresponding protein products. Dawkins, for example, likens an organism to a high mountain peak. He compares climbing the sheer precipice up the front side of the mountain to building a new organism by chance. He acknowledges that his approach up "Mount Improbable" will not succeed. Nevertheless, he suggests that there is a gradual slope up the backside of the mountain that could be climbed in small incremental steps. In his analogy, the backside climb up "Mount Improbable" corresponds to the process of natural selection acting on random changes in the genetic text. What chance alone cannot accomplish blindly or in one leap, selection (acting on mutations) can accomplish through the cumulative effect of many slight successive steps.[42]

Yet the extreme specificity and complexity of proteins presents a difficulty, not only for the chance origin of specified biological information (i.e., for random mutations acting alone) but also for selection and mutation acting in concert. Indeed, mutagenesis experiments cast doubt on each of the two scenarios by which neo-Darwinists envisioned new information arising from the mutation/selection mechanism.[43] For neo-Darwinism, new functional genes either arise from noncoding sections in the genome or from preexisting genes. Both scenarios are problematic.

In the first scenario, neo-Darwinists envision new genetic information arising from those sections of the genetic text that can presumably vary freely without consequence to the organism. According to this scenario, noncoding sections of the genome or duplicated sections of coding regions can experience a protracted period of "neutral evolution" during which alterations in nucleotide sequences have no discernible effect on the function of the organism.[44] Eventually, however, a new gene sequence will arise that can code for a novel protein. At that point, natural selection can favor the new gene and its functional protein product, thus securing the preservation and heritability of both.

This scenario has the advantage of allowing the genome to vary through many generations, as mutations "search" the space of possible base sequences. However, it has an overriding problem: the size of the combinatorial space (i.e., the number of possible amino acid sequences) and the extreme rarity and isolation of the functional sequences within that space of possibilities. Since natural selection can do nothing to help *generate* new functional sequences, but rather can only preserve such sequences once they have arisen, chance alone—random variation—must do the work of information generation—that is, of finding the exceedingly rare functional sequences within the set of combinatorial possibilities. Yet the probability of randomly assembling (or "finding," in the previous sense) a functional sequence is extremely small.

Cassette mutagenesis experiments performed during the early 1990s suggest that the probability of attaining (at random) the correct sequencing for

a short protein one hundred amino acids long is about 1 in 10^{65}.[45] This result
agreed closely with earlier calculations that Yockey had performed based on
the known sequence variability of cytochrome c in different species and
other theoretical considerations.[46] More recent mutagenesis research has
provided additional support for the conclusion that functional proteins are
exceedingly rare among possible amino acid sequences.[47] Axe has performed site directed mutagenesis experiments on a 150-residue protein-fold-
ing domain within a B-lactamase enzyme. His experimental method im-
proves on earlier mutagenesis techniques and corrects for several sources of
possible estimation error inherent in them. On the basis of these experiments, Axe has estimated the ratio of (a) proteins of typical size (150 residues) that perform a specified function via any folded structure to (b) the
whole set of possible amino acids sequences of that size. Based on his experiments, Axe has estimated his ratio to be 1 to 10^{77}. Thus, the probability
of finding a functional protein among the possible amino acid sequences
corresponding to a 150-residue protein is similarly 1 in 10^{77}.[48]

Other considerations imply additional improbabilities. First, new Cambrian animals would require proteins much longer than one hundred residues to perform many necessary specialized functions. Ohno (1996) has
noted that Cambrian animals would have required complex proteins such
as lysyl oxidase in order to support their stout body structures.[49] Lysyl oxidase molecules in extant organisms comprise over four hundred amino
acids. These molecules are both highly complex (nonrepetitive) and functionally specified. Reasonable extrapolation from mutagenesis experiments done on shorter protein molecules suggests that the probability of
producing functionally sequenced proteins of this length at random is so
small as to make appeals to chance absurd, even granting the duration of
the entire universe.[50] Yet, second, fossil data[51] and even molecular analyses
supporting deep divergence,[52] suggest that the duration of the Cambrian
explosion (between 5-10 x 10^6 and, at most, 7 x 10^7 years) is far smaller
than that of the entire universe (1.3-2 x 10^{10} years). Third, DNA mutation
rates are far too low to generate the novel genes and proteins necessary
to building the Cambrian animals, given the most probable duration of the
explosion as determined by fossil studies.[53] As Ohno notes, even a mutation rate of 10^{-9} per base pair per year results in only a 1 percent change
in the sequence of a given section of DNA in 10 million years. Thus, he
argues that mutational divergence of preexisting genes cannot explain the

origin of the Cambrian forms in that time.[54]

The selection/mutation mechanism faces another probabilistic obstacle. The animals that arise in the Cambrian exhibit structures that would have required many new *types* of cells, each of which would have required many novel proteins to perform their specialized functions. Further, new cell types require *systems* of proteins that must, as a condition of functioning, act in close coordination with one another. The unit of selection in such systems ascends to the system as a whole. Natural selection selects for functional advantage. But new cell types require whole systems of proteins to perform their distinctive functions. In such cases, natural selection cannot contribute to the process of information generation until *after* the information necessary to build the requisite *system* of proteins has arisen. Thus random variations must again do the work of information generation—and now not simply for one protein, but for many proteins arising at nearly the same time. Yet the odds of this occurring by chance alone are, of course, far smaller than the odds of the chance origin of a single gene or protein—so small, in fact, as to render the chance origin of the genetic information necessary to build a new cell type (a necessary but not sufficient condition of building a new body plan) problematic given even the most optimistic estimates for the duration of the Cambrian explosion.

Dawkins has noted that scientific theories can rely on only so much "luck" before they cease to be credible.[55] The neutral theory of evolution, which, by its own logic, prevents natural selection from playing a role in generating genetic information until after the fact, relies on entirely too much luck. The sensitivity of proteins to functional loss, the need for long proteins to build new cell types and animals, the need for whole new *systems* of proteins to service new cell types, the probable brevity of the Cambrian explosion relative to mutation rates—all suggest the immense improbability (and implausibility) of any scenario for the origination of Cambrian genetic information that relies on random variation alone unassisted by natural selection.

Yet the neutral theory requires novel genes and proteins to arise—essentially—by random mutation alone. Adaptive advantage accrues *after* the generation of new functional genes and proteins. Thus natural selection cannot play a role *until* new information-bearing molecules have independently arisen. Thus neutral theorists envisioned the need to scale the steep face of a Dawkins-style precipice of which there is *no* gradually sloping backside—a situation that, by Dawkins' own logic, is probabilistically untenable.

In the second scenario, neo-Darwinists envisioned novel genes and proteins arising by numerous successive mutations in the preexisting genetic text that codes for proteins. To adapt Dawkins's metaphor, this scenario envisions gradually climbing down one functional peak and then ascending another. Yet mutagenesis experiments again suggest a difficulty. Recent experiments show that even when exploring a region of sequence space populated by proteins of a single fold and function, most multiple-position changes quickly lead to loss of function. Yet to turn one protein into another with a completely novel structure and function requires specified changes at many sites. Indeed, the number of changes necessary to produce a new protein greatly exceeds the number of changes that will typically produce functional losses. Given this, the probability of escaping total functional loss during a random search for the changes needed to produce a new function is extremely small—and this probability diminishes exponentially with each additional requisite change.[56] Thus Axe's results imply that, in all probability, random searches for novel proteins (through sequence space) will result in functional loss long before any novel functional protein will emerge.

Blanco, Angrand and Serrano have come to a similar conclusion. Using directed mutagenesis, they have determined that residues both in the hydrophobic core and on the surface of the protein play essential roles in determining protein structure. By sampling intermediate sequences between two naturally occurring sequences that adopt different folds, they found that the intermediate sequences "lack a well defined three-dimensional structure." Thus, they conclude that it is unlikely that a new protein fold via a series of folded intermediates sequences.[57]

Although this second neo-Darwinian scenario has the advantage of starting with functional genes and proteins, it also has a lethal disadvantage: any process of random mutation or rearrangement in the genome would in all probability generate nonfunctional intermediate sequences before fundamentally new functional genes or proteins would arise. Clearly, nonfunctional intermediate sequences confer no survival advantage on their host organisms. Natural selection favors *only* functional advantage. It cannot select or favor nucleotide sequences or polypeptide chains that do not yet perform biological functions, and still less will it favor sequences that efface or destroy preexisting function.

Evolving genes and proteins will range through a series of nonfunctional intermediate sequences that natural selection will not favor or preserve but will, in all probability, eliminate.[58] When this happens, selection-driven evolution will cease. At this point, neutral evolution of the genome (unhinged from selective pressure) may ensue, but as we have seen, such a process must overcome immense probabilistic hurdles, even granting cosmic time.

Whether one envisions the evolutionary process beginning with a noncoding region of the genome or a preexisting functional gene, the functional specificity and complexity of proteins impose very stringent limitations on the efficacy of mutation and selection. In the first case, function must arise first, before natural selection can act to favor a novel variation. In the second case, function must be continuously maintained in order to prevent deleterious (or lethal) consequences to the organism and to allow further evolution. Yet the complexity and functional specificity of proteins implies that both these conditions will be extremely difficult to meet. Therefore, the neo-Darwinian mechanism appears to be inadequate to generate the new information present in the novel genes and proteins that arise with the Cambrian animals.

Novel Body Plans

The problems with the neo-Darwinian mechanism run deeper still. In order to explain the origin of the Cambrian animals, we must account not only for new proteins and cell types but also for the origin of new body plans. Within the past decade developmental biology has dramatically advanced our understanding of how body plans are built during ontogeny. In the process it has also uncovered a profound difficulty for neo-Darwinism.

Significant morphological change in organisms requires attention to timing. Mutations in genes that are expressed late in the development of an organism will not affect the body plan. Mutations expressed early in development, however, could conceivably produce significant morphological change.[59] Thus events expressed early in the development of organisms have the only realistic chance of producing large-scale macroevolutionary change.[60] As John and Miklos explain, macroevolutionary change requires alterations in the very early stages of ontogenesis.[61]

Yet recent studies in developmental biology make clear that mutations expressed early in development typically have deleterious effects.[62] For example, when early-acting body plan molecules, or morphogens such as *bicoid* (which helps to set up the anterior-posterior head-to-tail axis in *Drosophila*), are perturbed, development shuts down.[63] The resulting embryos die. Moreover, there is a good reason for this. If an engineer modifies the length of the piston rods in an internal combustion engine without modifying the crankshaft accordingly, the engine won't start. Similarly, processes of development are tightly integrated spatially and temporally such that changes early in development will require a host of other coordinated changes in separate but functionally interrelated developmental processes downstream. For this reason mutations will be much more likely to be deadly if they disrupt a functionally deeply embedded structure such as a spinal column than if they affect more isolated anatomical features such as fingers.[64]

This problem has led to what J. F. McDonald has called "a great Darwinian paradox."[65] He notes that genes that are observed to vary within natural populations do not lead to major adaptive changes, while genes that could cause major changes—the very stuff of macroevolution—apparently do not vary. In other words, mutations of the kind that macroevolution doesn't need (namely, viable genetic mutations in DNA expressed late in development) do occur, but those that it does need (namely, beneficial body plan mutations expressed early in development) apparently don't occur.[66] According to Darwin, natural selection cannot act until favorable variations arise in a population.[67] Yet there is no evidence from developmental genetics that the kind of variations required by neo-Darwinism—namely, favorable body plan mutations—ever occur.

Developmental biology has raised another formidable problem for the mutation/selection mechanism. Embryological evidence has long shown that DNA does not wholly determine morphological form,[68] suggesting that mutations in DNA alone cannot account for the morphological changes required to build a new body plan.

DNA helps direct protein synthesis.[69] It also helps to regulate the timing and expression of the synthesis of various proteins within cells. Yet DNA alone does not determine how individual proteins assemble themselves into larger systems of proteins; still less does it solely determine how cell types, tissue types and organs arrange themselves into body plans.[70] Instead, other factors—such as the three-dimensional structure and organization of the cell membrane and cytoskeleton and the spatial architecture of the fertilized egg—play important roles in determining body plan formation during embryogenesis.

For example, the structure and location of the cytoskeleton influence the patterning of embryos. Arrays of microtubules help to distribute the essential proteins used during development to their correct locations in the cell. Of course, microtubules themselves are made of many protein subunits. Nevertheless, like bricks that can be used to assemble many different structures, the tubulin subunits in the cell's microtubules are identical to one another. Thus, neither the tubulin subunits nor the genes that produce them account for the different shape of microtubule arrays that distinguish different kinds of embryos and developmental pathways. Instead, the structure of the microtubule array itself is determined by the location and arrangement of its subunits, not the properties of the subunits themselves. For this reason it is not possible to predict the structure of the cytoskeleton of the cell from the characteristics of the protein constituents that form that structure.[71]

Two analogies may help further clarify the point. At a building site, builders will make use of many materials: lumber, wires, nails, drywall, piping and windows. Yet building materials do not determine the floor plan of the house or the arrangement of houses in a neighborhood. Similarly, electronic circuits are composed of many components, such as resistors, capacitors and transistors. But such lower-level components do not determine their own arrangement in an integrated circuit. Biological symptoms also depend on hierarchical arrangements of parts. Genes and proteins are made from simple building blocks—nucleotide bases and amino acids—arranged in specific ways. Cell types are made of, among other things, systems of specialized proteins. Organs are made of specialized arrangements of cell types and tissues. And body plans comprise specific arrangements of specialized organs. Yet, clearly, the properties of individual proteins (or indeed the lower-level parts in the hierarchy generally) do not fully determine the organization of the higher-level structures and organizational patterns.[72] It follows that the genetic information that codes for proteins does not determine these higher-level structures either.

These considerations pose another challenge to the sufficiency of the neo-Darwinian mechanism. Neo-Darwinism seeks to explain the origin of new information, form and structure as a result of selection acting on randomly arising variation at a very low level within the biological hierarchy, namely, within the genetic text. Yet major morphological innovations depend on a specificity of arrangement at a much higher level of the organizational hierarchy, a level that DNA alone does not determine. Yet if DNA is not wholly responsible for body plan morphogenesis, then DNA sequences can mutate indefinitely, without regard to realistic probabilistic limits, and still not produce a new body plan. Thus the mechanism of natural selection acting on random mutations in DNA cannot *in principle* generate novel body plans, including those that first arose in the Cambrian explosion.

Of course, it could be argued that, while many single proteins do not by themselves determine cellular structures or body plans, proteins acting in concert with other proteins or suites of proteins could determine such higher-level form. For example, it might be pointed out that the tubulin subunits are assembled by other helper proteins—gene products—called microtubule associated proteins (MAPS). This might seem to suggest that genes and gene products alone do suffice to determine the development of the three-dimensional structure of the cytoskeleton.

Yet MAPS, and indeed many other necessary proteins, are only part of the story. The location of specified target sites on the interior of the cell membrane also helps to determine the shape of the cytoskeleton. Similarly, so does the position and structure of the centrosome that nucleates the microtubules that form the cytoskeleton. While both the membrane targets and the centrosomes are made of proteins, the location and form of these structures are not wholly determined by the proteins that form them. Indeed, centrosome structure and membrane patterns *as a whole* convey three-dimensional structural information that helps determine the structure of the cytoskeleton and the location of its subunits.[73] Moreover, the centrioles that compose the centrosomes replicate independently of DNA replication.[74] The daughter centriole receives its form from the overall structure of the mother centriole, not from the individual gene products that constitute it.[75] In ciliates, microsurgery on cell membranes can produce heritable changes in membrane patterns, even though the DNA of the ciliates has not been altered.[76] This suggests that membrane patterns (as opposed to membrane constituents) are impressed directly on daughter cells. In both cases, form is transmitted from parent three-dimensional structures to daughter three-dimensional structures directly and is not wholly contained in constituent proteins or genetic information.[77]

In each new generation the form and structure of the cell arises as the result of *both* gene products and preexisting three-dimensional structure and organization. Cellular structures are built from proteins, but proteins find their way to correct locations in part because of preexisting three-dimensional patterns and organization inherent in cellular structures. Preexisting three-dimensional form present in the preceding generation (whether inherent in the cell membrane, the centrosomes, the cytoskeleton or other features of the fertilized egg) contributes to the production of form in the next generation. Neither structural proteins alone nor the genes that code for them are sufficient to determine the three-dimensional shape and structure of the entities they form. Gene products provide necessary, but not sufficient, conditions for the development of three-dimensional structure within cells, organs and body plans.[78] But if this is so, then natural selection acting on genetic variation alone cannot produce the new forms that arise in the history of life.

Self-Organizational Models

Of course, neo-Darwinism is not the only evolutionary theory for explaining the origin of novel biological form. Kauffman (1995) doubts the efficacy of the mutation/selection mechanism. Nevertheless, he has advanced a self-organizational theory to account for the emergence of new form and presumably the information necessary to generate it. Whereas neo-Darwinism attempts to explain new form as the consequence of selection acting on random mutation, Kauffman suggests that selection acts not mainly on random variations but on emergent patterns of order that self-organize via the laws of nature.[79]

Kauffman illustrates how this might work with various model systems in a computer environment. In one, he conceives a system of buttons connected by strings. Buttons represent novel genes or gene products; strings represent the law-like forces of interaction that obtain between gene products—that is, proteins. Kauffman suggests that when the complexity of the system (as represented by the number of buttons and strings) reaches a critical threshold, new modes of organization can arise in the system "free"—that is, naturally and spontaneously—after the manner of a phase transition in chemistry.[80]

Another model that Kauffman develops is a system of interconnected lights. Each light can flash in a variety of states—on, off, twinkling and so forth. Since there is more than one possible state for each light, and many lights, there are a vast number of possible states that the system can adopt. Further, in his system, rules determine how past states will influence future states. Kauffman asserts that as a result of these rules the system will, if properly tuned, eventually produce a kind of order in which a few basic patterns of light activity recur with greater-than-random frequency. Since these actual patterns of light activity represent a small portion of the total number of possible states in which the system can reside, Kauffman seems to imply that self-organizational laws might similarly result in highly improbable biological outcomes—perhaps even sequences (of bases or amino acids) within a much larger sequence space of possibilities.

Do these simulations of self-organizational processes accurately model the origin of novel genetic information? It is hard to think so.

First, in both examples, Kauffman presupposes, but does not explain, significant sources of preexisting information. In his buttons-and-strings system, the buttons represent proteins, themselves packets of CSI and the result of preexisting genetic information. Where does this information come from? Kauffman doesn't say, but the origin of such information is an essential part of what needs to be explained in the history of life. Similarly, in his light system, the order that allegedly arises for "free" actually arises only if the programmer of the model system "tunes" it in such a way as to keep it from either (a) generating an excessively rigid order or (b) developing into chaos.[81] Yet this necessary tuning involves an intelligent programmer selecting certain parameters and excluding others—that is, inputting information.

Second, Kauffman's model systems are not constrained by functional considerations and thus are not analogous to biological systems. A system of interconnected lights governed by preprogrammed rules may well settle into a small number of patterns within a much larger space of possibilities. But because these patterns have no function, and need not meet any functional requirements, they have no specificity analogous to that present in actual organisms. Instead, examination of Kauffman's model systems shows that they do not produce sequences or systems characterized by *specified* complexity, but instead by large amounts of symmetrical order or internal redundancy interspersed with aperiodicity or (mere) complexity.[82] Getting a law-governed system to generate repetitive patterns of flashing lights, even with a certain amount of variation, is clearly interesting, but not biologically relevant. On the other hand, a system of lights flashing the title of a Broadway play would model a biologically relevant self-organizational process, at least if such a meaningful or functionally specified sequence arose without intelligent agents previously programming the system with equivalent amounts of CSI. In any case, Kauffman's systems do not produce *specified* complexity and thus do not offer promising models for explaining the new genes and proteins that arose in the Cambrian.

Even so, Kauffman suggests that his self-organizational models can specifically elucidate aspects of the Cambrian explosion. According to Kauffman, new Cambrian animals emerged as the result of "long jump" mutations that established new body plans in a discrete rather than gradual fashion.[83] He also recognizes that mutations affecting early development are almost inevitably harmful. Thus he concludes that body plans, once established, will not change and that any subsequent evolution must occur within an established body plan.[84] And indeed, the fossil record does show a curious (from a neo-Darwinian point of view) top-down pattern of appearance, in which higher taxa (and the body plans they represent) appear first, only later to be followed by the multiplication of lower taxa representing variations within those original body designs.[85] Further, as Kauffman expects, body plans appear suddenly and persist without significant modification over time.

But here again Kauffman begs the most important question, which is: what produces the new Cambrian body plans in the first place? Granted, he invokes long-jump mutations to explain this, but he identifies no specific self-organizational process that can produce such mutations. Moreover, he concedes a principle that undermines the plausibility of his own proposal. Kauffman acknowledges that mutations that occur early in development are almost inevitably deleterious. Yet developmental biologists know that these are the only kind of mutations that have a realistic chance of producing large-scale evolutionary change—that is, the big jumps that Kauffman invokes. Though Kauffman repudiates the neo-Darwinian reliance on random mutations in favor of self-organizing order, in the end, he must invoke the most implausible kind of random mutation in order to provide a self-organizational account of the new Cambrian body plans. Clearly, his model is not sufficient.

Punctuated Equilibrium

Of course, still other causal explanations have been proposed. During the 1970s the paleontologists Eldredge and Gould (1972) proposed the theory of evolution by punctuated equilibrium in order to account for a pervasive pattern of "sudden appearance" and "stasis" in the fossil record.[86] Though advocates of punctuated equilibrium were mainly seeking to describe the fossil record more accurately than earlier gradualist neo-Darwinian models had done, they also proposed a mechanism—known as species selection—by which the large morphological jumps evident in the fossil record might have been produced. According to punctuationalists, natural selection functions more as a mechanism for selecting the fittest species rather than the most fit individual among a species. Accordingly, on this model, morphological change should occur in larger, more discrete intervals than it would given a traditional neo-Darwinian understanding.

Despite its virtues as a descriptive model of the history of life, punctuated equilibrium has been widely criticized for failing to provide a mechanism sufficient to produce the novel form characteristic of higher taxonomic groups. For one thing, critics have noted that the proposed mechanism of punctuated evolutionary change simply lacked the raw material on which to work. As Valentine and Erwin note, the fossil record fails to document a large pool of species prior to the Cambrian explosion. Yet the proposed mechanism of species selection requires just such a pool of species on which to act. Thus, they conclude that the mechanism of species selection probably does not resolve the problem of the origin of the higher taxonomic groups.[87] Further, punctuated equilibrium has not addressed the more specific and fundamental problem of explaining the origin of the new biological information (whether genetic or epigenetic) necessary to produce novel biological form. Advocates of punctuated equilibrium might assume that the new species (on which natural selection acts) arise by known microevolutionary processes of speciation (such as founder effect, genetic drift or bottleneck effect) that do not necessarily depend on mutations to produce adaptive changes. But in that case the theory lacks an account of how the specifically *higher* taxa arise. Species selection will only produce more fit species. On the other hand, if punctuationalists assume that processes of genetic mutation can produce more fundamental morphological changes and variations, then their model becomes subject to the same problems as neo-Darwinism. This dilemma is evident in Gould (2002:710) insofar as his attempts to explain adaptive complexity inevitably employ classical neo-Darwinian modes of explanation.[88]

Structuralism

Another attempt to explain the origin of form has been proposed by the structuralists, such as Gerry Webster and Brian Goodwin.[89] These biologists, drawing on the earlier work of D'Arcy Thompson,[90] view biological form as the result of structural constraints imposed on matter by morphogenetic rules or laws. For reasons similar to those discussed above, the structuralists have insisted that these generative or morphogenetic rules do not reside in the lower-level building materials of organisms, whether in genes or proteins. Webster and Goodwin further envisioned morphogenetic rules or laws operating ahistorically, similar to the way in which gravitational or electromagnetic laws operate.[91] For this reason, structuralists see phylogeny as of secondary importance in understanding the origin of the higher taxa, though they think that transformations of form can occur. For structuralists, constraints on the arrangement of matter arise not mainly as the result of historical contingencies—such as environmental changes or genetic mutations—but instead because of the continuous ahistorical operation of fundamental laws of form—laws that organize or inform matter.

While this approach avoids many of the difficulties currently afflicting neo-Darwinism (in particular those associated with its "genocentricity"), critics (such as Maynard Smith)[92] of structuralism have argued that the structuralist explanation of form lacks specificity. They note that structuralists have been unable to say just where laws of form reside—whether in the universe or in every possible world or in organisms as a whole or in just some part of organisms. Further, according to structuralists, morphogenetic laws are mathematical in character. Yet structuralists have yet to specify the mathematical formulas that determine biological forms.

Others have questioned whether physical laws could in principle generate the kind of complexity that characterizes biological systems.[93] Structuralists envision the existence of biological laws that produce form in much the same way that physical laws produce form. Yet the forms that physicists regard as manifestations of underlying laws are characterized by large amounts of symmetric or redundant order, by relatively simple patterns such as vortices or gravitational fields or magnetic lines of force. Indeed, physical laws are typically expressed as differential equations (or algorithms) that almost by definition describe recurring phenomena—patterns of compressible "order," not "complexity," as defined by algorithmic information theory.[94] Biological forms, by contrast, manifest greater complexity and derive in ontogeny from highly complex initial conditions—that is, nonredundant sequences of nucleotide bases in the genome and other forms of information expressed in the complex and irregular three-dimensional topography of the organism or the fertilized egg. Thus the kind of form that physical laws produce is not analogous to biological form—at least not when compared from the standpoint of (algorithmic) complexity. Further, physical laws lack the information content to specify biological systems. As Polanyi and Yockey have shown, the laws of physics and chemistry allow, but do not determine, distinctively biological modes of organization.[95] In other words, living systems are consistent with but not deducible from physical-chemical laws.[96]

Of course, biological systems do manifest some recurring patterns, processes and behaviors. The same type of organism develops repeatedly from similar ontogenetic processes in the same species. Similar processes of cell division recur in many organisms. Thus one might describe certain biological processes as being law governed. Even so, the existence of such biological regularities does not solve the problem of the origin of form and information, since the recurring processes described by such biological laws (if there be such laws) only occur as the result of preexisting stores of (genetic or epigenetic) information, and these information-rich initial conditions impose the constraints that produce the recurring behavior in biological systems (e.g., processes of cell division recur with great frequency in organisms but depend on information-rich DNA and protein molecules). In other words, distinctively biological regularities depend on preexisting biological information. Thus appeals to higher-level biological laws presuppose but do not explain the origination of the information necessary to morphogenesis.

Structuralism faces a difficult dilemma, in principle. On the one hand, physical laws produce very simple redundant patterns that lack the complexity characteristic of biological systems. On the other hand, distinctively biological laws—if there are such laws—depend on preexisting information-rich structures. In either case, laws are not good candidates for explaining the origination of biological form or the information necessary to produce it.

Cladism: An Artifact of Classification?

Some cladists have advanced another approach to the problem of the origin of form, specifically as it arises in the Cambrian explosion. They have argued that the problem of the origin of the phyla is an artifact of the classification system and therefore does not require explanation. G. E. Budd and S. E. Jensen, for example, argue that the problem of the Cambrian explosion resolves itself if one keeps in mind the cladistic distinction between "stem" and "crown" groups. Since crown groups arise whenever new characters are added to simpler, more ancestral stem groups during the evolutionary process, new phyla will inevitably arise once a new stem group has arisen. Thus for Budd and Jensen what requires explanation is not the crown groups corresponding to the new Cambrian phyla but the earlier more primitive stem groups that presumably arose deep in the Proterozoic period. Yet since these earlier stem groups are by definition less derived, explaining them will be considerably easier than explaining the origin of the Cambrian animals de novo. In any case, for Budd and Jensen the explosion of new phyla in the Cambrian period does not require explanation. As they put it, "given that the early branching points of major clades is an inevitable result of clade diversification, the alleged phenomenon of the phyla appearing early and remaining morphologically static is not seen to require particular explanation."[97]

While superficially plausible, perhaps, Budd and Jensen's attempt to explain away the Cambrian explosion begs crucial questions. Granted, as new characters are added to existing forms, novel morphology and greater morphological disparity will likely result. But what causes new characters to arise? And how does the information necessary to produce new characters originate? Budd and Jensen do not specify. Nor can they say how derived the ancestral forms are likely to have been, and what processes might have been sufficient to produce them. Instead, they simply assume the sufficiency of known neo-Darwinian mechanisms.[98] Yet this assumption is now problematic. In any case, Budd and Jensen do not explain what causes the origination of biological form and information.

Convergence and Teleological Evolution

More recently, Conway Morris has suggested another possible explanation, based on the tendency for evolution to converge on the same structural forms during the history of life. Conway Morris cites numerous examples of organisms that possess very similar forms and structures, even though such structures are often built from different material substrates and arise (in ontogeny) by the expression of very different genes. Given the extreme improbability of the same structures arising by random mutation and selection in disparate phylogenies, Conway Morris argues that the pervasiveness of convergent structures suggests that evolution may be in some way "channeled" toward similar functional or structural endpoints. Such an end-directed understanding of evolution, he admits, raises the controversial prospect of a teleological or purposive element in the history of life. For this reason, he argues that the phenomenon of convergence has received less attention than it might have otherwise. Nevertheless, he argues that just as physicists have reopened the question of design in their discussions of anthropic fine-tuning, the ubiquity of convergent structures in the history of life has led some biologists (e.g., Denton) to consider extending teleological thinking to biology. And, indeed, Conway Morris himself intimates that the evolutionary process might be "underpinned by a purpose."[99]

Conway Morris, of course, considers this possibility in relation to a very specific aspect of the problem of organismal form, namely, the problem of explaining why the same forms arise repeatedly in so many disparate lines of decent. But this raises a question. Could a similar approach shed explanatory light on the more general causal question that has been addressed in this review? Could the notion of purposive design help provide a more adequate explanation for the origin of organismal form generally? Are there reasons to consider design as an explanation for the origin of the biological information necessary to produce the higher taxa and their corresponding morphological novelty?

The remainder of this review will suggest that there are such reasons. In so doing, it may also help explain why the issue of teleology or design has reemerged within the scientific discussion of biological origins[100] and why some scientists and philosophers of science have considered teleological explanations for the origin of form and information despite strong methodological prohibitions against design as a scientific hypothesis.[101]

First, the possibility of design as an explanation follows logically from a consideration of the deficiencies of neo-Darwinism and other current theories as explanations for some of the more striking "appearances of design" in biological systems. Neo-Darwinists such as Ayala, Dawkins, Mayr and Lewontin have long acknowledged that organisms appear to have been designed.[102] Of course, neo-Darwinists assert that what Ayala calls the "obvious design" of living things[103] is only apparent since the selection/mutation mechanism can explain the origin of complex form and organization in living systems without an appeal to a designing agent. Indeed, neo-Darwinists affirm that mutation and selection—and perhaps other similarly undirected mechanisms—are fully sufficient to explain the appearance of design in biology. Self-organizational theorists and punctuationalists modify this claim but affirm its essential tenet. Self-organization theorists argue that natural selection acting on self-organizing order can explain the complexity of living things—again, without any appeal to design. Punctuationalists similarly envision natural selection acting on newly arising species with no actual design involved.

Clearly, the neo-Darwinian mechanism does explain many appearances of design, such as the adaptation of organisms to specialized environments that attracted the interest of nineteenth-century biologists. More specifically, known microevolutionary processes appear quite sufficient to account for changes in the size of Galapagos finch beaks that have occurred in response to variations in annual rainfall and available food supplies.[104]

But does neo-Darwinism, or any other fully materialistic model, explain all appearances of design in biology, including the body plans and information that characterize living systems? Arguably, biological forms—such as the structure of a chambered nautilus, the organization of a trilobite, the functional integration of parts in an eye or molecular machine—attract our attention in part because the organized complexity of such systems seems reminiscent of our own designs. But this review has argued that neo-Darwinism does not adequately account for the origin of all appearances of design, especially when we consider animal body plans and the information necessary to construct them as especially striking examples of the appearance of design in living systems. Indeed, Dawkins and Bill Gates have noted that genetic information bears an uncanny resemblance to computer software or machine code.[105] For this reason, the presence of CSI in living organisms, and the discontinuous increases of CSI that occurred during events such as the Cambrian explosion, appears at least suggestive of design.

Does neo-Darwinism or any other purely materialistic model of morphogenesis account for the origin of the genetic and other forms of CSI necessary to produce novel organismal form? If not, as this review has argued, could the emergence of novel information-rich genes, proteins, cell types and body plans have resulted from actual design rather than from a purposeless process that merely mimics the powers of a designing intelligence? The logic of neo-Darwinism, with its specific claim to have accounted for the appearance of design, would itself seem to open the door to this possibility. Indeed, the historical formulation of Darwinism in dialectical opposition to the design hypothesis,[106] coupled with neo-Darwinism's inability to account for many salient appearances of design, including the emergence of form and information, would seem logically to reopen the possibility of actual (as opposed to apparent) design in the history of life.

A second reason for considering design as an explanation for these phenomena follows from the importance of explanatory power to scientific theory evaluation and from a consideration of the potential explanatory power of the design hypothesis. Studies in the methodology and philosophy of science have shown that many scientific theories, particularly in the historical sciences, are formulated and justified as inferences to the best explanation.[107] Historical scientists, in particular, assess or test competing hypotheses by evaluating which hypothesis would, if true, provide the best explanation for some set of relevant data.[108] Those with greater explanatory power are typically judged to be better, more probably true, theories. Darwin used this method of reasoning in defending his theory of universal common descent.[109] Moreover, contemporary studies on the method of "inference to the best explanation" have shown that determining which among a set of competing possible explanations constitutes the best depends on judgments about the causal adequacy or "causal powers" of competing explanatory entities.[110] In the historical sciences, uniformitarian or actualistic canons of method suggest that judgments about causal adequacy should derive from our present knowledge of cause and effect relationships.[111] For historical scientists "the present is the key to the past" means that present experience-based knowledge of cause and effect relationships typically guides the assessment of the plausibility of proposed causes of past events.

Yet it is precisely for this reason that current advocates of the design hypothesis want to reconsider design as an explanation for the origin of biological form and information. This review, and much of the literature it has surveyed, suggests that four of the most prominent models for explaining the origin of biological form fail to provide adequate causal explanations for the discontinuous increases of CSI that are required to produce novel morphologies. We have repeated experience of rational and conscious agents—in particular ourselves—generating or causing increases in complex specified information, both in the form of sequence-specific lines of code and in the form of hierarchically arranged systems of parts.

In the first place, intelligent human agents—in virtue of their rationality and consciousness—have demonstrated the power to produce information in the form of linear sequence-specific arrangements of characters. Indeed, experience affirms that information of this type routinely arises from the activity of intelligent agents. A computer user who traces the information on a screen back to its source invariably comes to a *mind*—that of a software engineer or programmer. The information in a book or in inscriptions ultimately derives from a writer or scribe—from a mental rather than a strictly material cause. Our experience-based knowledge of information flow confirms that systems with large amounts of specified complexity (especially codes and languages) invariably originate from an intelligent source from a mind or personal agent. As Henry Quastler put it, the "creation of new information is habitually associated with conscious activity."[112] Experience teaches this obvious truth.

Further, the highly specified hierarchical arrangements of parts in animal body plans also suggest *design,* again because of our experience of the kinds of features and systems that designers can and do produce. At every level of the biological hierarchy, organisms require specified and highly improbable arrangements of lower-level constituents in order to maintain their form and function. Genes require specified arrangements of nucleotide bases; proteins require specified arrangements of amino acids; new cell types require specified arrangements of systems of proteins; body plans require specialized arrangements of cell types and organs. Organisms not only contain information-rich components (such as proteins and genes), but they comprise information-rich arrangements of those components and the systems that comprise them. Yet we know, based on our present experience of cause and effect relationships, that design engineers—possessing purposive intelligence and rationality—have the ability to produce information-rich hierarchies in which both individual modules and the arrangements of those modules exhibit complexity and specificity—information so defined. Individual transistors, resistors and capacitors exhibit considerable complexity and specificity of design; at a higher level of organization, their specific arrangement within an integrated circuit represents additional information and reflects further design. Conscious and rational agents have, as part of their powers of purposive intelligence, the capacity to design information-rich parts and to organize those parts into functional information-rich systems and hierarchies. Further, we know of no other causal entity or process that has this capacity. Clearly we have good reason to doubt that mutation and selection, self-organizational processes or laws of nature, can produce the information-rich components, systems and body plans necessary to explain the origination of morphological novelty such as that which arises in the Cambrian period.

There is a third reason to consider purpose or design as an explanation for the origin of biological form and information: purposive agents have just those necessary powers that natural selection lacks as a condition of its causal adequacy. At several points in the previous analysis, we saw that natural selection lacked the ability to generate novel information precisely because it can only act *after* new functional CSI has arisen. Natural selection can favor new proteins and genes, but only after they perform some function. The job of generating new functional genes, proteins and systems of proteins therefore falls entirely to random mutations. Yet without functional criteria to guide a search through the space of possible sequences, random variation is probabilistically doomed. What is needed is not just a source of variation (i.e., the freedom to search a space of possibilities) or a mode of selection that can operate after the fact of a successful search, but instead a means of selection that (a) operates during a search—before success—and that (b) is guided by information about or knowledge of a functional target.

Demonstration of this requirement has come from an unlikely quarter: genetic algorithms. Genetic algorithms are programs that allegedly simulate the creative power of mutation and selection. Dawkins and Kuppers, for example, have developed computer programs that putatively simulate the production of genetic information by mutation and natural selection.[113] Nevertheless, as shown elsewhere, these programs only succeed by the illicit expedient of providing the computer with a "target sequence" and then treating relatively greater proximity to *future* function (i.e., the target sequence), not actual present function, as a selection criterion.[114] As David Berlinski has argued, genetic algorithms need something akin to a "forward looking memory" in order to succeed.[115] Yet such foresighted selection has no analogue in nature. In biology, where differential survival depends on maintaining function, selection cannot occur before new functional sequences arise. Natural selection lacks foresight.

What natural selection lacks, intelligent selection—purposive or goal-directed design—provides. Rational agents can arrange both matter and symbols with distant goals in mind. In using language the human mind routinely "finds" or generates highly improbable linguistic sequences to convey an intended or *pre*conceived idea. In the process of thought, functional objectives precede and constrain the selection of words, sounds and symbols to generate functional (and indeed meaningful) sequences from among a vast ensemble of meaningless alternative combinations of sound or symbol.[116] Similarly, the construction of complex technological objects and products, such as bridges, circuit boards, engines and software, result from the application of goal-directed constraints.[117] Indeed, in all functionally integrated complex systems where the cause is known by experience or observation, design engineers or other intelligent agents applied boundary constraints to limit possibilities in order to produce improbable forms, sequences or structures. Rational agents have repeatedly demonstrated the capacity to constrain the possible to actualize improbable but initially unrealized future functions. Repeated experience affirms that intelligent agents (minds) uniquely possess such causal powers.

Analysis of the problem of the origin of biological information, therefore, exposes a deficiency in the causal powers of natural selection that corresponds precisely to powers that agents are uniquely known to possess. Intelligent agents have foresight. Such agents can select functional goals *before* they exist. They can devise or select material means to accomplish those ends from among an array of possibilities and then actualize those goals in accord with a *pre*conceived design plan or set of functional requirements. Rational agents can constrain combinatorial space with distant outcomes in mind. The causal powers that natural selection lacks—almost by definition—are associated with the attributes of consciousness and rationality—with purposive intelligence. Thus, by invoking design to explain the origin of new biological information, contemporary design theorists are not positing an arbitrary explanatory element unmotivated by a consideration of the evidence. Instead, they are positing an entity possessing precisely the attributes and causal powers that the phenomenon in question requires as a condition of its production and explanation.

Conclusion

An experience-based analysis of the causal powers of various explanatory hypotheses suggests purposive or intelligent design as a causally adequate—and perhaps the most causally adequate—explanation for the origin of the complex specified information required to build the Cambrian animals and the novel forms they represent. For this reason, recent scientific interest in the design hypothesis is unlikely to abate as biologists continue to wrestle with the problem of the origination of biological form and the higher taxa.

Bibliography

Adams, M. D., et al. "The Genome Sequence of *Drosophila melanogaster*." *Science* 287 (2000): 2185-95.

Aris-Brosou, S., and Z. Yang. "Bayesian Models of Episodic Evolution Support a Late Precambrian Explosive Diversification of the Metazoa." *Molecular Biology and Evolution* 20 (2003): 1947-54.

Arthur, W. *The Origin of Animal Body Plans.* Cambridge: Cambridge University Press, 1997.

Axe, D. D. "Extreme Functional Sensitivity to Conservative Amino Acid Changes on Enzyme Exteriors." *Journal of Molecular Biology* 301, no. 3 (2000): 585-96.

———. "Estimating the Prevalence of Protein Sequences Adopting Functional Enzyme Folds." *Journal of Molecular Biology* (2004).

Ayala, F. "Darwin's Revolution." In *Creative Evolution?!* edited by J. Campbell and J. Schopf. Boston: Jones & Bartlett, 1994.

Ayala, F., A. Rzhetsky and F. J. Ayala. "Origin of the Metazoan Phyla: Molecular Clocks Confirm Paleontological Estimates." *Proceedings of the National Academy of Sciences USA* 95 (1998): 606-11.

Becker, H., and Wolf-Ekkehard Loennig. "Transposons: Eukaryotic." In *Nature Encyclopedia of Life Sciences.* Vol. 18. London: Nature Publishing, 2001.

Behe, Michael. "Experimental Support for Regarding Functional Classes of Proteins to Be Highly Isolated from Each Other." In *Darwinism: Science or Philosophy?* edited by J. Buell and V. Hearn. Richardson, Tex.: Foundation for Thought and Ethics, 1992.

———. *Darwin's Black Box.* New York: Free Press, 1996.

———. "Irreducible Complexity: Obstacle to Darwinian Evolution." In *Debating Design: From Darwin to DNA,* edited by W. A. Dembski and M. Ruse. Cambridge: Cambridge University Press, 2004.

Benton, M., and F. J. Ayala. "Dating the Tree of Life." *Science* 300 (2003): 1698-1700.

Berlinski, David. "On Assessing Genetic Algorithms." Public lecture at the "Science and Evidence of Design in the Universe" Conference. Yale University, New Haven, Connecticut, November 4, 2000.

Blanco, F., I. Angrand, and L. Serrano. "Exploring the Confirmational Properties of the Sequence Space Between Two Proteins with Different Folds: An Experimental Study." *Journal of Molecular Biology* 285 (1999): 741-53.

Bowie, J., and R. Sauer. "Identifying Determinants of Folding and Activity for a Protein of Unknown Sequences: Tolerance to Amino Acid Substitution." *Proceedings of the National Academy of Sciences, U.S.A.* 86 (1989): 2152-56.

Bowring, S. A. et al. "Calibrating Rates of Early Cambrian Evolution." *Science* 261 (1993): 1293-98.

————. "A New Look at Evolutionary Rates in Deep Time: Uniting Paleontology and High-Precision Geochronology." *GSA Today* 8 (1998a): 1-8.

————. "Geochronology Comes of Age." *Geotimes* 43 (1998b): 36-40.

Bradley, Walter L. "Information, Entropy and the Origin of Life." In *Debating Design: From Darwin to DNA,* edited by William A. Dembski and Michael Ruse. Cambridge: Cambridge University Press, 2004.

Brocks, J. J., et al. "Archean Molecular Fossils and the Early Rise of Eukaryotes." *Science* 285 (1999): 1033-36.

Brush, S. G. "Prediction and Theory Evaluation: The Case of Light Bending." *Science* 246 (1989): 1124-29.

Budd, G. E., and S. E. Jensen. "A Critical Reappraisal of the Fossil Record of the Bilaterial Phyla." *Biological Reviews of the Cambridge Philosophical Society* 75 (2000): 253-95.

Carroll, R. L. "Towards a New Evolutionary Synthesis." *Trends in Ecology and Evolution* 15 (2000): 27-32.

Cleland, C. "Historical Science, Experimental Science, and the Scientific Method." *Geology* 29 (2001): 987-90.

————. "Methodological and Epistemic Differences Between Historical Science and Experimental Science." *Philosophy of Science* 69 (2002): 474-96.

Chothia, C., I. Gelfland and A. Kister. "Structural Determinants in the Sequences of Immunoglobulin Variable Domain." *Journal of Molecular Biology* 278 (1998): 457-79.

Morris, Conway S. "The Question of Metazoan Monophyly and the Fossil Record." *Progress in Molecular and Subcellular Biology* 21 (1998): 1-9.

————. "Early Metazoan Evolution: Reconciling Paleontology and Molecular Biology." *American Zoologist* 38 (1998): 867-77.

————. "Evolution: Bringing Molecules into the Fold." *Cell* 100 (2000): 1-11.

————. "The Cambrian 'Explosion' of Metazoans." In *Origination of Organismal Form: Beyond the Gene in Developmental and Evolutionary Biology,* edited by Gerd B. Muller and Stuart A. Newman. Cambridge, Mass.: MIT Press, 2003.

————. "Cambrian 'Explosion' of Metazoans and Molecular Biology: Would Darwin Be Satisfied?" *International Journal of Developmental Biology* 47, nos. 7-8 (2003): 505-15.

————. *Life's Solution: Inevitable Humans in a Lonely Universe.* Cambridge: Cambridge University Press, 2003.

Crick, Francis. "On Protein Synthesis." *Symposium for the Society of Experimental Biology* 12 (1958): 138-63.

Darwin, Charles. *On the Origin of Species.* London: John Murray, 1859.

————. "Letter to Asa Gray." In *Life and Letters of Charles Darwin,* edited by F. Darwin. Vol. 1. London: D. Appleton, 1896.

Davidson, E. *Genomic Regulatory Systems: Development and Evolution.* New York: Academic Press, 2001.

Dawkins, Richard. *The Blind Watchmaker.* London: Penguin, 1986.

————. *River Out of Eden.* New York: Basic Books, 1995.

————. *Climbing Mount Improbable.* New York: W. W. Norton, 1996.

Dembski, William A. *The Design Inference.* Cambridge: Cambridge University Press, 1998.

————. *No Free Lunch: Why Specified Complexity Cannot Be Purchased Without Intelligence.* Lanham, Md.: Rowman & Littlefield, 2002.

————. "The Logical Underpinnings of Intelligent Design." In *Debating Design: From Darwin to DNA,* edited by William A. Dembski and Michael Ruse. Cambridge: Cambridge University Press, 2004.

Denton, Michael. *Evolution: A Theory in Crisis.* London: Adler & Adler, 1986.

————. *Nature's Density.* New York: Free Press, 1998.

Eden, M. "The Inadequacies of Neo-Darwinian Evolution as a Scientific Theory." In *Mathematical Challenges to the Darwinian Interpretation of Evolution,* edited by P. S. Morehead and M. M. Kaplan. Wistar Institute Symposium Monograph, New York: Allen R. Liss, 1967.

Eldredge, N., and Stephen Jay Gould. "Punctuated Equilibria: An Alternative to Phyletic Gradualism." In *Models in Paleobiology,* edited by T. Schopf. San Francisco: W. H. Freeman, 1972.

Erwin, D. H. "Early Introduction of Major Morphological Innovations." *Acta Palaeontologica Polonica* 38 (1994): 281-94.

————. "Macroevolution Is More than Repeated Rounds of Microevolution." *Evolution & Development* 2 (2000): 78-84.

————. "One Very Long Argument." *Biology and Philosophy* 19 (2004): 17-28.

Erwin, D. H., J. Valentine, and D. Jablonski. "The Origin of Animal Body Plans." *American Scientist* 85 (1997): 126-37.

Erwin, D. H., J. Valentine, and J. J. Sepkoski. "A Comparative Study of Diversification Events: The Early Paleozoic Versus the Mesozoic." *Evolution* 41 (1987): 1177-86.

Foote, M. "Sampling, Taxonomic Description, and Our Evolving Knowledge of Morphological Diversity." *Paleobiology* 23 (1997): 181-206.

Foote, M. et al. "Evolutionary and Preservational Constraints on Origins of Biologic Groups: Divergence Times of Eutherian Mammals." *Science* 283 (1999): 1310-14.

Frankel, J. "Propagation of Cortical Differences in *Tetrahymena.*" *Genetics* 94 (1980): 607-23.

Gates, Bill. *The Road Ahead.* Boulder, Colo.: Blue Penguin, 1996.

Gerhart, J., and M. Kirschner. *Cells, Embryos, and Evolution.* London: Blackwell Science, 1997.

Gibbs, W. W. "The Unseen Genome: Gems Among the Junk." *Scientific American* 289 (2003): 46-53.

Gilbert, S. F., J. M. Opitz and R. A. Raff. "Resynthesizing Evolutionary and Developmental Biology." *Developmental Biology* 173 (1996): 357-72.

Gillespie, N. C. *Charles Darwin and the Problem of Creation.* Chicago: University of Chicago Press, 1979.

Goodwin, B. C. "What Are the Causes of Morphogenesis?" *BioEssays* 3 (1985): 32-36.

———. *How the Leopard Changed Its Spots: The Evolution of Complexity.* New York: Scribner's, 1995.

Gould, Stephen Jay. "Is Uniformitarianism Necessary?" *American Journal of Science* 263 (1965): 223-28.

———. *The Structure of Evolutionary Theory.* Cambridge, Mass.: Harvard University Press, 2002.

Grant, P. R. *Ecology and Evolution of Darwin's Finches.* Princeton, N.J.: Princeton University Press, 1999.

Grimes, G. W., and K. J. Aufderheide. *Cellular Aspects of Pattern Formation: The Problem of Assembly.* Monographs in Developmental Biology 22. Baseline, Switzerland: Karger, 1991.

Grotzinger, J. P., et al. "Biostratigraphic and Geochronologic Constraints on Early Animal Evolution." *Science* 270 (1995): 598-604.

Harold, F. M. "From Morphogenes to Morphogenesis." *Microbiology* 141 (1995): 2765-78.

———. *The Way of the Cell: Molecules, Organisms, and the Order of Life.* New York: Oxford University Press, 2001.

Hodge, M. J. S. "The Structure and Strategy of Darwin's Long Argument."
   British Journal for the History of Science 10 (1977): 237-45.

Hooykaas, R. "Catastrophism in Geology, Its Scientific Character in Relation
   to Actualism and Uniformitarianism." In *Philosophy of Geohistory (1785-
   1970)*, edited by C. Albritton. Stroudsburg, Penn: Dowden, Hutchinson &
   Ross, 1975.

John, B., and G. Miklos. *The Eukaryote Genome in Development and Evolution*. London: Allen & Unwin, 1988.

Kauffman, S. *At Home in the Universe*. Oxford: Oxford University Press,
   1995.

Kenyon, D., and G. Mills. "The RNA World: A Critique." *Origins & Design* 17,
   no. 1 (1996): 9-16.

Kerr, R. A. "Evolution's Big Bang Gets Even More Explosive." *Science* 261
   (1993): 1274-75.

Kimura, M. *The Neutral Theory of Molecular Evolution*. Cambridge: Cambridge University Press, 1983.

Koonin, E. "How Many Genes Can Make a Cell? The Minimal Genome Concept." *Annual Review of Genomics and Human Genetics* 1 (2000): 99-116.

Kuppers, B. O. "On the Prior Probability of the Existence of Life." In *The
   Probabilistic Revolution*, edited by L. Kruger et al. Cambridge, Mass.: MIT
   Press, 1987.

Lange, B. M. et al. "Centriole Duplication and Maturation in Animal Cells."
   In *The Centrosome in Cell Replication and Early Development*, edited by
   R. E. Palazzo and G. P. Schatten. Current Topics in Developmental Biol-
   ogy 49. San Diego: Academic Press, 2000.

Lawrence, P. A., and G. Struhl. "Morphogens, Compartments and Pattern:
   Lessons from Drosophila?" *Cell* 85 (1996): 951-61.

Lenior, T. *The Strategy of Life*. Chicago: University of Chicago Press, 1982.

Levinton, J. *Genetics, Paleontology, and Macroevolution*. Cambridge: Cambridge University Press, 1988.

———. "The Big Bang of Animal Evolution." *Scientific American* 267
   (1992): 84-91.

Lewin, Roger. "A Lopsided Look at Evolution." *Science* 241 (1988): 292.

Lewontin, R. "Adaptation." In *Evolution: A Scientific American Book*. San
   Francisco: W. H. Freeman, 1978.

Lipton, P. *Inference to the Best Explanation*. New York: Routledge, 1991.

Loennig, Wolf-Ekkehard. "Natural Selection." In *The Corsini Encyclopedia of*

Psychology and Behavioral Sciences. Vol. 3. Edited by W. E. Craighead and C. B. Nemeroff. 3rd edition. New York: John Wiley, 2001.

Loennig, Wolf-Ekkehard, and H. Saedler. "Chromosome Rearrangements and Transposable Elements." *Annual Review of Genetics* 36 (2002): 389-410.

Lovtrup, S. "Semantics, Logic and Vulgate Neo-Darwinism." *Evolutionary Theory* 4 (1979): 157-72.

Marshall, W. F., and J. L. Rosenbaum. "Are There Nucleic Acids in the Centrosome?" In *The Centrosome in Cell Replication and Early Development,* edited by R. E. Palazzo and G. P. Schatten. *Current Topics in Developmental Biology* 49. San Diego: Academic Press, 2000.

Maynard Smith, J. "Structuralism Versus Selection: Is Darwinism Enough?" In *Science and Beyond,* edited by S. Rose and L. Appignanesi. London: Basil Blackwell, 1986.

Mayr, E. Foreword to M. Ruse, *Darwinism Defended.* Boston: Pearson Addison Wesley, 1982.

McDonald, J. F. "The Molecular Basis of Adaptation: A Critical Review of Relevant Ideas and Observations." *Annual Review of Ecology and Systematics* 14 (1983): 77-102.

McNiven, M. A., and K. R. Porter. "The Centrosome: Contributions to Cell Form." In *The Centrosome,* edited by V. I. Kalnins. San Diego: Academic Press, 1992.

Meyer, Stephen C. *Of Clues and Causes: A Methodological Interpretation of Origin of Life Studies.* Doctoral diss., Cambridge: University of Cambridge, 1991.

———. "DNA by Design: An Inference to the Best Explanation for the Origin of Biological Information." *Rhetoric & Public Affairs* 1, no. 4 (1998): 519-55.

———. "The Scientific Status of Intelligent Design: The Methodological Equivalence of Naturalistic and Non-Naturalistic Origins Theories." In *Science and Evidence for Design in the Universe.* Proceedings of the Wethersfield Institute. San Francisco: Ignatius Press, 2002.

———. "DNA and the Origin of Life: Information, Specification and Explanation." In *Darwinism, Design and Public Education,* edited by J. A. Campbell and Stephen C. Meyer. Lansing: Michigan State University Press, 2003.

———. "The Cambrian Information Explosion: Evidence for Intelligent Design." In *Debating Design: From Darwin to DNA,* edited by William A.

Dembski and Michael Ruse. Cambridge: Cambridge University Press, 2004.

Meyer, Stephen C., et al. "The Cambrian Explosion: Biology's Big Bang." In *Darwinism, Design and Public Education,* edited by J. A. Campbell and S. C. Meyer. Lansing: Michigan State University Press, 2003. (See also appendix C, "Stratigraphic First Appearance of Phyla Body Plans.")

Miklos, G. L. G. "Emergence of Organizational Complexities During Metazoan Evolution: Perspectives from Molecular Biology, Paleontology and Neo-Darwinism." *Mem. Ass. Australas. Palaeontols* 15 (1993): 7-41.

Monastersky, R. "Siberian Rocks Clock Biological Big Bang." *Science News* 144 (1993): 148.

Moss, L. *What Genes Can't Do.* Cambridge, Mass.: MIT Press, 2004.

Muller, Gerd B., and Stuart A. Newman. "Origination of Organismal Form: The Forgotten Cause in Evolutionary Theory." In *Origination of Organismal Form: Beyond the Gene in Developmental and Evolutionary Biology,* edited by Gerd B. Muller and Stuart A. Newman. Cambridge, Mass.: MIT Press, 2003.

Nanney, D. L. "The Ciliates and the Cytoplasm." *Journal of Heredity* 74 (1983): 163-70.

Nelson, P., and J. Wells. "Homology in Biology: Problem for Naturalistic Science and Prospect for Intelligent Design." In *Darwinism, Design and Public Education,* edited by J. A. Campbell and Stephen C. Meyer. East Lansing: Michigan State University Press, 2003.

Nijhout, H. F. "Metaphors and the Role of Genes in Development." *BioEssays* 12 (1990): 441-46.

Nusslein-Volhard, C., and E. Wieschaus. "Mutations Affecting Segment Number and Polarity in Drosophila." *Nature* 287 (1980): 795-801.

Ohno, S. "The Notion of the Cambrian Pananimalia Genome." *Proceedings of the National Academy of Sciences, U.S.A.* 93 (1996): 8475-78.

Orgel, L. E., and Francis H. Crick. "Selfish DNA: The Ultimate Parasite." *Nature* 284 (1980): 604-7.

Perutz, M. F., and H. Lehmann. "Molecular Pathology of Human Hemoglobin." *Nature* 219 (1968): 902-9.

Polanyi, Michael. "Life Transcending Physics and Chemistry." *Chemical and Engineering News* 45, no. 35 (1967): 54-66.

———. "Life's Irreducible Structure." *Science* 160 (1968): 1308-12, esp. p. 1309.

Pourquie, O. "Vertebrate Somitogenesis: A Novel Paradigm for Animal Segmentation?" *International Journal of Developmental Biology* 47, nos. 7-8 (2003): 597-603.

Quastler, Henry. *The Emergence of Biological Organization.* New Haven, Conn.: Yale University Press, 1964.

Raff, R. "Larval Homologies and Radical Evolutionary Changes in Early Development." In *Homology.* Novartis Symposium 222. Chichester, U.K.: John Wiley, 1999.

Reidhaar-Olson, J., and R. Sauer. "Functionally Acceptable Solutions in Two Alpha-Helical Regions of Lambda Repressor." *Proteins, Structure, Function, and Genetics* 7 (1990): 306-16.

Rutten, M. G. *The Origin of Life by Natural Causes.* Amsterdam: Elsevier, 1971.

Sapp, J. *Beyond the Gene.* New York: Oxford University Press, 1987.

Sarkar, S. "Biological Information: A Skeptical Look at Some Central Dogmas of Molecular Biology." In *The Philosophy and History of Molecular Biology: New Perspectives.* Edited by S. Sarkar. Dordrecht: Kluwer Academic, 1996.

Schutzenberger, M. "Algorithms and the Neo-Darwinian Theory of Evolution." In *Mathematical Challenges to the Darwinian Interpretation of Evolution,* edited by P. S. Morehead and M. M. Kaplan. Wistar Institute Symposium Monograph. New York: Allen R. Liss, 1967.

Shannon, Claude Elwood. "A Mathematical Theory of Communication." *Bell System Technical Journal* 27 (1948): 379-423, 623-56.

Shu, D. G., et al. "Lower Cambrian Vertebrates from South China." *Nature* 402 (1999): 42-46.

Shubin, N. H., and C. R. Marshall. "Fossils, Genes, and the Origin of Novelty." In *Deep Time,* edited by Douglas H. Erwin and Scott L. Wing. Lawrence, Kans.: Allen, 2000.

Simpson, G. "Uniformitarianism: An Inquiry into Principle, Theory, and Method in Geohistory and Biohistory." In *Essays in Evolution and Genetics in Honor of Theodosius Dobzhansky,* edited by M. K. Hecht and W. C. Steered. New York: Appleton-Century-Crofts, 1970.

Sober, E. *The Philosophy of Biology.* 2nd ed. San Francisco: Westview, 2000.

Sonneborn, T. M. "Determination, Development, and Inheritance of the Structure of the Cell Cortex." *Symposia of the International Society for Cell Biology* 9 (1970): 1-13.

Sole, R. V., P. Fernandez, and S. A. Kauffman. "Adaptive Walks in a Gene Network Model of Morphogenesis: Insight into the Cambrian Explosion." *International Journal of Developmental Biology* 47, nos. 7-8 (2003): 685-93.

Stadler, B. M. R., et al. "The Topology of the Possible: Formal Spaces Underlying Patterns of Evolutionary Change." *Journal of Theoretical Biology* 213 (2001): 241-74.

Steiner, M., and R. Reitner. "Evidence of Organic Structures in Ediacara-Type Fossils and Associated Microbial Mats." *Geology* 29, no. 12 (2001): 1119-22.

Taylor, S. V., et al. "Searching Sequence Space for Protein Catalysts." *Proceedings of the National Academy of Sciences, U.S.A.* 98 (2001): 10596-601.

Thaxton, Charles B., Walter L. Bradley and Roger L. Olsen. *The Mystery of Life's Origin: Reassessing Current Theories.* Dallas: Lewis & Stanley, 1992.

Thompson, D'Arcy W. *On Growth and Form.* 2nd ed. Cambridge: Cambridge University Press, 1942.

Thomson, K. S. "Macroevolution: The Morphological Problem." *American Zoologist* 32 (1992): 106-12.

Valentine, J. W. "Late Precambrian Bilaterians: Grades and Clades." In *Temporal and Mode in Evolution: Genetics and Paleontology 50 Years After Simpson,* edited by W. M. Fitch and F. J. Ayala. Washington, D.C.: National Academy Press, 1995.

———. *On the Origin of Phyla.* Chicago: University of Chicago Press, 2004.

Valentine, J. W., and D. H. Erwin. "Interpreting Great Developmental Experiments: The Fossil Record." In *Development as an Evolutionary Process,* edited by R. A. Raff and E. C. Raff. New York: Alan R. Liss, 1987.

Valentine, J. W., and D. Jablonski. "Morphological and Developmental Macroevolution: A Paleontological Perspective." *International Journal of Developmental Biology* 47 (2003): 517-22.

Wagner, G. P. "What Is the Promise of Developmental Evolution? Part II: A Causal Explanation of Evolutionary Innovations May Be Impossible." *Journal of Experimental Zoology* (Mol. Dev. Evol.) 291 (2001): 305-9.

Wagner, G. P., and P. F. Stadler. "Quasi-Independence, Homology and the Unity-C of Type: A Topological Theory of Characters." *Journal of Theoretical Biology* 220 (2003): 505-27.

Webster, Gerry, and Brian Goodwin. "A Structuralist Approach to Morphology." *Rivista di Biologia* 77 (1984): 503-10.

————. *Form and Transformation: Generative and Relational Principles in Biology.* Cambridge: Cambridge University Press, 1996.

Weiner, J. *The Beak of the Finch.* New York: Vintage, 1994.

Willmer, P. *Invertebrate Relationships: Patterns in Animal Evolution.* Cambridge: Cambridge University Press, 1990.

————. "Convergence and Homoplasy in the Evolution of Organismal Form." In *Origination of Organismal Form: Beyond the Gene in Developmental and Evolutionary Biology,* edited by G. B. Muller and S. A. Newman. Cambridge, Mass.: MIT Press, 2003.

Woese, C. 1998. "The Universal Ancestor." *Proceedings of the National Academy of Sciences, U.S.A.* 95 (1998): 6854-59.

Wray, G. A., J. S. Levinton and L. H. Shapiro. "Molecular Evidence for Deep Precambrian Divergences Among Metazoan Phyla." *Science* 274 (1996): 568-73.

Yockey, H. P. "A Calculation of the Probability of Spontaneous Biogenesis by Information Theory." *Journal of Theoretical Biology* 67 (1978): 377-98.

————. *Information Theory and Molecular Biology.* Cambridge: Cambridge University Press, 1992.

13

GENETIC ANALYSIS OF COORDINATE FLAGELLAR AND TYPE III REGULATORY CIRCUITS IN PATHOGENIC BACTERIA

SCOTT A. MINNICH AND STEPHEN C. MEYER

The rotary engine comprising the bacterial flagellum has been studied as a model for organelle development and assembly. Approximately thirty genes are involved in coding for its individual parts with another ten genes regulating expression and assembly and ten more for sensory perception or chemotaxis. These fifty genes constitute roughly 1 percent of the *Escherichia coli* or *Salmonella typhimurium* genome, a modest but significant information investment.

The expression of flagellar genes is tightly controlled and regulated in a sequential genetic hierarchy mirroring organelle assembly from the inner membrane to the outer cell surface. As such, the sequence is initiated by the master control operon, *flhDC*, or Class I genes. FlhDC forms a protein tetramer to transcriptionally activate the set of genes comprising the flagellar basal body-hook complex, collectively referred to as Class II genes. This complex includes several ring structures housing the motor, rod proteins equivalent to a drive shaft and the hook, which acts as a universal joint. Once this complex is assembled, a Class II flagellar-specific RNA polymerase sigma factor, FliA, activates the flagellar Class III genes, which include chemotaxis receptors and relays and the major flagellar component, flagellin, which forms the long filamentous propeller.

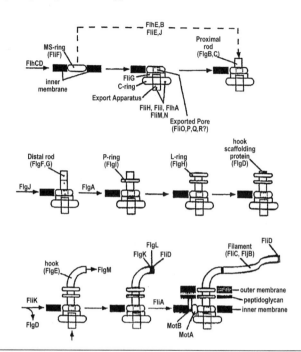

Figure 13.1. Flagellar gene regulatory cascade and assembly

Flagellin monomers are transported through the hollow core of the basal body hook structure and polymerize at the distal tip. This assembly process (see fig. 13.1) includes several key check points (feedback loops) that must be satisfied during the expression of each class of genes in the hierarchy, or the succeeding sets of genes are transcriptionally arrested. Additionally, a protein complex of approximately ten factors is positioned at the base of the flagellum to discriminate the order and number of basal body and flagellin components secreted through the flagellar core. This secretory complex is referred to as a type III secretory apparatus.[1]

Viewed as a whole, the flagellum is a true nanomachine of remarkable complexity in structure and assembly control. This macromolecular machine self-assembles and repairs, displays assembly control and processing, operates with two gears, is fueled by proton motive force, and the apparatus is "hard-wired" to a sensory apparatus that functions on short-term memory (chemotaxis). Rotor speeds for *E. coli* are estimated at 17,000 rpm but motors of some marine vibrios have been clocked upwards of 100,000 rpm.[2]

The Gram negative genus *Yersinia* includes three human and animal pathogens, *Y. enterocolitica, Y. pseudotuberculosis* and *Y. pestis* (etiologic agent of bubonic plague). Although each species causes a unique disease in the mammalian host, they share the same basic strategy for infection. In part, this strategy relies on a commonly shared set of genes localized on a large plasmid, referred to as pYV (plasmid for *Yersinia* virulence). This common plasmid encodes at least twelve antihost factors collectively referred to as Yops (for *Yersinia* outer proteins) that are now known to be injected directly into target host cells. All of the Yops are required for full virulence. The structure built to deliver these toxins into mammalian cells is likewise encoded by pYV (termed *ysc* and *lcr* genes). Expression of the Yop, Ysc, and Lcr genes is temperature dependent. These genes are not transcribed at temperatures of $\leq 30°C$, but when cells encounter a temperature of $37°C$, reflective of mammalian body temperature, their expression is activated. Secretion and injection of Yops is triggered by contact of the delivery system apparatus with a mammalian cell. In vitro expression of the *yop*, *ysc* and *lcr* genes is induced at $37°C$ but Yop secretion and increased production only occurs under limiting calcium ion concentration. In fact, in vitro, *Yersinia* shows calcium-dependent growth, a unique phenotype, and along with temperature, calcium limitation is a key environmental cue used by these organisms.[3]

Temperature also affects several other suits of genes in these organisms. These include flagellar synthesis for *Y. enterocolitica* and *Y. pseudotuberculosis* (*Y. pestis* is nonmotile), outer-membrane proteins, and surface proteins required for invasion of mammalian cells. Of these, motility shows the opposite temperature requirements compared to Yop synthesis.

In the early 1990s, several observations intimated the reciprocal temperature requirements of flagellar biosynthesis and Yop expression were more than coincidental effects. In the dissection of the mechanism of temperature regulation, R. R. Rohde, J. M. Fox and Scott Minnich isolated several spontaneous mutants that showed loss of motility and Yop synthesis, suggesting regulatory coordination of these two disparate systems.[4] G. Ramakrishnan, J. L. Zhao and A. Newton, and L. A. Sanders, S. Van Way and D. A. Mullin also found that the sequence of the *Caulobacter crescentus flbA* flagellar gene showed predicted sequence similarity to pYV-encoded LcrD.[5] Soon thereafter another set of *Caulobacter* flagellar genes showed sequence similarity to additional *ysc* and *lcr* genes.[6] The common denominator for this association identified both sets of genes as required for protein secretion in each process, flagellar assembly or Yop secretion. It was this association that led to the recognition that virulence protein secretion and flagellar protein assembly involved similar mechanisms now classed as type III secretion systems III (TTSS). More recently, yet another TTSS was identified in *Y. enterocolitica*. The function of this system, designated Ysa, is not clearly defined. However, it does contribute to virulence and is expressed in vitro under high salt/low temperature conditions, conditions that inhibit the expression of Yop and flagellar biosynthesis.[7]

Rationale for Coordinate Segregation of *Yersinia* TTSSs

Before the separate parallel nature of the flagellar and Yop systems was dissected, our laboratory proposed that the simplest explanation for temperature regulation of these systems was a direct overlap in function. Thus the flagellum in our view could be used not only as a propulsion machine but also as a dedicated highly efficient protein secretory machine for the Yops. Hence, the *Y. enterocolitica* and *Y. pseudotuberculosis* coordinate reciprocal regulation of motility and Yop secretion by temperature was due to alternative secretory function of the flagellar basal body dependent on habitat.

This hypothesis was testable by the following predictions. First, nonmotile *Y. pestis* would also have to contain a set of flagellar genes, some of which would be expressed. Second, it predicted that both *Y. enterocolitica* and *Y. pseudotuberculosis* would maintain expression of a subset of flagellar genes even at the nonpermissive temperature (37°C) for motility. Finally, it suggested that a subclass of mutants should be defective in both Yop secretion and motility.

Using genetic fusion with a reporter gene *(lacZ)* and northern blot assays, it was determined that at 37°C *Y. enterocolitica* continues to synthesize Class I and Class II flagellar genes. The loss of motility appears to be due to the immediate transcriptional arrest of Class III genes (twenty-five- to thirtyfold repression) on exposure to 37°C.[8] A subset set of flagellar gene transposon insertionally inactivated genes were mapped to flagellar genes and also failed to secrete Yops. Finally, *Y. pestis* does contain a complete set of flagellar genes, and using RT PCR, some are transcribed.[9]

Whereas all these predictions turned out to be true, subsequent work in a number of laboratories showed that the TTSS and flagellum are in fact separate and parallel systems. Further, it was shown by Kubori and coworkers that the TTSS system of *Salmonella typhimurium* assembles into a "needle-like" structure, remarkably similar to the basal body of the flagellum. This structure acts as a nanoinjector; its hollow core serves as a specific conduit to export virulence proteins.[10] These structures have since been isolated in a number of organisms including *Yersinia*.[11]

Given that the TTSS of the flagellum and virulence systems are separate parallel structures, several factors suggest that the relationship and regulatory parameters of the flagellar and virulence TTSS are still intimately related. Further, maintaining the segregational nature of their regulation may be critical for function. A key to this understanding was the observation that TTSS virulence proteins from disparate organisms could be cross-recognized for secretion. Schneewind and coworkers showed that TTSS proteins of plant pathogens were recognized and secreted by *Y. enterocolitica* and vice versa.[12] This suggested that the secretory signal of TTS proteins was conserved across genera. Based on this observation, we came full circle back to our original hypothesis of dual function of the flagellar basal body structure. Expression of multiple TTSSs in the same organism by mutually exclusive environmental conditions prevents cross-contamination of secreted proteins. That is to say, if flagellum biosynthesis were expressed simultaneously with the Yop TTSS, flagellin monomers may be exported out the needlelike structure as well as the flagellar basal body and vice versa in regard to Yops. Efficiency of both systems would suffer as a consequence.

Type Three Secretory Systems Display Promiscuity

To test the above hypothesis, a flagellin and two Yop genes (*fleB* and, *yopD* and *yopM* respectively) were cloned and fused to the inducible p*tac* promoter. By placing these plasmid constructs in various mutant strains lacking either a flagellum TTSS (*flhD-*, *flhB-*) or Yop TTSS (*yscE-*, pYV-), expression and secretion of FleB, YopM, and YopD could be examined under conditions when these proteins are not normally expressed. For example, p*tac* induction of FleB in an *flhD-* strain at the 37°C and limiting calcium ion concentration (conditions conducive to Yop expression and secretion) permit determination if FleB is exported outside the cell. Expression of FleB in the double mutant strain *flhD-* and pYV- or *yscE-* serves as a negative control. As can be seen in figure 13.2, it was found that FleB is indeed secreted by the Yop TTSS and even the Ysa TTSS (low temperature, high salt). Conversely, it has been shown that YopM is recognized by both the Ysa and flagellar TTSSs.

Practical Implications

The potential for cross-recognition between type III exported proteins of different systems in the same cell carries several implications. First, these observations explain why segregation of these systems by specific environmental cues is necessary. For example, expression of a flagellum under host conditions would result in loss of polarized secretion of Yop proteins into target host cells. Additionally, display of flagellin to macrophages by direct injection via the Ysc secretin would countermand the anti-inflammatory strategy used by the *Yersinia*. Flagellin is a potent cytokine inducer.[13] Further, because flagellin expression is controlled by such high expression promoters, it also suggests that flagellin, if expressed, may competitively interfere with virulence protein secretion. Indeed, this latter suggestion may explain why an important subset of major human pathogens, including *Y. pestis, Shigella* spp., *Bordetella pertussis* and recent isolates of *E. coli* O157:H7, have lost flagellar biosynthetic capacity altogether, even though they have the requisite flagellar genes. Each of these species has mutations in the flagellar master control operon *flhDC*, shutting down the entire flagellar regulon. For example, *Y. pestis* has a single T insertion in *flhD,* causing a frameshift mutation. Function of FlhD can be restored by a spontaneous five base pair insertion.[14] Based on these results, we predicted that recent atypical virulent strains of *E. coli* O157:H- (nonmotile) would contain a similar genetic lesion. Indeed, this clonal isolate carries a twelve-base pair in-frame deletion in *flhC.*[15] Further analysis showed this deletion involves a critical amino acid residue for FlhC function. Repair of this lesion restores full motility and H7 antigen expression.

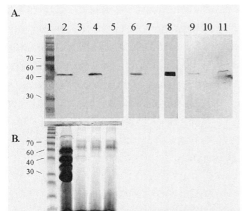

	Flagellum	Ysc
1. M. W. Std's		
2. *flhD*::Ω 37°C - Ca^{2+}	-	+
3. *flhD*::Ω 37°C + Ca^{2+}	-	-
4. *flhD*::Ω 25°C (neg. control)	-	-
5. *flhD*::Ω 37°C -pYV + Ca^{2+}	-	-
6. *flhB*::*lacZI* 37°C - Ca^{2+}	-	+
7. *flhB*::*lacZI* 37°C + Ca^{2+}	-	-
8. Wild-type 37°C - Ca^{2+}	-	+

Figure 13.2. (A) Western blot showing extracellular secretion of p*tac*-induced FleB. Lane designations are in the table to the right showing strain genotypes. These results show secretion of FleB occurs only when either the Yop or Ysa TTSS are expressed. (B) Stained gel from western blot represented in lanes 1-5 showing Yop secretion.

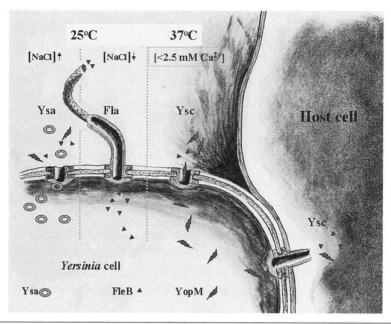

Figure 13.3. Schematic showing the three segregating environmental conditions required for Ysa, flagellar and Yop TTSS expression.

Examination of the *Y. pestis* flagellar system shows there are additional consequences in loss of motility (*flhD* phenotype). DNA sequence analysis shows that at least three other deletions are present in flagellar genes. Loss of expression of *flhDC* in this organism renders downstream genes silent and susceptible to further decay and permanent loss due to the improbability of repairing four lesions sequentially. B. M. Pruss and her coworkers showed that the *E. coli flhDC* operon regulates about thirty nonflagellar operons, including genes involved with amino acid synthesis and growth under various environmental conditions.[16] Using microarray analysis of *Y. enterocolitica* wild-type and *flhDC* mutants, similar results have been observed. *Y. pestis* displays multiple vitamin and amino acid requirements, some of which may be due to loss of *flhC*.

Philosophical Implications

To quote the original rendition of the U.S. Department of Energy's "Genomes to Life" website, "the molecular machines present in the simplest cells, produced by evolution, dwarf the engineering feats of the twentieth century."[17] The dissection of the complexity and sophistication of simple machines like the bacterial flagellum are indeed a testimony to the power of modern molecular biological techniques. Yet the elegant structural properties, efficiency and the highly controlled genetic programming to produce these machines was neither anticipated nor predicted. The potential applications of this knowledge are legion and have spawned a new discipline focused on nanotechnology.

In light of this new information, some scientists have questioned whether the mechanism of mutation, natural selection and time are sufficient to account for the origin of such machines. Michael Behe has proffered the concept of irreducible complexity using the flagellum as a paradigmatic example.[18] It is this very concept that has been the bread and butter of molecular geneticists, allowing them to identify genes in any given system by loss of function. Behe argues that natural selection and random mutation cannot produce the irreducibly complex bacterial flagellar motor with its roughly forty separate protein parts, since the motor confers no functional advantage on the cell unless all the parts are present. Natural selection can preserve the motor once it has been assembled, but it cannot detect anything to preserve until the motor has been assembled and performs a function. If there is no function, there is nothing to select. Given that the flagellum requires nearly fifty genes to function, how did these arise? Contrary to popular belief, we have no detailed account for the evolution of any molecular machine. The data from *Y. pestis* presented here seems to indicate that loss of one constituent in the system leads to the gradual loss of others. For progression to work, each gene product must maintain some function as it is adapted to another.

To counter this argument, particularly as it applies to the flagellum, others have used the TTSS. Since the secretory system that forms part of the flagellar mechanism can also function separately, Miller [18, 19] has argued that natural selection could have "co-opted" the functional parts from the TTTS and other earlier simple systems to produce the flagellar motor.[19] And indeed the TTSS contains eight to ten proteins that are also found in the forty protein bacterial flagellar motor. Miller thus regards the virulence secretory pump of the *Yersinia* Yop system as a Darwinian intermediate, case closed.

This argument seems only superficially plausible in light of some of the findings presented in this paper. First, if anything, TTSSs generate more complications than solutions to this question. As shown here, possessing multiple TTSSs causes interference. If not segregated, one or both systems are lost. Additionally, the other thirty proteins in the flagellar motor (that are not present in the TTSS) are unique to the motor and are not found in any other living system. From whence were these protein parts co-opted? Even if all the protein parts were somehow available to make a flagellar motor during the evolution of life, the parts would need to be assembled in the correct temporal sequence similar to the way an automobile is assembled in a factory. Yet, to choreograph the assembly of the parts of the flagellar motor, present-day bacteria need an elaborate system of genetic instructions as well as many other protein machines to time the expression of those assembly instructions.

Arguably, this system is itself irreducibly complex. In any case, the co-option argument tacitly presupposes the need for the very thing it seeks to explain—a functionally interdependent system of proteins. Finally, the phylogenetic analyses of the gene sequences show that the flagella proteins arose first and those of the TSS came later.[20] In other words, if anything, the TSS (less complex) evolved from the flagellum (more complex).

Molecular machines display a key signature or hallmark of design, namely, irreducible complexity. In all irreducibly complex systems in which the cause of the system is known by experience or observation, intelligent design or engineering played a role in the origin of the system. Given that neither standard neo-Darwinism nor co-option has adequately accounted for the origin of these machines or the appearance of design that they manifest, we might now consider the design hypothesis as the best explanation for the origin of irreducibly complex systems in living organisms. That we have encountered systems that tax our own capacities as design engineers justifiably leads us to question whether these systems are the product of undirected, unpurposed chance and necessity. Indeed, in any other context we would immediately recognize such systems as the product of very intelligent engineering. Although some may argue this is merely an argument from ignorance, we regard it as an inference to the best explanation,[21] given what we *know* about the powers of intelligence as opposed to strictly natural or material causes.

We know that intelligent designers can and do produce irreducibly complex systems. We find such systems within living organisms. We have good reason to think that these systems defy the creative capacity of the selection/mutation mechanism. The real problem may not be determining the best explanation of the origin of the flagellum. Rather, it may be amending the methodological strictures that prevent consideration of the most natural and rational conclusion—albeit one with discomfiting philosophical implications.

PART V

EVER-INCREASING SPHERES
OF INFLUENCE

INTELLIGENT DESIGN AND THE DEFENSE OF REASON

NANCY PEARCEY

—

I had the honor of paying tribute to Phillip Johnson when I wrote the foreword to his 2002 book *The Right Questions*. There I analyzed his reframing of the scientific debate over Darwinian evolution. Here I want to focus on the ways he has provided strategic direction for engaging the cultural and philosophical implications of evolutionary thought as well.

People often wonder why Johnson, a lawyer, would get involved in the evolution issue in the first place. When asked that question in an interview, he explained that in the past he was thoroughly grounded in secular rationalism, having earned degrees at Harvard and the University of Chicago. In his career as a Berkeley law professor, the prevailing opinion among those who considered themselves enlightened was that religion was mere wish fulfillment.

Upon reaching his mid-thirties, however, Johnson began to consider the claims of Christianity. "I wanted to know whether the fundamentals of the Christian worldview were fact or fantasy," he told the interviewer. To answer that question, "Darwinism is a logical place to begin." Why? "Because if Darwinism is true, Christian metaphysics is fantasy."[1] As he was to write later in *The Wedge of Truth,* religion is something that "scientific intellectuals usually relegate to the private sphere, where illusory beliefs are acceptable 'if they work for you.' "[2]

As Johnson goes on to explain, the most significant impact of naturalistic evolution does not consist in the details of mutation and natural selection but in a change in the definition of truth (epistemology). It reinforces the so-called fact-value split that divides human experience into two separate and incompatible domains. On one hand, reliable knowledge is assumed to be a matter of scientific "facts," which are objective, rational and value free. But what does that mean for "values"? They have been relegated to the realm of human subjectivity, where they are, technically speaking, not a matter of true or false at all. Values may be personally meaningful, may be part of our cultural tradition, but ultimately they express something about ourselves only, not about objective reality. This change in the dominant definition of truth explains why naturalistic evolution has contributed to a wide-ranging cultural revolution that continues unabated to this day.

Symbiotic Relationship

I had the opportunity to speak on this topic at Stanford University for a Veritas Forum conference in May 2005. On the podium before me was Michael Behe, making the case against Darwinian evolution, along with Richard Rorty, one of the best-known proponents of postmodernism. As a result, I proposed to address how the two are related—how Darwinian naturalism *gives rise to* postmodernism.

Drawing a link between Darwinism and postmodernism may initially seem a bit of a stretch. We have grown accustomed to the idea that the academic world is split into what C. P. Snow famously called "two cultures,"[3] with the humanities sharply divided off from the sciences. Today this bifurcation reaches down to the elementary grades, so that in subjects like the humanities, teachers have tossed out their red pencils and act as though correct spelling or grammar were forms of oppression imposed by those in power. Ironically, however, down the hallway in the science classroom, only one view is tolerated. For example, Darwinian evolution is not open to question; students are not exposed to the evidence against it and invited to judge for themselves whether or not it is true. Science is treated as public truth that all are required to accept, regardless of their private beliefs.

"Two Cultures" Reflect the Fact-Value Split
> *Humanities:*
> human values

> *Science:*
> objective facts

In short, the sciences still hold the ideal of objective truth, while the humanities treat truth as a matter of personal values—subjective and relativistic.[4] My argument is that these two approaches, though apparently widely divergent, are actually logically connected: It is because science has largely accepted Darwinian naturalism that the humanities are in such a muddle of subjectivism and postmodernism.

This thesis will not be difficult to support, because Richard Rorty has already done it for me. All I have to do is demonstrate why he is right. In many of his writings, Rorty has explicitly stated that Darwinism is the basis of postmodernism. As he put it in the *New Republic,* some philosophers continue to seek a transcendent, universal, capital-T Truth—one "not subject to chance." Other philosophers, however, have found ways of "keeping faith with Darwin" (an interesting phrase) by recognizing that all truth claims "are as much products of chance as are tectonic plates and mutated viruses." He goes on: "The idea that one species of organism is, unlike all the others, oriented not just toward its own increased prosperity but toward Truth, is . . . un-Darwinian."[5]

In other words, if we are organisms that arose by adaptation to the environment, then our brains are products of evolutionary forces. Ideas arise by random variations in the brain, just like Darwin's random variations in nature. Thus Rorty treats all the great formative ideas of Western culture as evolutionary accidents: Just as "a cosmic ray scrambles the atoms in a DNA molecule" to produce a mutation, so too the ideas of Aristotle or St. Paul or Newton could be "the results of cosmic rays scrambling the fine structure of some crucial neurons in their respective brains." These ideas have had great staying power not because they reflect reality in some way but only because they are useful. They help people organize their experience and get ahead in the struggle for existence.[6]

In short, concepts are not true or false; they are merely problem-solving tools that serve human goals and purposes. Rorty considers himself a disciple of John Dewey, who drew an analogy to silverware: You don't ask if a spoon or a fork is *true*. That would be a category mistake. You ask only which one works for eating soup. Ideas are simply mental tools; we judge them by how well they work in getting us what we want. Thus by "keeping faith with Darwin," Rorty concludes, we end up with a form of postmodernism very much like that of Heidegger, Derrida and Foucault.[7]

The Fact-Value Split

There is a certain irony in Rorty's position: though postmodernism rejects the very concept of objective truth, there is *one* idea that it continues to treat as unquestioned truth—namely, Darwinism itself. Evolution is treated as an objective fact and not merely a human construction, because unless it is true, there is no reason to accept postmodernism. Thus a symbiotic relationship exists between the two. We might say that in our day the "two cultures" look like this:

A Symbiotic Relationship
> *Humanities:*
> postmodern relativism
> (Ideas are tools for human purposes.)

> _____

> *Science:*
> naturalistic evolution
> (The mind evolved by natural selection.)

Today most Westerners have absorbed this two-track epistemology. Alan Bloom, author of the well-known book *The Closing of the American Mind,* in another essay wrote, "Every school child knows that values are relative. [They] are not based on facts but are mere individual subjective preferences."[8] In other words, the fact-value split has become so deeply ingrained in the American mind that it is something "every school child knows."

So does every college student. Philosophy professor Peter Kreeft of Boston College says the students coming into his classroom are "perfectly willing to believe in objective truth in science, or even in history sometimes, but certainly not in ethics or morality."[9] Notice the split mentality: Students arrive on campus already convinced that science is about objective facts, but morality is about subject values. Moreover, what they learn in the college classroom typically reinforces that dichotomy. For example, one college economics textbook states the fact-value split as unquestioned dogma: "Facts are objective, that is, they can be measured, and their truth tested. . . . Value judgments, on the other hand, are subjective, being matters of personal preference. . . . Such preferences are based on personal likes and feelings, rather than on facts and reasons."[10]

Francis Schaeffer placed this two-track epistemology at the center of his analysis of Western thought, using the imagery of two stories in a building.[11] In the lower story are science and reason, which are thought to yield "public truth," binding on everyone. Above it is an upper story of religion and morality, the arts and humanities, which are thought to be the realm of "private truth," derived from personal experience. When people say, "That may be true for you, but it's not true for me," they are expressing an upper-story definition of truth.

Values Reduced to Existential Decisions

Value:

subjective preferences

Fact:

objective verifiable truths

No major civilization at any other period of history has ever held such a divided conception of truth. Until the rise of late-modern Western culture, virtually every culture has assumed that the universe is possessed of both a physical order and a moral/spiritual order. Individuals were thought to have an obligation to bring their lives into harmony with that objective order.

How did this unified view of reality unravel? Many historians identify the crucial turning point as the rise of Darwinism. As one philosophy textbook puts it, "Until 1859 [when Darwin's *Origin of Species* was published], the fundamental unity of knowledge was assumed by virtually all serious writers in America." That is, there was a conviction of a single universal order established by God, encompassing both the natural and the moral order. "What the controversy over evolution did was to shatter this unity of knowledge," reducing religion and morality to "noncognitive subjects."[12] To use Schaeffer's imagery, they were relegated to the upper story.

Similarly, a legal historian says Darwinism led to a naturalistic view of knowledge, causing a shift "from religion as *knowledge* to religion as *faith*." Because "there was no longer any function for God to carry out in the world, He was, at best, a gratuitous philosophical concept derived from personal need." Individuals are free to continue holding to some form of religion, so long as they recognize that it has become "private, subjective, and artificial."[13] In short, if God's existence does not serve any cognitive function in explaining the world, then the only function left is an emotional one. Religion becomes something that can be tolerated for people who need that kind of crutch.

Today those who are blunt put religion on the same level as fairy tales.

Writing in the *New York Times* the Darwinist philosopher Daniel Dennett said, "We don't believe in ghosts or elves or the Easter Bunny—or God."[14] This explains why controversies over Darwinism keep bubbling up in school districts across the country. The public senses intuitively that when naturalistic evolution is taught in the science classroom, then a naturalistic view of religion and morality will be taught in history, social studies, English and all the rest of the curriculum.

The Public-Private Divide

Because ideas inevitably interact with social reality, it is not surprising that the fact-value split has a counterpart in the way modern societies are organized. Western culture has become sharply divided between the public and the private sphere.

"Modernization brings about a novel dichotomization of social life," explains sociologist Peter Berger. "The dichotomy is between the huge and immensely powerful institutions of the public sphere," by which he means the state, large corporations, academia, and so forth, "and the private sphere"—the realm of family, church and personal relationships.[15] We might diagram it like this:

The Dichotomization of Social Life

> *Private sphere:*
> family, church, personal relationships

> *Public sphere:*
> state, corporations, academia

Notice how this social dichotomy correlates with the split in the realm of ideas. The public sphere is the place for public truth—for facts that can be verified scientifically. And what about values? They have been relegated to the private realm of family and personal life.

This correlation became strikingly clear during the 2004 presidential campaign.[16] My purpose here is not to talk about politics per se. However, during times of cultural upheaval, underlying worldview themes often bubble to the surface and become easier to recognize. As a result, these examples provide lasting object lessons in the way many Americans tend to think.

The defining feature of the 2004 election was a "morality gap," said

Thomas Byrne Edsall in the *Atlantic Monthly*. In the past, the left-right division in American politics was over economic issues. It was an accepted axiom that people vote their pocketbooks. But today the cutting edge issues have to do with sex and reproduction: abortion, homosexual marriage, embryonic stem-cell research and so on. "Whereas elections once pitted the party of the working class against the party of Wall Street," Edsall concludes, "they now pit voters who believe in a fixed and universal morality against those who see moral issues, especially sexual ones, as elastic and subject to personal choice."[17]

In short, the issue is not the *content* of morality (i.e., which actions are right or wrong) so much as the *truth status* of moral claims. Is morality a universal normative standard or only a matter of subjective preference? The question at the heart the cultural conflict in America is over epistemology—the nature of truth.

For example, at the 2004 Democratic National Convention, Ron Reagan, son of the former president, made a widely publicized remark about opponents of embryonic stem-cell research. "Their belief is just that—an article of faith—and they are entitled to it," he said. "But it does not follow that the theology of a few should be allowed to forestall the health and well-being of the many."

What moral theory is expressed here? Put another way, which side of the "morality gap" is Reagan on? Notice that people are invited to believe whatever they want—they're even "entitled to it"—so long as they are willing to hold it as a subjective "article of faith," not something objectively true that should be allowed to guide scientific research. The assumption is that morality and religion are not universal truths, as they were traditionally thought to be, but merely personal "values," and as such they belong strictly in the private sphere.

In the Washington, D.C., area (where I live), public policy groups are constantly strategizing how to get past the gatekeepers. Yet the most important gatekeeper is not any group of people; it is the definition of truth. The dominant epistemology determines which ideas are taken seriously in the public arena and which are simply dismissed out of hand. Thus the fact-value grid functions as the most powerful gatekeeper today. It removes religion and morality from the realm of objectivity and relegates them to the realm of subjective values.

Once that happens, arguments on the detail level have no traction. In principle, private values do not belong at the table of public discourse. This explains why the late Christopher Reeve could say, "When matters of public policy are debated, *no religions should have a seat at the table.*"[18] He was not weighing whether particular religious-based viewpoints are right or wrong; instead he insisted that they do not belong at the table in the first place. Why not? Because private preferences should not be allowed to shape public policy.

Hidden Agendas

Of course, most politicians are not so blunt. They realize that it can be politically risky to attack religion directly or debunk it as false. But that is precisely why the language of values proves so useful—because it takes religion out of the realm of true and false altogether. That way politicians can reassure voters that of course they "respect" everyone's religious belief—while at the same time denying that it has any relevance to the public realm, where we make decisions about what we're really going to do as a society. To quote Phillip Johnson, the fact-value split "allows the metaphysical naturalists to mollify the potentially troublesome religious people, by assuring them that science does not rule out 'religious *belief*' (so long as it does not pretend to be *knowledge*)."[19]

In the 2004 presidential debates, John Kerry employed this strategy repeatedly. On embryonic stem-cell research: "I really respect the feeling that's in your question." On abortion: "I cannot tell you how deeply I respect the belief about life and when it begins." He made it clear, however, that he would not allow such considerations to affect the way he actually voted. The strategy is that first you placate religious people by telling them how much you respect their "beliefs" and "feelings"—but then remind them that private feelings are not something we can impose on others in terms of public policy.

To get a handle on this, imagine that you present your position on some issue and the other person responds, "Oh that's just science, that's just facts, don't impose it on me." Of course, no one says that. But they *do* say, "That's just your religion, don't impose it on me." What's the difference? Science is thought to be public truth, binding on everyone, while religion is defined as private feelings relevant only to those who believe it.

Of course, labels like *science* or *reason* are often used to mask what is actually a substantive philosophical position. In another example from the 2004 campaign, *Newsweek* columnist Eleanor Clift criticized President Bush for allowing religion to inform public policy in matters like abortion, while she praised Kerry for keeping faith out of politics. "Voters have a choice," she concluded, "between a president who governs by *belief* and a challenger who puts his faith in *rational decision-making*."[20]

What is the implication here? Obviously, that Christianity is *not* rational. But notice that Clift is also presenting the liberal position as though it were not any particular ideology, but only a rational weighing of the facts. The article was titled "Faith Versus Reason," as though liberal views were purely a product of reason. This is a common rhetorical trick: Christian views are dismissed as biased and irrational, while secular views are presented as unbiased and objective, derived from facts and reason. But this is sheer bluff. In reality, the liberal position on abortion and bioethics is an expression of utilitarianism and pragmatism, based on a cost-benefit analysis.

The lesson is that worldviews do not come neatly labeled. No one says the conflict is a *utilitarian, pragmatic* standard of ethics versus a *normative, transcendent* standard. Instead, public rhetoric typically casts the conflict as science versus religion, facts versus faith. That kind of language should be a tip-off to start looking for implicit worldview assumptions. The underlying conflict is between two belief systems, two philosophies, two worldviews. By unmasking hidden worldviews we can level the playing field, debunking the double standard that allows secular views—but not religious ones—a seat at the table in public discourse.

It is evident from these examples that the challenge to Christianity is much more radical than it was in the past. Secularists used to argue that religion is *false*, which meant Christians could at least engage them in discussions about reasons, logic and evidence. But today secularists are much more likely to argue that religion does not have the status of a testable truth claim at all—that it doesn't even belong at the table. The famous physicist Wolfgang Pauli once told a colleague, "Your theory is so bad, it's not even wrong." That is, it's not even in ballpark of possible answers. That is a good metaphor for the way Christianity is treated in mainstream Western culture today. The fact-value grid functions as a gatekeeper to rule it out of the ballpark of possible candidates in the public debate.

The Body Machine

Having traced the way the fact-value split is reflected socially in the public-private dichotomy, let's see how the same division has been applied to the human person. René Descartes, often considered the first modern philosopher, divided the human person along the lines of a radical dualism. He treated the human body as nothing but a glorified machine, running by scientific laws. But he treated the human mind as an autonomous power of choice that in a sense *uses* the body in an instrumental way—almost the way you use a car to take you where you want to go.

Descartes's Radical Dualism
> *Mind:*
> free, autonomous, self-determining

> *Body:*
> mechanical, determined, instrumental

For Descartes, religion and morality apply to the realm of freedom in the upper story, while science gives us knowledge of the lower story. Over the centuries, as science became increasingly committed to materialism and determinism, the two stories came to stand in tension and even contradiction with one another. In our own day, they represent two pictures of the universe that "are really at war" with one another (in the words of philosopher John Searle).[21]

As an example, consider Steven Pinker, a leader in the field of cognitive science and author of the bestselling book *How the Mind Works*. Pinker's worldview could be called evolutionary naturalism—nature is all there is. Things traditionally thought to be transcendent to nature, like spirit or soul or mind, are illusions. Pinker argues that our brains are nothing but computers, complex data-processing machines.

At the same time, he acknowledges that morality depends on the idea that humans are *more* than machines—that we are capable of making undetermined, free choices. Here is his dilemma, then: When working in the lab, Pinker adopts what he calls "the mechanistic stance," treating humans as complex mechanisms. But "when those discussions wind down for the day," he writes, " we go back to talking about each other as free and dignified human beings."

In other words, when he goes home to his family and friends, his scientific naturalism does not work as a viable philosophy. You can't treat your wife like a complex data-processing machine or program your kids like little computers. So in ordinary life, Pinker admits that he has to switch to a completely contradictory paradigm. Here's how he puts it: "A human being is simultaneously a machine and a sentient free agent, depending on the purposes of the discussion."[22]

This is a fatal internal contradiction, and Schaeffer provocatively called it a secular leap of faith. Thinkers like Pinker embrace evolutionary naturalism as their professional ideology (in the lower story). But it does not fit their real-life experience. So what do they do? They take a leap of faith to the upper story, where they affirm a completely contradictory set of ideas like moral freedom and human dignity—*even though these things have no basis within their own intellectual system.*

Pinker comes close to calling it a leap as well; he labels it mysticism. "Consciousness and free will seem to suffuse the neurobiological phenomena at every level," he writes. "Thinkers seem condemned either to denying their existence or to wallowing in mysticism."[23] That is, either you can try to be consistent with evolutionary naturalism in the lower story—in which case you have to deny the existence of things like consciousness and free will. Or else you can affirm their existence, even though they have no basis within your intellectual system—which is sheer mysticism. An irrational leap.

Rodney Brooks of MIT provides another example. In his book *Flesh and Machines,* he argues that a person is nothing but an automaton—"a big bag of skin full of biomolecules" interacting by the laws of physics and chemistry. It is not easy to think this way, he admits. But "when I look at my children, I can, when I force myself, . . . see that they are machines."

And yet: "That is not how I treat them. . . . They have my unconditional love, the furthest one might be able to get from rational analysis." If this sounds incoherent, Brooks admits as much: "I maintain two sets of inconsistent beliefs."[24] In other words, he has to take a secular leap of faith as well.

Marvin Minsky, also at MIT, is famous for his punchy phrase that the human brain is nothing but "a three-pound computer made of meat." But he too takes a secular leap of faith. In *The Society of Mind* he writes, "The physical world provides no room for freedom of will." And yet "that concept is essential to our models of the mental realm. [And so] we are virtually forced to maintain that belief, *even though we know it's false.*"[25]

This is an astonishing statement. Because of ordinary, undeniable human experience, people are *forced* to affirm certain things—like moral freedom—even when they claim to "know" these ideas are false, based on their naturalistic philosophy. This is the tragedy of the postmodern age. The things that matter most in life, the things that make us truly human, have been reduced to nothing but useful fictions. Convenient falsehoods. These thinkers are forced to hang their entire hopes for dignity and meaning on an upper-story realm that they themselves regard as noncognitive and ultimately unreal.

Of course, the very fact that these scientists have to make a leap of faith ought to tell them something—it means that evolutionary naturalism is not an adequate worldview. After all, the purpose of a worldview is to explain the *world*. And if it *fails* to explain some part of the world, then there is something wrong with that worldview. The only way these scientists can account for human nature *as they themselves experience it* is by accepting an outright contradiction.

Critics of Christianity often dismiss it as irrational, but it's really the other way around. Christianity does *not* require this kind of self-contradictory leap of faith. As a system of thought, it begins with a personal God—a God who is a personal agent. Then the fact that humans are personal agents makes perfect sense. Every worldview is limited to the categories allowed by its starting assumptions. *If* you begin with matter operating by blind, mechanical forces, then logically humans are nothing but machines—complex mechanisms. But *if* you begin with a transcendent personal agent, that gives a rationally consistent basis to explain the full range of human experience. The very things that are so problematic for Darwinian naturalism—like free will and moral responsibility—are accounted for simply and elegantly within the Christian worldview. There is no need to resort to an irrational upper-story leap in order to account for them.

Treating the Body Like a Machine

Let's turn to the way this same Cartesian dualism affects ethical issues. In the 1970s ethicist Paul Ramsey noticed that dualism had become the underlying worldview in abortion, genetic engineering and all the other life issues.[26] Pro-life groups used to think the battle was over getting people to agree that the fetus is human: Surely, it was thought, they would realize then that abortion is morally wrong. But today abortion advocates are perfectly willing to agree that the fetus is *physiologically* human. However, that fact is regarded as irrelevant to its moral status, nor does it warrant legal protection. The deciding factor is "personhood," typically defined in terms of autonomy or the power of choice.

The Two-Story Approach to Life Issues
Personhood:
warrants legal protection

Physiologically human:
irrelevant to moral status

For example, during the 2004 presidential campaign, John Kerry surprised everyone by agreeing that "life begins at conception." How, then, could he support abortion? Because, as he explained in an interview with Peter Jennings, the fetus is "not the form of life that takes [on] *personhood*" as we have defined it.[27] In personhood theory, if there is no power of choice, no autonomous self, then there is no "person," and the body is merely a machine that can be treated in a utilitarian manner, to be discarded or used for research and experimentation.

This is the logic now used to justify euthanasia. The most significant debate over Terri Schiavo (in terms of clashing worldviews) was overlooked by most of the media. On *Court TV,* Bill Allen, a bioethicist from the University of Florida, was asked point blank, "Do you think Terri is a person?" He replied: "No, I do not. I think having awareness is an essential criterion for personhood."[28]

Now, Christian ethicists agree that there is no moral obligation to prolong the dying process, but Terri was not dying. So the heart of the issue is a theory of "personhood" in which simply being part of the human race is not enough to accord any intrinsic moral worth. Instead, a set of *additional* criteria must be met—a certain level of autonomy, the ability to make choices and so on. Anyone who lacks full cognitive abilities is considered a nonperson, a category that includes the fetus, the newborn and the mentally impaired. Many ethicists have begun to argue that "nonpersons" can be used for harvesting organs or other utilitarian purposes. Those who favored letting Terri die included some, like Dr. Ronald Cranford, who have openly defended denying food and water even to disabled people who are conscious and partly mobile, like the case of a Washington man who could operate an electric wheelchair.[29]

The problem is that there is no way to establish a normative definition of "personhood." The traditional criterion was based on biology—anyone who was biologically human qualified for the moral status and legal protection due a human person. Humans are much more than biology, of course, but that criterion provided an objective, empirically detectable marker. Once the concept of personhood is detached from biology, however, it becomes nonempirical, nonscientific and ultimately subjective. There are no universally detectable criteria, and in the end each ethicist ends up proposing his or her own definition. Personhood becomes a purely relativistic, upper-story concept.

This suggests a way to turn the tables in the abortion debate. Critics of the pro-life position often charge that it is religious and subjective, that it is based on mere faith that life begins at conception. Yet the beginning of life is a biological fact. Biologically speaking, a unique individual life begins once the full genetic component comes together. Thus the pro-life position is actually based on scientific, empirically knowable facts. By contrast, arguments for abortion rest on the concept of "personhood," a nonempirical, nonscientific philosophical concept that is ultimately private and subjective.

Postmodern Sexuality

The same dualistic view of the human being underlies the liberal approach to sexuality. The body is treated as simply an instrument that is *used* by the autonomous self for its own goals and purposes. A sex education video widely used in public schools defines sex as "something done by two adults to give each other pleasure."[30] There is no hint that the body has its own teleology and moral dignity that calls for respect.

In fact, the cutting edge today is the idea that gender is a social construction—and therefore it can be *de*constructed. People "don't want to fit into *any* boxes—not gay, straight, lesbian, or bisexual ones. . . . [T]hey want to be free to change their minds," says a magazine for homosexuals. "It's as if we're seeing a challenge to the old modernist way of thinking 'This is who I am, period,' and a movement toward a postmodern version, 'This is who I am right now.' "[31] Sexual identity is being redefined as fluid and changing.

"This is seen as liberating, a way to take control of one's own identity, rather than accepting the one that has been culturally 'assigned,'" writes Gene Edward Veith.

At some colleges, students no longer have to check 'M' or 'F' on their health forms. Instead they are asked to 'describe your gender identity history.' "[32] The body has become an instrumental tool that can be used by the autonomous self any way it chooses, in a purely utilitarian calculus of personal pleasure.

Postmodern Sexuality
> *Autonomous self:*
> uses the body any way it chooses
>
> ---
>
> *Physical body:*
> morally neutral mechanism for pain or pleasure

The irony is that Christianity is often criticized for having a low view of the body because of biblical sexual morality. But it actually has a much *higher* view than today's secular utilitarian view. The doctrine of the incarnation—that God himself took on a human body—set Christianity apart from all the world-denying philosophies of the ancient world (such as Gnosticism), and it still does today. It is Christianity that actually gives a basis for according a high dignity to embodied existence.

Recovering Total Truth

Having traced the dichotomy in Western thought in several contexts, let's return to the epistemological questions with which we began. As we have seen, Richard Rorty is quite right to argue that postmodernism is an implication of Darwinism. If you put Darwinian evolution in the lower story, in the realm of "facts," then you will end up with postmodern relativism in the upper story of "values."

Of course, there is a fatal inconsistency in this formulation. If pushed to its logical conclusion, it undercuts itself. For if Darwinism means that the ideas in our minds are not true but only useful, then the same principle applies to the idea of Darwinism itself—and why should the rest of us give it any credence? Darwinian naturalism is self-refuting. Darwin himself wrote repeatedly about this conundrum, calling it his "horrid doubt." For example, he once wrote, "With me, the horrid doubt always arises whether the convictions of man's mind, which has been developed from the mind of the lower animals, are of any value or at all trustworthy."[33] But of course, Darwin's own theory was a "conviction of man's mind," so he was cutting off the branch he was sitting on.[34]

There is a remarkable passage in which Rorty acknowledges that the notion of capital-T Truth comes only from Christianity. In *Contingency, Irony, and Solidarity,* he contrasts the older idea that truth is "out there" (something objective that is found or discovered) with the postmodern idea that truth is created (a social construction). He then writes, "The suggestion that truth is *out there* is a legacy of an age in which the world was seen as the creation of a being who had a language of his own."[35] Rorty's point is that truth applies only to descriptions. The world itself is not true or false; only descriptions of it can be true or false. Thus to recover the concept of objective truth, it is not enough to affirm that God exists. God must also be a Being who speaks, who communicates his own description of the world—a God of revelation. As Rorty goes on to say, "The very idea that the world or the self has an intrinsic nature [that can be objectively known] . . . is a remnant of the idea that the world is a divine creation, the work of someone who had something in mind, who Himself spoke some language in which He described His own project."[36]

Philosophers talk about an objective concept of truth as a "God's-eye view" of the world, because only a transcendent God would be capable of knowing the world as it truly is. As human beings, our experience is constrained to a tiny slot in time and history; obviously it is impossible for us to transcend our context to discover a universal, objective, capital-T Truth. But the audacious claim of Christianity is that God himself *has* spoken—that he has not only created us but that he has also communicated with us, giving us a glimpse into his own God's-eye view. Within the Christian worldview this makes perfect sense—that the God who made humans capable of speech is a God who himself speaks and communicates to us.

Thus Christianity offers a way to heal the split in the Western mind—a way out of two-track dichotomy, a way to recover a unified vision of truth. Here's how one historical account puts it:

> Christianity posited a single reality, with some kind of rational coherence integrating it. . . . A single omnipotent and all-wise God had created the universe, including human beings, who shared to some extent in the rationality behind creation. *Given this creation story,* it followed the knowledge, too, comprised a single whole.[37]

The Darwinian creation story leads to an upper story of postmodern relativism, and ultimately undercuts itself. But Christianity offers a rationally coherent, logically consistent worldview that is comprehensive enough to explain the full range of human experience without a crippling leap of faith. Christianity does not offer merely religious truth—truth about an isolated *part* of life. It lays claim to be truth about every aspect of reality, thus offering the basis for a consistent and unified truth that applies to all aspects of our experience of the world. In that sense it is *total Truth*.

PHILLIP JOHNSON WAS RIGHT

THE RIVALRY OF NATURALISM AND NATURAL LAW

J. BUDZISZEWSKI

—

The critics of Darwinism attack it as bad science. Fair enough; viewed strictly as science, it is very bad indeed. But as Phillip Johnson has always emphasized, it is more than science: It is the great origins myth of our time. Myth matters because Marx was wrong. Men are not moved to act by mere things—by the forces and relations of production—but by the devices and desires of their hearts.

The noblest versions of the Darwinian origins myth have a tragic grandeur hardly matched in the literature of man. As C. S. Lewis tells the story:

> The drama . . . is preceded . . . by . . . the infinite void . . . endlessly, aimlessly moving to bring forth it knows not what. Then by some millionth, millionth chance—what tragic irony!—the conditions at one point of space and time bubble up into . . . organic life. At first everything seems to be against the infant hero . . . just as everything always was against the seventh son or ill-used step-daughter. . . . But . . . with incalculable sufferings . . . it spreads, it breeds, it complicates itself. . . . There comes forth a little, naked, shivering, cowering biped, shuffling, not yet fully erect, promising nothing: the product of another millionth, millionth chance. . . . But these were only growing pains. In the next act he has become true Man. He learns to master Nature. . . . See him in the last act, though not the last scene, of this great mystery. A race of demi-gods now rule the planet. . . . Eugenics have made certain that only demi-gods will now be born: psycho-analysis that none of them shall lose or smirch his divinity: economics that they shall have to hand all that demi-gods require. Man has ascended his throne . . . become God. . . . And now, mark well the final stroke. . . . All this time Nature, the old enemy who only seemed to be de-

feated, has been gnawing away, silently, unceasingly, out of the reach of human power. The Sun will cool—all suns will cool—the whole universe will run down. Life . . . will be banished without home of return from every cubic inch of infinite space. All ends in nothingness.[1]

The elixir which opens the mind to these mysteries is Nothingness. On a temperament like that of the young Lewis, this elixir produces a sense of exaltation. On a more typical temperament, like that of the young Aldous Huxley, its effect is bathetic. As he explains:

> I had motives for not wanting the world to have meaning; consequently, assumed it had none, and was able without any difficulty to find satisfying reasons for this assumption. . . . For myself, as no doubt for most of my contemporaries, the philosophy of meaninglessness was essentially an instrument of liberation. The liberation we desired was simultaneously liberation from a certain political and economic system and liberation from a certain system of morality. We objected to the morality because it interfered with our sexual freedom.[2]

The Bloodhound Gang gives the bathetic version of the myth a more concise and memorable formulation; we "ain't nothin' but mammals," so we should behave like the rest of the animals shown on the Discovery Channel.[3]

Is it possible for the Darwinian elixir to have a third effect, neither pathetic nor bathetic, but moral? Is it possible to get a Somethingness from the drink of Nothingness? Phillip Johnson says no, and convincingly traces the deadly and disintegrating effects of the Darwinian origins myth in law, education, and culture.[4] But amazingly, some moral thinkers say yes.

What Darwin discovered, these thinkers say, is the very profile of human nature. And hasn't human nature been the organizing idea of Western moral reflection since Aristotle—the axis around which it revolves?

Yes and no. Let us investigate. What we find in the end is that Phil Johnson was right.

The Ground of Ethics

In ethics, there are two ways to take human nature seriously. The first way is to regard nature as the design of a supernatural intelligence; you take it seriously because you take God seriously. The other is to regard nature (in a physical or material sense) as all there is. Here you ascribe to matter—or to some process, property or aspect of matter—the ontological position that theists ascribe to God himself. Natural lawyers follow the first way; naturalists follow the second. Similar name—radically different meaning.

Nature means something different to naturalists than it does to natural law-yers. It has to. Naturalists cannot view it as a design because on their view there isn't anyone whose design it might be. What is, just is. If you accept the principle of sufficient reason, this is rather unsatisfactory, for no one seriously maintains that the universe had to be just the way it is. There might have been fewer stars or more. There might have been creatures like us, or there might not. There might not have been a universe at all. Nature, then, is a contingent being, not a necessary being like God, and contingent beings need causes. Naturalists reject this line of reasoning, or at least limit it. They might concede that each thing in nature needs a cause but deny that the entire ensemble of things needs a cause. This exception seems suspiciously arbitrary.

It's easy to see how the first approach can ground ethics. If God himself is the good—the uncreated source of all being, all meaning and all value in created things—then inasmuch as his intentions are reflected in our own de-sign, in human nature, these intentions are normative for us.[5] Consider, for example, the inclination to associate in families. This is not the same as a mere desire to do so; indeed we have conflicting desires, and some people would rather be alone. It would be more accurate to say that we are made for family life, that fitness for family life is one of our design criteria. For hu-mans, then, the familial inclination is a *natural* inclination. When we follow this inclination, we are not acting in the teeth of our design but in accord with our design. Family is not a merely apparent good for us but a real one, and the rules and habits necessary to its flourishing belong to the natural law. Next consider the universal testimony of conscience against murder. This is more than a matter of guilty feelings; indeed no one always feels re-morse for doing wrong, and some people never do. Nevertheless, the wrong of deliberately taking innocent human life is acknowledged at all times and everywhere, and this too belongs to the natural law. Notice that both exam-ples concern design. The former concerns the design of the inclinations as apprehended by the intellect. The latter concerns the design of the intellect itself, for we are so made that there are certain moral truths, things we *can't not know*.[6]

How the other view could ground ethics is hard to see. If material nature is all there is, then how could actions have nonmaterial properties like right and wrong? Another puzzle is how there could be true moral "law" without a lawgiver. Perhaps it would be like the "law" of gravity—a pattern we cannot help but enact, a force to which we cannot help but yield. But in that case, "you ought to" would mean the same thing as "you do." Stones do not deliberate about whether they "ought" to fall.

Is Nature Pointless?

Some naturalists concede the point, or as we must say here, the pointlessness. William Provine declares that "No purposive principles exist in nature. . . . No inherent moral or ethical laws exist, nor are there absolute guiding principles for human society. The universe cares nothing for us and we have no ultimate meaning in life."[7] Richard Dawkins opines, "The universe that we observe has precisely the properties we should expect if there is, at bottom, no design, no purpose, no evil and no good, nothing but blind, pitiless indifference."[8] According to E. O. Wilson, "Human behavior—like the deepest capacities for emotional response which drive and guide it—is the circuitous technique by which human genetic material has been and will be kept intact. Morality has no other demonstrable ultimate function."[9] Wilson and Michael Ruse write:

> Our belief in morality is merely an adaptation put in place to further our reproductive ends. . . . [E]thics as we understand it is an illusion fobbed off on us by our genes to get us to co-operate (so that human genes survive). . . . Furthermore the way our biology enforces its ends is by making us think that there is an objective higher code to which we are all subject.[10]

On the subject of conscience, Robert Wright chimes in, "It's amazing that a process as amoral and crassly pragmatic as natural selection could design[!] a mental organ that makes us feel as if we're in touch with higher truths. Truly a shameless ploy."[11]

From views like this, it is only a small step to the opinion that a truly authentic morality would be Promethean, setting itself *against* the shameless ploy. That's what Richard Dawkins thinks. First he sets the stage: "We are survival machines, robot vehicles blindly programmed to preserve the selfish molecules known as genes." Then he issues the call to arms: "Let us understand what our own selfish genes are up to, because we may then at least have the chance to upset their designs, something that no other species has ever aspired to."[12] *Écrasez l'infâme!* It is all very stimulating, but of course if we really are really "blindly programmed" by our genes, then the call to revolt is worse than futile. One might as well expect a typewriter to revolt against the keys.

Perhaps Dawkins is setting his hopes on cultural evolution, for later he suggests that higher-level genetic programs are "open" and do not settle every detail of the way we live. Yet this is hardly a promising gambit, for his discussion of culture merely exchanges one form of determinism for another. As he sees things, our bodies are blindly programmed to preserve the self-replicating molecules called genes, and our cultures are blindly programmed to preserve the self-replicating ideas called "memes." If we take him at his word, then presumably the idea of revolt is merely another of the replicators. In this case he rails against blind destiny only because he is blindly destined so to rail.

Further complicating the story is that from time to time, the very writers who say that naturalism destroys morality have sometimes propounded the view that it *implies* a morality. Wilson, for example, believes that we are *morally* obligated to preserve all extant living species. The reasoning seems to be that (1) whatever is, is lovable; (2) the preservation of whatever is, is right; and (3) if we fail to pay sufficient homage to whatever is, there will be retribution. This is not quite how Wilson puts it. Here is how he frames the idea in a newspaper column adapted from his 2002 book *The Future of Life:*

> "Don't mess with Mother Nature." The lady is our mother all right, and a mighty dispensational force as well. After evolving on her own for more than three billion years, she gave birth to us a mere million years ago, an eye blink in evolutionary time. Ancient and vulnerable, she will not tolerate the undisciplined appetite of her gargantuan infant much longer.

Could it be that he is speaking poetically and does not intend his words to be taken in a moral sense? On the contrary:

> The issue, like all great decisions, is moral. Science and technology are what we can do; morality is what we agree we should or should not do. The ethic from which moral decisions spring is a norm or standard of behavior in support of a value, and value in turn depends on purpose. Purpose, whether personal or global, whether urged by conscience or graven in sacred script, expresses the image we hold of ourselves and our society. A conservation ethic is that which aims to pass on to future generations the best part of the nonhuman world. To know this world is to gain a proprietary attachment to it. To know it well is to love and take responsibility for it.[13]

The foregoing passage is rather cloudy. For starters, what does Wilson mean by "moral"? Is there an "ought" in there—is he saying anything more than "I have feelings of love, awe and fear toward nonhuman nature, and I want you to have them too"? Plainly, we can *elicit* such feelings on the part of other people without recourse to an "ought." For example, I might get you to share my fear of environmental disaster by conjuring an image of it. But can we recommend such feelings *as moral* without recourse to an "ought"? To sharpen the point: We see that Wilson might regard people who fail to share his fear as deficient in *imagination,* but it is hard to see how he could regard them as deficient in *duty.* Duty doesn't look any more like a property of matter than right and wrong.[14] From time to time Wilson notices the difficulty. On such occasions he waves his hands and refers vaguely to "emergent" properties of matter—properties that appear only when matter is complexly organized. But this is sleight of hand, because he has no idea how complexly organized matter could give rise to such properties either. Finding a property that he cannot account for, he calls it "emergent" and says that he has explained it.

Evolutionary Ethics

A heterogeneous movement, variously styled "evolutionary ethics" and "evolutionary psychology," shares the goal of providing a naturalistic basis for moral judgments, but it tries to be more systematic. This new naturalist fashion comes in three overlapping varieties.

The variety closest in spirit to Wilson's own work tries to demonstrate that a moral sense has evolved among human beings because it confers a selective advantage. Consider, for example, the human tendency to help out other people, even at some cost to oneself. At first it might seem that a genetic predisposition for such behavior could never have evolved by natural selection because unselfishness spends resources for nothing; every selfless act reduces the likelihood of passing on the genes that have made one act selflessly. But if the ancestors of human beings already lived in family groups, maybe not.[15] Under those circumstances, those most likely to receive aid would be relatives, and for each degree of relationship, there is a certain likelihood that the relative is carrying *the same* gene. So, even though an act of self-sacrifice reduces the likelihood that I will pass on my *own* copy of the gene, it increases the likelihood that my relatives will pass on theirs. If my act helps a sufficient number of such relatives, then the proliferation of the gene in question is assisted even more than it would have been by selfishness. This is called "kin selection."[16]

If kin selection really happens, then it might explain the tendency to help out other people. It might even explain why we approve of the tendency. The problem is that it can't explain whether we *ought* to approve of it. After all, the fact that we developed one way rather than another is an accident. We might have turned out like guppies, who eat their young instead of helping them. Someone might reply, "That we *might have* turned out differently is no concern of ours. The fact is that we didn't. Besides, natural selection has determined not only that we are the way we are, but that we're happy about the way we are. We don't need a justification for being pleased!" Not so fast. We may be pleased about our tendency to render aid, but we are not so pleased about its limits. As a matter of fact, many of our tendencies *dis*please us; consider how appalled we are by our propensity for territorial aggression. Now our tendency to territorial aggression and our propensity to be appalled by it must both belong to the genome. What sense could there be, then, in judging between them? Genes provide no basis for judging between gene and gene. The basis of morality must lie elsewhere.

The second variety of evolutionary ethics tries to show that by considering how we came to be, we will learn more about how we are. According to this view, Darwinism reveals the universal, persistent features of human nature. Why it should do so is very strange because Darwinism is not a predictive theory. It does not proceed by saying "According to our models, we should expect human males to be more interested in sexual variety than human females; let's find out if this is true." Rather it proceeds by saying "Human males seem to be more interested in sexual variety than human females; let's cook up some models about how this might have come to pass." In other words, the theory *discovers nothing*. It depends entirely on what we know already and proceeds from there to a purely conjectural evolutionary history.

These conjectures are made to order. You can "explain" fidelity, and you can "explain" infidelity. You can "explain" monogamy, and you can "explain" polygamy. Best of all (for those who devise them), none of your explanations can be disconfirmed—because all of the data about what actually happened are lost in the mists of prehistory. In the truest sense of the word, they are myths—but with one difference. The dominant myths of most cultures encourage adherence to cultural norms. By contrast, the myths of evolutionary ethicists encourage cynicism about them. Robert Wright is remarkably candid about this effect:

Our generosity and affection have a narrow underlying purpose. They're aimed either at kin, who share our genes, at nonkin of the opposite sex who can help package our genes for shipment to the next generation, or at nonkin of either sex who seem likely to return the favor. What's more, the favor often entails dishonesty or malice; we do our friends the favor of overlooking their flaws, and seeing (if not magnifying) the flaws of their enemies. Affection is a tool of hostility. We form bonds to deepen fissures. . . .

It is safe to call this a cynical view of behavior. So what's new? There's nothing revolutionary about cynicism. Indeed, some would call it the story of our time—the by now august successor to Victorian earnestness.[17]

An evolutionary ethicist of this second sort does not claim that Darwinism itself provides the foundation for ethics. What it does tell us, he thinks, is the general features of human nature that ethics must come to terms with. Wright's ethics, for example, are utilitarian; he holds that "the fundamental guidelines for moral discourse are pleasure and pain." Given utilitarian ethics, here is how he explains the usefulness of Darwinism:

Of course, happiness is great. There's every reason to seek it. There's every reason for psychologists to try to instill it, and no reason for them to mold the kinds of people natural selection "wants." But therapists will be better equipped to make people happy once they understand what natural selection *does* "want," and how, with humans, it tries to get it. What burdensome mental appliances are we stuck with? How, if at all, can they be defused? And at what cost—to ourselves and to others? Understanding what is and isn't pathological from natural selection's point of view can help us confront things that are pathological from our point of view.[18]

If we ask Wright why he *does* favor utilitarianism, he gives the intriguing answer that once Darwinism gets loose in the world, we find that it becomes harder and harder to find principles on which everyone will agree. All the old ones have been destroyed. "In a post-Darwinian world" which "for all we know is godless," minimalism rules, fewer principles are better than more. Utilitarianism, of course, has only one—pleasure good, pain bad. Does this *prove* the goodness of pleasure and the badness of pain? No, but we do regard pleasure as good and we do regard pain as bad. "Who could disagree with that?" Wright asks. Like most utilitarians, he is convinced that even people who do not call themselves utilitarians are utilitarians at heart.[19]

The argument is less transparent than it seems to be. In the first place, the kind of minimalism that is likely to strike people as plausible depends on what kind of people they are. In cynical times when they are well-fed, the "one plausible principle" may seem to be "Pleasure is good." But in violent times when they are afraid, the one plausible principle may seem to be "Death is bad." In fact, this was the very principle propounded by Thomas Hobbes in 1641, in very violent times. Another problem is that minimalism won't get you very far. Hobbes thought his one plausible principle was very powerful, but he confused consensus that death is bad with consensus that death is the *greatest* bad. Though most people do think death is bad, most also think that there are some things worse than death. For that reason, even if they agree that death is to be avoided, they will not agree that death is to be avoided *above all things,* as Hobbes would have them do.

Utilitarianism runs into similar problems. People may agree with Wright that happiness is good, yet they may not agree with him that happiness is the same as pleasure (most of the Western tradition has denied it). Or they may agree with him that pleasure is good, yet they may not agree with him that pleasure should be pursued as a *goal* (the Western tradition has maintained that pleasure is best enjoyed as a byproduct of pursuing other ends; the search for pleasure dries up the springs of pleasure). Or they may agree with him that pleasure should be pursued as a goal, yet they may not agree with him that *aggregate* pleasure should be pursued as a goal, as utilitarianism requires (if torturing one innocent soul would make everyone else much happier, then concern for the aggregate pleasure would require torturing him or her).

For all that, it is easy to see why naturalists find utilitarianism attractive. I asked earlier if material nature is all there is, how could actions have nonmaterial properties like right and wrong? Confronted with this question the naturalist has only two ways to proceed. Straightforwardly deny moral properties, or try reducing them to nonmoral properties—which is a more roundabout way to deny them. The only puzzle is why a naturalist would want to do either of these things.

The common method of reduction is to explain moral properties in terms of desire. This move has four steps.[20] *Step one* is to say that the right is nothing but what brings about the good. This is called consequentialism. To consequentialists, a maxim like "It is wrong to do evil that good may result" means nothing, because apart from results they have no concept of evil or good. *Step two* is to say that the good is nothing but the desirable. This is the only unproblematic step in the argument. *Step three* is to say that the desirable is nothing but what we actually desire. This definition renders it impossible to make sense of perverse desires, desires we wish we had but don't or desires we wish we didn't have but do. John Stuart Mill tied himself in knots over the problem.[21] *Step four* is to say what it is that we actually desire. According to utilitarians like Wright, this is pleasure. You may think you desire many things—love, skill, friendship, achievement, salvation—but according to utilitarians, you're wrong. They say you desire nothing except as either a part of pleasure or a means to pleasure; hence, the only thing you ultimately desire is pleasure itself. For example, you may think you want dinner, but what you really want is the pleasure of feeling full; knowledge, but what you really want is the pleasure of feeling knowledgeable; love, but what you really want is the pleasure of feeling loved; or God, but what you really want is the pleasure of feeling—well, whatever God would make you feel. It follows that if it were possible to have the pleasures without the things, then that would be just as good. Eat, purge and eat again.

Natural Law and Naturalism

The third and most paradoxical variety of evolutionary ethics proclaims that natural law and naturalism are not at odds after all—that they are getting at the same thing. A dash of Darwin, as it were, makes Thomas Aquinas more powerful and precise. Yes, yes, we must do away with Thomas's silly superstition that a God is somehow behind things and that nature is designed—but he is better off without it anyway.

This kind of evolutionary ethics has been especially popular among conservatives who think they believe in natural law theory but don't notice the sleight of hand. The most vigorous exponent of this "Darwinian" natural law is Larry Arnhart.[22] Arnhart uses the expressions "natural right" and "natural law" interchangeably. Although he borrows liberally from the other two varieties of evolutionary ethics, his approach requires more detailed attention.

The structure of Arnhart's theory is easy to explain. He makes three of the four moves that utilitarians do, differing only as to the fourth.[23]

1. He *tacitly* supposes that the right is nothing but what brings about the good, so he is a consequentialist. This critical move is not defended. The unwary reader often joins in the silent assumption that the end justifies the means before realizing what is happening. What is astonishing here is that historically the natural law tradition has been invoked *against* consequentialism in all of its forms. Yes, the tradition says that good is to be done and

that evil is to be avoided, but it has also insisted that some acts are *intrinsically* good and evil aside from all consideration of their consequences. This Arnhart denies, as a consequentialist must. For him there can't be such a thing as an intrinsically evil act—not even rape or murder. His understanding of the virtue of prudence is that there are no inviolable rules; *everything* depends on circumstances because circumstances determine results. This utterly obliterates the distinction between the right, pursued by prudence, and the expedient, pursued by craft. Within his theory, one can distinguish between socially approved expedience and socially disapproved expedience, but this is not the same thing.

2. He *explicitly* declares that the good is nothing but the desirable; in fact, he asserts and defends the claim repeatedly. Not that it helps much to do so, because this is the only step that is not problematic.

3. He *tacitly* supposes that the desirable is nothing but what we actually desire. Again there is no justification and no discussion of cases which do not seem to fit. For example, what about a sadomasochist who strongly desires "bondage and discipline" but loathes himself for this desire and strongly desires not to be burdened by it? On Arnhart's account, we would have to conclude *both* that bondage and discipline are desirable for that person, *and* that freedom from such desire is desirable for him. This seems incoherent. A more straightforward view of the matter is that such a person *recognizes that what is subjectively desired is not objectively desirable.* It is for precisely this reason that he desires liberation from this burden.

Arnhart does mention one difficult case: When I think I want something, but then discover that it wasn't what I wanted after all. Unfortunately, the case is equivocal, and Arnhart does not analyze it. Consider two instances in which I might wish to say that something wasn't what I wanted after all. Instance one: I want to be drunk. As soon as I succeed, I throw up. I tell myself, *I guess that's not what I wanted after all.* This is probably the sort of thing that Arnhart has in mind. Unfortunately, it isn't really true that I didn't know what I wanted. I really did want to be drunk, and I knew it—but I changed my mind. Instance two: I have a longing for "that unnamable something, desire for which pierces us like a rapier at the smell of a bonfire, the sound of wild ducks flying overhead, the title of *The Well at the World's End,* the opening lines of Kubla Khan, the morning cobwebs in late summer, or the noise of falling waves."[24] Trying to understand what it is that I want, I form one hypothesis after another: "What I *really* want is beauty," "What I *really* want is the remote and mysterious," "What I *really* want is ecstatic union with the rest of nature." Pursuing each of these things in turn, I find to my dismay that none of them actually satisfies the longing. Eventually I realize that what I long for is not to be found within the created order at all. What I am longing for is God. This case is different than the other one because until the end, I *really don't* know what it is that I want. Unfortunately, Arnhart has no resources to analyze a case like this because he does not acknowledge the reality of God. The closest his classification comes to such longing for God is the "desire for religious understanding," which of course is not the same thing.

4. Not until the step of stating just what it is that we desire does the structure of Arnhart's theory differ significantly from that of utilitarianism. The utilitarian acknowledges only one human desire—pleasure. Arnhart acknowledges twenty, though why Darwinism is needed to discover them is not explained: The desire for a complete life, for parental care, for sexual identity, for sexual mating, for familial bonding, for friendship, for social ranking, for justice as reciprocity, for political rule, for war, for health, for beauty, for wealth, for speech, for practical habituation, for practical reasoning, for practical arts, for aesthetic pleasure, for religious understanding, and for intellectual understanding. No doubt it is better to recognize twenty desires than the "one big desire" of utilitarianism. In the context of Arnhart's theory, however, the list presents difficulties of its own.

The first great peculiarity is that for Arnhart, the general human desires simply *are* the natural laws; there are no others. The natural law tradition has always denied this. Speaking for the mainstream of the tradition, Thomas Aquinas thought that a good summary of the natural law is found in the Decalogue, or Ten Commandments, which of course are found in divine law too.[25] The prohibition of murder, for example, is one of the "general" precepts[26] that Thomas Aquinas calls "the same for all both as to rectitude and as to knowledge," meaning that it is both right for all and known to all. General principles brook no exceptions, no matter the circumstances.[27] Arnhart denies that there are any such principles—other than the desires themselves. As he says in his discussion of prudence, "The natural *desires* of human beings constitute a universal norm for morality and politics, but there are *no universal rules* for what should be done in particular circumstances."[28]

The reason that the natural *desires* constitute a universal norm for morality and politics is that for Arnhart the right is nothing but what brings about the good, the good is nothing but the desirable, and the desirable is nothing but what we actually desire; therefore, the right is what causes what we want. Together with the list of desires itself, this entails some very strange conclusions. War, for example, is one of the desires on the list; war is therefore a universal norm for morality and politics. Notice what Arnhart's theory does *not* say here. It does *not* say that war is sometimes an unfortunate necessity for securing justice, as the just-war doctrine of the natural law tradition declares. Rather, it implies that war is *good and right in itself*—simply because we do in fact desire war. Arnhart's actual discussion of war softens the point (there is, in fact, a great deal of softening of points in his book), but it follows inescapably from his premises.

Yet another oddity of Arnhart's list is the tension in his theory between general and exceptional desires. In the opening section of the second chapter, Arnhart affirms, "I reject skeptical and solipsistic relativism, which asserts that there are no standards of ethical judgment beyond the impulses of unique individuals." Later in the chapter, in explaining the "big twenty," he remarks, "In the case of each desire, I speak of what human beings 'generally' desire, because I am speaking of general tendencies or proclivities that are true for all societies but not for all individuals in all circumstances."[29] But although in one sense his theory is based on general desires (for he does in fact generalize about the desires), his fundamental equation between the right, the good, the desirable and what is actually desired pulls him helplessly in the other direction. For by the logic of the argument, the pursuit of what is *generally* desired is right only for the *generality* of people, those who actually experience them as desires. Should there be any people with abnormal desires, they must be viewed as standing outside of our morality; they have their own morality. This is necessarily the case, because what is right *for them* is what brings about the good *for them,* which is the desirable *for them,* which is what *they* actually desire.

This implication becomes strangely clear in chapter eight, which is devoted to psychopaths—those who "lack the social desires that support the moral sense in normal people." Such people, says Arnhart, are "moral strangers." Most of us would simply say that they lack the desire to do right. Because Arnhart *reduces* right to desire, however, he cannot speak this way. In his view, if desire is different for them *then right must be different for them too.* Arnhart says they have "no moral obligation" to conform to what our "moral sense" demands. If we use force and fear to restrain them, it is not because they are doing wrong, for given their desires, they are doing right. It is merely that, given our own quite different desires, we too are doing right in restraining them.

Once this is understood, we can see that many of Arnhart's statements about his theory are misleading. Consider for example the sentence quoted a few paragraphs earlier, "I reject skeptical and solipsistic relativism, which asserts that there are no standards of ethical judgment beyond the impulses of unique individuals." It would be more accurate to say that he accepts solipsistic relativism based on the impulses of unique individuals *and* affirms standards of judgment beyond the impulses of these individuals. On the one hand, psychopaths have a morality of their own which our morality cannot touch; on the other hand, the rest of us are not psychopaths. Nor are psychopaths the only ones to get their own morality. By the logic of the case, *all* whose desires are significantly different than the rest of us get their own moralities. If the foundational principles of the natural law are "the same for all both as to rectitude and as to knowledge," then Arnhart's theory does not affirm the natural law but rather rejects it.

In fact, nothing in Arnhart's theory is quite as it appears. One of his most vigorously argued theses is that slavery violates natural right. He devotes all of chapter seven to the subject, warmly endorsing Lincoln's remark that "if slavery is not wrong, nothing is wrong." I have no reason to doubt Arnhart's sincerity. However, his theory cannot support his conclusion.

The reason slavery violates natural right, according to Arnhart, is that it "frustrates the desire to be free from exploitation"—put another way, the desire to enjoy justice as reciprocity (desire 8). But if the right is nothing but what brings about the good, which is the desirable, which is what is actually desired, then the fact that the slaves and the masters desire different things is an insuperable obstacle to the conclusion that Arnhart wants to draw. He tries to get over the obstacle by emphasizing the social desires that might lead nonslaves to sympathize with the slaves' desire for justice. The difficulty, of course, is that not all nonslaves do sympathize with slaves. I believe I am right to say that members of the master class have not generally been known to do so.

The truth is that slavery represents a protracted state of war between the master class and the slave class—and Arnhart seems to forget that he has included war on his list of the twenty general human desires. Although the practice of slavery may frustrate the desire of the slaves for reciprocity, it satisfies the desire of the masters for war, and Arnhart's theory provides no principled basis to judge between them. As he states in another context, "When individuals or groups compete with one another, we must either find some common ground of shared interests, or we must allow for an appeal to force or fraud to settle the dispute."[30] In slavery, however, there are no shared interests; the interest of the masters is to continue ruling, and the interest of the slaves is to escape. I agree with Arnhart that slavery is against the natural law; I am glad that he reaches a different conclusion than his theory requires. That does not change what it requires.

The quotation in the previous paragraph is not from either the chapter on war or the chapter on slavery, but from the chapter on men and women. From its context, this too is highly revealing. Arnhart criticizes Darwin for giving two conflicting accounts of "the relationship between male and female norms in the moral economy of human life":

> In one account, [Darwin] defends a moral realism that combines typically male norms such as dominance and courage and typically female norms such a nurturance and sympathy, which he presents as complementary and interdependent inclinations of the human moral constitution. In the other account, he defends a moral utopianism that subordinates the male norms to the female norms, and he expands female sympathy into a disinterested sentiment of universal humanitarianism.[31]

But Arnhart also gives two conflicting accounts. The account that he purports to defend is that typically male and typically female norms are complementary. The account that actually emerges from his theory is that these two sets of norms are substantially—though not entirely—at war. It is hard to see why else he would conclude his section on male and female complementarity with a paragraph explaining that "deep conflicts of interest between individuals or groups can create moral tragedies in which there is no universal moral principle or sentiment to resolve the conflict."[32] This is, by the way, the same paragraph in which Arnhart offers the comment quoted above, concerning disputes that can be settled only by force and fraud. Perhaps the clearest example is the conflict between the female desire for a faithful spouse and the male desire for sexual variety, which is settled, apparently, by fraud.

To defend the idea of sexual complementarity, Arnhart argues (correctly, I think) that even though human males characteristically have a greater desire than females do for a variety of sexual partners, they are actually more satisfied by monogamous marriage than by a life of promiscuous abandon. He does *not* say, however, that males are most satisfied in *faithful* marriage, and this is not the conclusion that emerges from his account of male desire. The Arnhart male will want to be married, but he will also want to cheat now and then—provided that he can get away with it. In the interests of his desire for a stable relationship, he will discipline his desire for sexual variety—but not so thoroughly that he becomes faithful. Men will desire to cheat occasionally—and because Arnhart takes desire as the measure of morality, he is logically compelled to conclude that for men such cheating is right. Does he say this in so many words? No, but nothing else could follow from his premises. What of the opening to the chapter, where Arnhart says that his theory "allows us to recognize and condemn cultural practices that frustrate the natural desires of women"?[33] The statement is not wholly false; Arnhart's theory does allow us to criticize the practice of female circumcision, a subject to which he devotes a number of pages. But his discussion of female circumcision seems little more than a diversion. After all, cheating husbands also "frustrate the natural desires of women," but against the occasional furtive adultery, Arnhart has nothing to say.

The strangest implication of the "big twenty" is that in Arnhart's determined attempt to make natural law safe for atheists, he is at war with his own theory. Numbers nineteen and twenty on his list of human desires are religious and intellectual understanding—the desire to understand the world "through supernatural revelation" and the desire to understand it "through natural reason." There is no priority here; the two desires are entirely distinct and equally general. If Arnhart means what he said earlier, that "the natural desires of human beings constitute a universal norm for morality and politics," then the implication would seem to be clear: Natural law instructs us to pursue them both. Unfortunately, not only does Arnhart's discussion obscure the point, but by the time the book concludes he is saying something quite different. Here are his words.

> Moved by their desire to understand, human beings will seek the uncaused ground of all causes. This will lead some human beings to a religious understanding of God. It will lead others to an intellectual understanding of nature. Yet, in either case, the good is the desirable. And perhaps the greatest human good, which would satisfy the deepest human desire, would be to understand human nature within the natural order of the whole.[34]

Instead of being urged to seek both kinds of understanding, suddenly we are urged to seek one or the other. They are no longer presented as either equal or distinct; natural reason is given priority over supernatural revelation and seems to want to absorb it. This does not wash: If the right is nothing but what brings about the good, the good is nothing but the desirable, the desirable is nothing but what we desire, and we desire *both* supernatural revelation and what reason can learn on its own, then Arnhart's own theory is instructing him to lay aside his atheism and pursue supernatural revelation, but he isn't listening. As Pascal once wrote of cases like this, the heart has its reasons, whereof the mind knows nothing.

Conclusion

From all that has been said, we may conclude that "Darwinian" natural law is entirely at odds with what has traditionally been called natural law. It differs not only in content (no precepts) and structure (consequentialist) but in basic ontology (no lawgiver and therefore no law). In these respects it affirms precisely those tendencies of thought that the natural law tradition has always sought to oppose. If any contemporary scientific movement holds promise for the furtherance of the natural law tradition, it is not the stale dogma of natural selection but the theory of intelligent design.

Phil Johnson was right after all.

A TAXONOMY OF TELEOLOGY

Phillip Johnson, the Intelligent Design Community and Young-Earth Creationism

MARCUS ROSS AND PAUL NELSON

I cannot believe that "Nature" was unknown before Rousseau's time; or method before Descartes; or the experimental system before Bacon; or anything that's self-evident before someone or other. Only, someone had to "make a song" about it!

Paul Valéry, *Analects*

In 1991, with the publication of *Darwin on Trial,* Phillip Johnson "made a song" about the role of naturalism in biology. Immediately, the main thesis of *Darwin on Trial*—its catchiest tune, so to speak—lodged in the thinking of many people, as catchy tunes do. Johnson argued that the authoritative place of neo-Darwinian evolution in modern culture was supported not by the evidence but by the scientific community's prior philosophical commitment to naturalism.[1] This thesis played out as a bothersome, even infuriating jingle for many in the scientific community; as a witty, irreverent *divertimento* to others in that community—but to young-earth creationists, exiles from modern science, Johnson's argument about naturalism was a powerfully evocative melody from their distant homeland. In the last two decades of the twentieth century, *creation* and *creationism* had become bywords in science, as textbook examples (even for many theists) of non- or anti-science. To become known as a "creationist" in the sciences could be career-imperiling.[2] Yet here stood a Berkeley professor, saying that in one very important respect, at least, those exiles and outcasts from science were right. Johnson even claimed to be a "creationist" of some sort himself. But how could that be, given that he also denied defending creation science and said he was unconcerned to reconcile the Bible with scientific evidence? How could Johnson be a creationist when he plainly wasn't, well, a creationist?

In this chapter, we will answer this seemingly paradoxical question and will argue that Johnson's naturalism thesis deeply shifted the debate—a debate that until *Darwin on Trial* most onlookers saw divided between the polar camps of "evolution" versus "creation science" (meaning young-earth creationism). From the perspective of mainstream science, all sensible people accepted "evolution"—that is, the common descent of life on earth via undirected causes such as natural selection—while a relict population of biblical literalists, self-identified as "scientific creationists," clung stubbornly to their pre-Darwinian views. Creationists occasionally made political trouble by persuading state legislatures to enact so-called "balanced treatment" laws, mandating the teaching of creation science whenever evolution was taught. The courts inevitably overturned those laws, however, and in 1987, the U.S. Supreme Court declared creation science to be a religious belief, effectively banishing it from discussion in science classes.

And there the matter might have remained. Johnson, however, glimpsed something that others had missed. To borrow a metaphor from biological classification, we can say that Johnson discovered the popular taxonomy of theories of origins was wrong. In that classification those who accepted creation held the view of six-day special creation and a young earth, while others accepted "evolution," a 4.5 billion-year-old earth and an even older universe. This classification took as its diagnostic markers the most widely promoted *narratives* of creation and evolution. One story, whose biblical roots were obvious, described the special creation of all life in a few days, on a young planet whose surface was later destroyed by a global flood. The other story—the *scientific* account, for most people—began with the origin of the universe in the big bang, continued through galactic, stellar and planetary evolution to the first stirrings of life on earth. Over billions of years, all other living things descended, via common ancestry and forces including natural selection, until a few million years ago, *Homo* walked across the plains of East Africa. These stories contradict each other. Make your choice.

But Johnson's analysis in *Darwin on Trial* begins by jettisoning this familiar polarity. Setting aside the usual diagnostic markers, Johnson dissects creation and evolution by first inspecting what might be called their *epistemological anatomy*. " 'Evolution' contradicts 'creation,' " he wrote, "only when it is explicitly or tacitly defined as *fully naturalistic evolution*—meaning evolution that is not directed by any purposeful intelligence."[3] The fundamental differences between the two theories, Johnson argued, did not stem from any particular historical narrative but rather from *what kinds of causes would be allowed in scientific explanation and what would count as evidence.* Epistemology—namely, what can be known empirically, and what counts as a scientific explanation—is what truly cuts the origins issue at its joints.

At first glance this analysis seems to get it all wrong. Theistic evolutionists—those who accept a 4.5 billion-year-old earth and relatedness of all organisms in a tree of life (through divine purpose)—are sorted into the same group as young-earth creationists, with whom they appear to share only the theological premise of "divine purpose." But Johnson presses on:

> Persons who believe that the earth is billions of years old, and that simple forms of life evolved gradually to become more complex forms including humans, are "creationists" if they believe that a supernatural Creator not only initiated this process but in some meaningful sense *controls* it in furtherance of a purpose. As we shall see, "evolution" (in contemporary scientific usage) excludes not just creation-science but creationism in the broad sense.[4]

To be sure, the narratives of theistic and naturalistic evolution bear many similarities, just as a dolphin can look remarkably like a shark (as ocean swimmers have discovered, to their relief or terror). Both are highly mobile aquatic predators with a streamlined, fusiform shape, similar dorsal and pectoral fins, and so forth. But these are convergences: similarities that mislead about genuine relationships. What we have with theistic and naturalistic evolution, then, is a case of convergence in narratives. Although theistic evolution may resemble its naturalistic counterpart, if the former theory is genuinely theistic, it is profoundly distinct from naturalistic evolution. Any naturalistically grounded theory cannot allow for inferences to divine design, whatever the evidence may indicate. Whether God created suddenly, as in a young-earth narrative, or did so over long spans of time, as theistic evolutionists think, are questions that cannot be entertained by science. Given naturalism, the questions do not arise because they simply cannot arise— again, whatever the evidence.

The consequences of this reframing of the origins controversy, from a choice between two very different *narratives* to the question of which *epistemology* science should adopt, are still unfolding. But already, many years after *Darwin on Trial,* Johnson's approach has revolutionized the debate.

The salient feature of that revolution has been the rapid emergence of the intelligent design (ID) community. As its inhabitants quickly learn, the ID community can be a bewildering place to live. One might imagine introducing a dolphin to a tree shrew and then taking the two of them to meet a koala.[5] "Gentlemen, despite your differences, would you mind standing together for the group photograph? Why? You're all mammals, of course." When faced with taxonomic confusion and a bewildering variety of appearances, the good systematist does not allow himself to be distracted by superficial similarities, no matter how compelling they may be, nor to be put off by apparent dissimilarities. Dolphins and sharks look similar; in fact, they are very different kinds of organisms. ID theorists look very different (young-earth, old-earth, theistic evolution, etc.), yet these apparently dissimilar viewpoints actually belong together at a deep level.

This was one of Phillip Johnson's key insights, and it stemmed from his discovery that *naturalism*—that is, not the detailed narrative of evolution, but its underlying *epistemology*—had become the strongest commitment of modern science since Darwin's time. The evolutionary narrative changed from one year to the next, sometimes wildly so, depending on the latest discoveries or academic fashions; the naturalistic commitment was a constant, so deep that in most cases it was entirely tacit. In the following section, then, we wish to refine Johnson's taxonomy of viewpoints about origins, with two goals in mind: (1) to show how previous classifications of various origins positions fail, and (2) to throw light on what unites young-earth creationism and intelligent design—but also to explain how and why these positions differ. Dolphins and koalas are both mammals, but koalas aren't going to be finding any meals ten meters below the surface of the ocean.

In considering the second goal, we will also answer our opening question about how, or in what sense, Johnson could be a "creationist," when he clearly wasn't one. We also hope to disentangle the confusion that surrounds the perceived relationship between the ID and young-earth creation communities. Uncertainty about what differentiates young-earth creationism from intelligent design has resulted in murkiness (and, frankly, some deliberate mischief on the part of ID critics has played a role).[6] We will argue that young-earth creation and ID proponents define their positions differently, have different goals, and employ different standards of method. Furthermore, both young-earth creationism and ID consider themselves distinguishable from each other, and both agree concerning the basic nature of the distinction: the level of authority (if any) given to the Bible in model construction. We will also argue that previous attempts to classify design-based positions on origins suffer from three major shortcomings: a strict but unsupportable science-nonscience demarcation, the use of ambiguous classification criteria, and assumptions of theological uniformity among teleological positions. We will then discuss a "nested hierarchy of design," a classification system that categorizes teleological positions according to the strength of claims regarding the reality, detectability, source, method and timing of design. This system results in an accurate and robust classification of numerous positions, while simultaneously avoiding the philosophical and theological pitfalls of previous methods. Ultimately, the nested hierarchy of design classification enables construction of accurate definitions for a suite of teleological positions.

What Name Should We Use for You?

ID theorists can expect to be called "creationists" at some point or another in their academic career, whether they like it or not. The descriptions and terms used for the various teleological (i.e., design-based) perspectives on origins have caused much confusion in the scientific, philosophical and popular literature. (We dare not even contemplate the op-ed pages). Phrases such as *creationism in disguise, neo-creationism* and *stealth creationism* abound. Even the term *creationism* can be ambiguous. Notwithstanding their rhetorical value, such vague or crossover terms can cause those who interact with ID and young-earth creation proponents to expect that both groups agree philosophically and theologically, when in fact they differ significantly. In public forums, such as debates, panel discussions or school board meetings, failure to recognize distinctions between these and other teleological positions becomes a barrier to constructive dialogue.

How do you see (and what do you call) yourself? If we listen, just after a controversial lecture on origins, we can hear many voices in the foyer of the biology building: "Don't call me a creationist—that's become a term of abuse." "Well, *I'm* a creationist, and proud of it." "I'm a design theorist." "I'm a theistic evolutionist." By looking at how ID and young-earth creation proponents view both themselves and each other, one quickly learns that they don't hold, and don't consider themselves to hold, equivalent positions. The Discovery Institute's Center for Science and Culture, the primary research and publication organ of ID, defines ID as "hold[ing] that certain features of the universe and of living things are best explained by an intelligent cause, not an undirected process such as natural selection."[7] Access Research Network, a meeting place for the ID community, defines ID as "the view that nature shows tangible signs of having been designed by a preexisting intelligence."[8] Conspicuously missing from these definitions are any references to religious texts, such as the Bible.

Young-earth-creation paleontologist Kurt Wise defined "Young Age Creationism" (for our purposes the same as young-Earth creationism) as "maintain[ing] that God created the entire universe during a six-day Creation Week about six thousand years ago."[9] While not giving an age for the earth, Paul Nelson and John Mark Reynolds provide four other characteristics of young-earth creationism:

1. An open philosophy of science (characterized by an openness to all possible modes of causation).

2. All basic types of organisms were directly created by God during the creation week of Genesis 1—2

3. The curse of Genesis 3:14-19 profoundly affected every aspect of the natural economy.

4. The flood of Noah was a historical event, global in extent and effect.[10]

Both ID and young-earth creation proponents eschew terms like "intelligent design creationism," considering them to be pejoratives designed to blur the distinctions between the groups. The Discovery Institute states that ID can be distinguished from young-earth creationism in five ways, two of which are of particular importance here:

1. ID is based on science, whereas young-earth creationism is based on sacred texts.

2. The religious implications of ID are unconnected to ID itself.

How do you see (and describe) others? Bill Dembski (1999) differentiates ID from young-earth creationism primarily because "intelligent design nowhere attempts to identify the intelligent cause responsible for the design in nature, nor does it prescribe in advance the sequence of events by which this intelligent cause had to act."[11] Thus the distinctions between ID and young-earth creationism drawn by ID proponents themselves center on the nonauthority (in science) of sacred texts and an official agnosticism about the nature and methods employed by the designer(s).

The reaction of young-earth creation proponents to ID has been mixed.

Henry Morris, coauthor of *The Genesis Flood* and founder of the Institute for Creation Research, has written in sharp opposition to ID, stating that the design argument "has been tried in the past and has failed, and it will fail today. The reason it won't work is because it is not the Biblical method."[12] Answers in Genesis, another leading creationist organization, has been more measured in its response. Carl Weiland, writing the AiG's official position on ID, outlined perceived strengths and weaknesses of ID from a young-earth creationism perspective, concluding: "Where we can be natural allies, [and] if this can occur without compromising our Biblical stance in any way, we want to be."[13] Wise defines ID as "a theory and movement that seeks to develop a secular method of identifying and defending design in the universe."[14] In each instance, young-earth creation proponents distinguish themselves from ID mainly by the place they believe biblical authority ought to have in scientific model construction.

It should by now be clear that ID and young-earth creationism consider themselves as distinct groups. The groups have different philosophies of science, methodologies and aims. As such, those interacting with them should be mindful of these differences. Utilization of crossover terminology (e.g., "neo-creationism" and "intelligent design creationism") is both inappropriate and misleading.

Despite their differences, there are still a number of "homologies" between ID and young-earth creationism. So much, in fact, that some young-earth creation proponents (such as ourselves) have found a home in the ID movement. To understand this fact, it is helpful to develop a classification scheme that can both (1) accurately define each position, and (2) provide a framework to understand the relationship between them. But before doing this, it would be prudent to look at previous attempts at classification.

Taxonomies That Don't Work
The most recent attempts to classify various positions of origins are those of Eugenie Scott and Donald Wise. These authors attempt to classify origins positions through one or more gradational characters. Scott's article "The Creation/Evolution Continuum" classifies various origins positions in terms of how literal an interpretation of the Bible is taken. All differences between each position are a matter of degree, and the continuum has "few sharp boundaries."[15] Wise combined the literal interpretation criterion with how much control God has in science in his "belief spectrum."[16] Both of

these classification schemes suffer from three major shortcomings: (1) a strict science-nonscience demarcation, (2) the use of ambiguous classification criteria, and (3) assumptions of theological uniformity among teleological positions.

Science-nonscience demarcation. Both the continuum and spectrum assert that there is a clear method of reliably distinguishing science from nonscience, a method of demarcation. Briefly, demarcation is the attempt to draw a distinction, in this case between science and nonscience, two things based on one or more characters, in this case science and nonscience. While the continuum makes a literal interpretation of the Bible the key for distinguishing various positions, a close reading of the article's text gives deeper insight into the nature of the classification. Regarding the interaction of science and the Bible for "Flat Earthers," Scott states, "the earth is flat because the Bible says it is flat. Scientific views are of secondary importance." For young-earth creationists, Scott notes that they "reject modern physics, chemistry, and geology concerning the age of the earth." As for the differences between evolutionary creationists and theistic evolutionists, they "lie not in science, but in theology."[17]

In essence, then, the continuum and spectrum (with its more obvious "Bible" and "science" axes) are identical. Both assert that there is a demarcation between the Bible and science. The admixture of the Bible (no scientific content) on one hand, and science (no biblical content) on the other, results in any one of the positions. Figure 16.1, a composite of the continuum and spectrum views, illustrates this point. But, to justify this scheme it must be shown that the Bible and science are mutually exclusive. It follows that if the Bible is nonscience, then the Bible can neither now nor ever have provided any framework for scientific investigation. Neither can it aid in generating any testable hypothesis. If we are to demarcate science and the Bible, then a scientist simply cannot use the Bible to gain meaningful insight while in the pursuit of scientific knowledge.

Yet the history of science testifies firmly against any such demarcation. The belief that the Bible provides information on the reproductive nature of plant and animal life led Karl Linne to construct the modern discipline of biological systematics. William Paley constructed his views on natural history based on his beliefs about the Bible and the nature of God, and his ideas resulted in an empirical investigation into the natural world that continues to this day. He also believed that certain observations in nature, such as the magnificent design of the human eye, pointed directly to the nature, character and power of God.

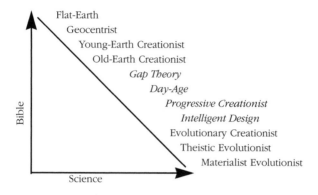

Figure 16.1. Composite continuum/spectrum classification

Conversely, Darwin often utilized a blended biblical-Platonic view, widely accepted at the time by natural theologians, as a foil in *On the Origin of Species* (see especially his discussions on immutability and biogeography), indicating that he believed that such ideas could indeed be empirically evaluated. To assume that there is a demarcation between the Bible and science would mean that Linne and Paley were not scientists (along with Newton and a host of others), and that many of Darwin's arguments in the *Origin of Species* do not count as scientific discourse.

In looking at demarcation attempts in the philosophy of science, assertions of a science-nonscience demarcation have fared poorly. Karl Popper tried to demarcate science from nonscience using the criterion of falsification. He argued that if a statement is testable through empirical investigation, it is thus falsifiable and counts as science. The falsification criterion of Popper fails on one level because theories do not exist in isolation. Rather, theories are usually multifaceted, with a number of auxiliary assumptions and hypotheses surrounding it. These auxiliary assumptions and hypotheses primarily exist to protect the central theory from the very thing that is supposed to make it scientific: falsification.

Furthermore, Popper's falsification relied on the independence of hypotheses and observations. But such a distinction is impossible since all observations entail theories to explain them. For example, observations of minerals in thin sections are dependent on various optical theories of microscopy that help scientists understand their observations. Without a distinction between theories and observation, Popper's falsification has no philosophical basis.

In the past, attempts by courts to distinguish science from creation science using the falsification criterion (such as Justice Overton's opinion on *McLean* v. *Arkansas Board of Education*) have been severely criticized.[18] In particular, falsification failed as a criterion because it can actually be met by creation science. Creation science passes the criterion of testability (and therefore falsification) since, for example, Steve Austin and his colleagues postulate a diluvial origin of Neoproterozoic to late Cretaceous sedimentary rocks.[19] This hypothesis can be empirically evaluated by looking at the inferred environment of deposition for relevant geologic units. So not only has falsification failed philosophically, it has also failed in practice by showing its inability to legally demarcate traditional science from creation science.

"Literal" interpretation of the Bible. A second problem for the continuum and spectrum views is, What does it mean to take the Bible "literally" as opposed to "nonliterally"? Here again we face a problem of demarcation. From the standpoint of the continuum, if we take the Bible entirely "literally," then we would be flat Earthers (the spectrum ends at young-earth creationism). Scott claims, "The strictest creationists are Flat Earthers."[20] Granted, flat Earthers would likely say that they take the Bible literally; indeed they might claim to take the Bible more literally than any other position represented on the continuum. But how is *literal* judged, and does the flat-earth position actually represent the most literal position on the continuum?

An interesting dilemma follows from this question. According to the continuum, young-earth and old-earth creationists take the Bible less literally than do flat Earthers. But young-Earth and old-Earth creationists might jointly claim that a flat-earth interpretation is actually taking the Bible *nonliterally.* How can this be? One charge might be that a flat-earth interpretation ignores grammatical and linguistic devices employed by the original writer. If a particular passage cited as support for a flat Earth has a poetic literary structure, then perhaps a nonliteral interpretation is actually literal with respect to the author's intent.

The book of Revelation will, perhaps ironically, clarify. John, in Revelation 7:1 writes, "After this I saw four angels standing at the four corners of the earth, holding back the four winds of the earth, so that no wind would blow on the earth or on the sea or on any tree" (NASB). The flat-Earth perspective may consider this passage to argue strongly for its case. But Revelation is a book written in a particular style known as apocalyptic. One of the main characteristics of this genre is the use of highly symbolic language. To complicate matters further, John indicates that what he is relating to his readers came from a vision (Rev 1:10; 4:1-2), so we can expect that the language used to describe scenes and events will also be symbolic. The "four corners" is typically understood to be the cardinal directions, not *literal* corners. It is literary phrase indicating "from everywhere." In light of these kinds of stylistic devices, saying that John actually meant that he saw a flat Earth would be like assuming that meteorologists are geocentrists because they tell us when the sun will rise and set.

Assumptions of theological unity. Third, the continuum and spectrum views fail because both assume theological unity among all positions. This problem is expressed in two ways. First, the continuum categorizes ID as being a form of old-earth creationism, located between progressive creationism and evolutionary creationists. But the diversity of Christian positions among ID proponents undercuts this argument. Though most individuals in the ID community are old-earth creationists of some form or another (e.g., Steven Meyer and William Dembski), the group includes theistic evolutionists (e.g., Michael Behe) and young-Earth creationists (e.g., Paul Nelson and John Mark Reynolds) that readily identify themselves as part of the ID community. Though individual ID proponents may integrate an old-earth creationist position with ID (e.g., Dembski, *Intelligent Design*), this is not an official position of ID. As it stands, such diversity among Christian beliefs on origins within the ID movement itself disqualifies ID as a subcategory of old-earth creationism.

Second, both the continuum and spectrum views consider that all positions not labeled materialist evolutionism/secular humanism to be derived from a Christian belief system, and that the designer is invariably God. While it is true that the vast majority of creationists and ID proponents are Christians, some are not. Some creationists (young-Earth and other types) are Jewish or Muslim. ID also includes non-Christian adherents such as Michael Denton and David Berlinski. Denton is a particularly illuminating case. His views on the origin and diversity of life are based on a decidedly neo-Platonic view of the universe, asserting that protein structures conform to "ideal forms" necessitated by a designer intrinsic to the universe.[21] To cloud matters further, the Raelian movement (which departs from theism altogether) has officially endorsed ID on its website[22] and identifies the designer as alien scientists who manufactured life. Supporters of directed panspermia, which submits that life was seeded on this planet by a dying alien race, could likewise be viewed as an ID-type position. Since each of these non-Christian positions can locate themselves within creationism or ID, the continuum and spectrum views fail to accurately describe the relationships among teleological positions.

The Nested Hierarchy of Design

The philosophical and theological problems encountered by the continuum and spectrum views can be avoided. A classification system that defines positions using a suite of discrete characters based on the presence or absence of particular design claims yet avoids demarcation arguments and naive theological assumptions yields positive results. The "nested hierarchy of design" (see fig. 16.2) is such a system. It is constructed similar to cladograms in biology and paleontology, and the various characters used in this system can be numerically coded.

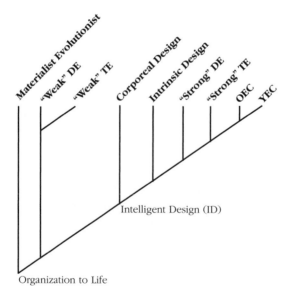

Figure 16.2. Nested hierarchy of design

The nested hierarchy of design is meant to classify teleological positions based on the relative strength of design claims. It is not intended to distinguish which of the positions classified can be referred to as "scientific" positions; hence it avoids the pitfalls of demarcation. Through the nested hierarchy of design, we can recognize the variety of theological positions represented among teleological positions. In fact, theological claims are employed to better resolve the relationship between the teleological positions classified here.

To classify the various teleological positions, the characters are defined as follows:

- Teleos (A): Real design does not (0) or does (1) exist in the biotic and/or abiotic universe.
- Detectable (B): Design is not recognizable or empirically detectable (0), or it is (1).
- Agency: The nature of the designing agent is
  (C) corporeal (0) or noncorporeal (1).
  (D) intrinsic to the universe (0) or transcendent to it (1).
  (E) deistic (0) or theistic (1).
- Biological complexity (F): continuous (0) or discontinuous (1) ancestry.
- Age (G): The age of earth is 4.5 billion years (0) or 6,000 to 10,000 years (1).

Table 16.1 is a character matrix (again, like cladistics) of eight teleological positions, with materialist evolutionist as an out-group. Note that in this analysis, there is a distinction between two types of deistic and theistic evolution, "strong" and "weak." These descriptors denote the relative strength of the design inference being made by adherents; they are not theological judgments. Note also the large gap between these views in the nested

Table 16.1. Character Matrix for Teleological Positions. (An X indicates that the position takes no stance on the character.)

	A	B	C	D	E	F	G
materialist evolutionist	0	0	X	X	X	0	0
"weak" deistic evolution	1	0	1	1	0	0	0
"weak" theistic evolution	1	0	1	1	1	0	0
corporeal design	1	1	0	X	X	0/1	0
intrinsic design	1	1	1	0	X	0	0
"strong" deistic evolution	1	1	1	1	0	0	0
"strong" theistic evolution	1	1	1	1	1	0	0
old-earth creation	1	1	1	1	1	1	0
young-earth creation	1	1	1	1	1	1	1

hierarchy of design, as compared to those in the continuum and spectrum views. The relationships elucidated by the nested hierarchy of design explain why most theistic evolutionists typically ally themselves with materialist evolutionists in common opposition to young-earth creationism and ID, which includes the smaller number of theistic evolutionists. This rejection of ID is based on a rejection of real design itself by materialists and by a rejection of detectability of real design by most theistic evolutionists.

Formal Definitions

We can now define various teleological positions based on the traits of the nodes that include the philosophy expressed by that position. Based on the above, the following definitions for ID and young-earth creationism can be advanced:

- Intelligent design: a teleological position that asserts recognition and empirical detectability of real design in the abiotic and/or biotic universe.

- Young-earth creationism: a teleological position that asserts recognition and detectability of real design in the abiotic and biotic universe by a transcendent, theistic being who has causally acted both during and after its initial formation, having designed discontinuous biological complexity approximately 6,000 years ago.

By looking at the structure of the nested hierarchy of design, we can now understand more fully the manner in which ID and young-earth creationism relate. ID occupies a node that contains all those teleological views that allow for the empirical detection of real design. That is, ID is philosophically minimalistic. In contrast, young-earth creationism occupies but one branch (like a taxon on a cladogram) and is defined by the successive accumulation of distinct philosophical and theological concepts as one moves up the diagram.

If we follow Johnson's logic that "creationists" are simply defined as those who "believe that a supernatural Creator . . . initiated [and] *controls*" the universe, then "creationists" become a "polyphyletic" assemblage.[23] That is, they are a group that cannot all be connected to a single node. So while "strong" theistic evolutionists, old-earth creationists and young-earth creationists are creationists, so too are "weak" theistic evolutionists. The first group all share a node on the nested hierarchy of design, while the last is on its own, separate branch in the diagram.

Though problematic to biologists and paleontologists, polyphyly need cause no consternation here. After all, this diagram does not entail common ancestry among the positions. Rather, this situation exists because, at least in the issues of creation and evolution, there are complexities involved in assessing one's position, and certain concepts carry more gravitas than others. If the classification were purely theological, then the situation would be quickly resolved.

Conclusion

Having explored the relationship between ID and young-earth creationism, we return to our three problems. We have answered the last two, the failure of previous classification methods and the relationship of ID and young-earth creationism, by constructing and interpreting the nested hierarchy of design. In doing so, we have also solved the riddle of our first query: How is it that Phillip Johnson could claim to be a creationist when he clearly wasn't one? The answer lies in the new way that Johnson looked at what it means to be a creationist. In focusing attention on the underlying principle of naturalism, rather than the distraction of the narratives of creation and evolution, a new taxonomy arises. In it we see the philosophical and theological concepts of any position laid out clearly and succinctly. Johnson is a creationist all right—just not a young-earth creationist.

COMPLEXITY, CHAOS AND GOD

WESLEY D. ALLEN AND HENRY F. SCHAEFER III

—

This chapter is dedicated to our friend Phillip E. Johnson. Our experiences with Phil attest to his marked influence over two scientists of different generations, so we will tell our stories separately, beginning with Henry (Fritz) and ending with Wesley.

Fritz: Phil joined the faculty at the University of California, Berkeley, in time for the fall quarter of 1967. I joined the Berkeley faculty a bit less than two years later, in June 1969. Thus Phil and I were faculty colleagues at Berkeley for a bit more than eighteen years, that is, until August 1987, when I moved to the University of Georgia.

Generally speaking, there is little interaction between the Boalt School of Law and the Department of Chemistry at Berkeley. The two departments share some frontage on Gayley Road, but that is about all they have in common. Thus our meeting was related to singular events in both of our lives. Namely, I became a Christian in 1973, and Phil became a Christian in 1977. Shortly after the latter date, we met in my office as a part of a group of Berkeley Christian faculty who met for lunch once a week. It did not take long for the rest of us (David Cole from the Department of Biochemistry was the leader of the group) to realize that Phil possessed a brilliant mind and an uncanny degree of clarity on intellectual issues. In my experience, Phil was one of the two individuals closest to "genius" among the many faculty I observed in eighteen years at Berkeley (the other being Professor William H. Miller in chemistry). Phil and I quickly became good friends, discovering that we both had a contrarian streak. As I occasionally tell my children, I have never aspired to be just like everyone else.

As I left Berkeley for the University of Georgia in August 1987, Phil departed for a sabbatical at University College, London, to study insurance law. However, he was quickly redirected, I think by the Almighty. Phil's diversion was laid out to me in a densely typed two-page letter from London, dated November 4, 1987. Phil writes:

> I have just read a fascinating book called *Evolution: A Theory in Crisis,* by Michael Denton. *Please* get hold of this book and tell me what you think of it. I would also like to read any reviews that have appeared. As far as I can judge, the book is an outstanding exposition of the problem. For a long time, I have been convinced that the evidence for evolution is nowhere near as strong as we have been told.

Fortunately, I had read the Denton book and was able to send Phil a copy of a review that appeared in the *Journal of the American Scientific Affiliation.* Phil's book *Darwin on Trial* was published less than four years later, and the rest is history.

I still return to the University of California, Berkeley, several times a year, and I always enjoy meeting with Phil. He has shown us a new dimension of courage and faith in recovering from his stroke of 2000, and we are pleased to honor him in some small way.

Wesley: While I received my Ph.D. in the Department of Chemistry at the University of California, Berkeley, in 1987, I did not have any direct interactions with Phil Johnson. It was only after I joined the faculty at Stanford University in 1988 that Phil's work began to have an affect on me. Determined to be a visible Christian faculty member in a difficult spiritual environment, I began to voraciously study a host of issues at the interface of science and Christianity in order to have maximum intellectual ammunition at my disposal. I was most surprised the day I found *Darwin on Trial* on the shelves at the Stanford Bookstore. I had originally swallowed the standard evolutionary models promoted in the textbooks I had been exposed to all the way back to junior high school, and without much analysis I had once syncretized Darwinian evolution with Christian theism. While my thinking had already begun to change as I assimilated quantum mechanics and statistical mechanics, Phil Johnson's writings gave the clearest possible expression to the deficiencies of the Darwinian worldview. His compelling arguments provided essential input as my own worldview matured. On each occasion that I have met Phil in the ensuing years, he has further inspired me to have the audacity to contend for truth. This legacy will continue to endure among numerous scientists of my generation.

Introduction

A complex, chaotic world undergoing very rapid and bewildering change naturally engenders angst in the human soul. To quote David Wells from his 1995 book *God in the Wasteland:*

> The central issue with which Our Time must now reckon [is the loss of its center]. The world is now filled with so many competing interests, so many rival values, so many gods, religions, and worldviews, so much activity, so many responsibilities, and so many choices that the older symphony of meaning has given way to the *random tumult* of the marketplace, to a perpetual assault on all of the senses. . . . We may now have everything, but none of it means anything anymore.[1]

The older symphony of meaning referred to here is the Judeo-Christian tradition, which grounded Western culture for many centuries and indeed nurtured the development of modern science.[2] As seen in a disturbing advertisement that popped up on a website a few years ago for an occultic group, some revelers not only reject the older symphony of meaning but also reject *any* symphony of meaning: "Tired of the same old boring religions? Does Jesus not return your phone calls? Does the goddess seem to be gathering nothing but mold? Are you tired of eating philosophical vomit? Then try a bit of good old-fashioned chaos! *Nothing is True. Everything is Permitted.*"

In an increasingly post-Christian culture, there might seem to be little room for a personal, providential God. Indeed, how could a world of such apparent complexity and chaos possibly be intricately superintended? Our approach to this issue will single out a more precise question, that is, whether there is any connection among philosophical relativism, the Christian faith and the *science* of complexity and chaos. Should we associate the very word *chaos* with the notion that "nothing is true, everything is permitted"? We contend that it is first critical to gain a proper scientific understanding of complexity and chaos theory, after which a host of deep and surprising philosophical and theological implications become apparent. Far from establishing philosophical relativism, we discover avenues by which classic Christian theism may be in coexistence with the realism advanced by complexity and chaos theory. Within the constraints of this short essay, we sketch out this line of inquiry for a broad audience. Part one: In presenting key elements of the science of complexity and chaos, we visit complexity vis-à-vis cosmos, the comprehensiveness of complexity, essential ingredients of complexity, the etymology of chaos, confusion over chaos, computers and chaos, mathematical characteristics of chaos, and the logistic equation as a class on chaos. Part two: In surveying the resulting philosophical and theological implications of complexity and chaos, we consider the uncertain prominence of chaos, the fundamental physics of complexity and chaos, the tension between classical reductionism and complexity theory, life and the metaphysics of complexity, the apparent demise of determinism, humankind confined to epistemic humility, paradoxes of free will and determinism, chance and divine providence, and the connection of complexity theory to Darwinism and the origin of life.

The Science of Complexity and Chaos

A striking transformation began in the structure of scientific research in the latter part of the twentieth century, one that will continue to accelerate in the coming decades. This transformation is the increasing interdisciplinary nature of modern science, even in the face of an information explosion unprecedented in human history and even as the burgeoning requirements for technical expertise push toward increasing specialization. In brief, the biggest, most important scientific problems and questions of human society cannot be adequately addressed by any single discipline, and only by pooling the insights and capabilities of myriad fields of chemistry, physics, mathematics, biology, and computer science can we hope to achieve real solutions. This realization requires us to stretch our understandings and actively build bridges between disciplines. For example, quantum physicists working on chemical problems must engage with those outside the field in showing that the detailed mechanisms of chemical reactions they discover provide the foundations for atmospheric modeling, and equally for pharmacology and drug design. A recent, large-scale research initiative of the U.S. Department of Energy coined SciDAC (Scientific Discovery through Advanced Computing) emphasizes collaborations of this type in solving problems ranging from climate modeling to astrophysics to combustion chemistry and dynamics. In recent years something called complexity theory, with attendant concepts such as mathematical chaos, has arisen as one means of collecting a cornucopia of scientific disciplines under a big tent.

This interdisciplinary tent claims to give key insights to some of the most vexing questions of science and society, as we intend to illustrate here in our broad introduction.

Two of the most ardent and populist proponents of complexity theory are Peter Coveney and Roger Highfield, who coauthored a 1995 book titled *Frontiers of Complexity: The Search for Order in a Chaotic World.*[3] These authors eloquently lay out their case and give us several notable quotes serving as springboards for discussion. They introduce their book as follows:[4]

> When viewed in profound close-up, the universe is an overwhelming and unimaginable number of particles dancing to a melody of fundamental forces. All about us and within us, molecules and atoms collide, vibrate, and spin. Gusts of nitrogen and oxygen molecules are drawn into our lungs with each breath we take. Lattices of atoms shake and jostle within the grains of sand between our toes. Armies of enzymes labor to turn chemicals into living energy for our cells. . . . Now a new branch of science is attempting to demonstrate why the whole universe is greater than the sum of its many parts, and how all its components come together to produce overarching patterns. This effort to divine order in a chaotic cosmos is the new science of complexity. It is weaving remarkable connections between the many and varied efforts of researchers working at its frontiers, across an astonishingly wide range of disciplines.[4]

The rich play on words "divine order in a chaotic cosmos" is juxtaposed with an eighteenth-century statement from philosopher Immanuel Kant, namely "God has put a secret art into the forces of Nature so as to enable it to fashion itself out of chaos into a perfect world system."[5] In these two short quotes we encounter many fundamental questions. Is there underlying order in chaotic systems? Can a true cosmos indeed exhibit chaos, or must chaos be supplanted to achieve a "perfect world system"? Is the "secret art" of natural forces humanly discernible in complex systems? Can a system "fashion itself" into a more ordered state; that is, is self-organization a property inherent in nature? Has God's propitious hand guided the creation of the ubiquitous and exquisite macroscopic complexity of the universe out of the primeval simplicity of atoms and molecules?

The Greek word *cosmos,* popularized in the Carl Sagan television series and book of the same name,[6] may be defined as "the universe, regarded as an orderly, harmonious whole." In contrast, *chaos* in Greek mythology refers to the primeval emptiness of the universe before the beginning of time, or the abyss of the underworld. In Genesis 1:2 we find that in the beginning "the earth was without form, and void, and darkness was upon the face of the deep." While the word *chaos* is not used in the Septuagint in this passage, Christian creeds have often contained such a translation. For example, in the baptismal covenant of the United Methodist Church, the second and third verses of Genesis are summarized as "when nothing existed but chaos, you swept across the dark waters and brought forth light." Moreover, in Thomas Burnet's *Telluris Theoria Sacra* from 1681, the history of the earth is depicted in seven stages: the chaotic liquid, the pristine earth, earth during the flood, modern earth, earth during the conflagration to come, earth during the millennium, and earth's ultimate fate as a star.[7] Therefore, in their original meanings, *cosmos* and *chaos* are essentially antithetical to one another. This historic antithesis is by no means retained in the modern scientific meaning of *chaos,* which now is properly considered to be a specific mathematical property present in many systems described by complexity theory.

The ideas of complexity theory have been invoked in a large number of scientific disciplines, including weather predictions, systems of global telecommunications, artificial intelligence, structural properties and hardening of cement, annealing of metals, magnetic properties of metal alloys, crystal growth, the formation of cell walls and intracellular structures, the physics of avalanches, earthquake predictions, the social behavior of ant and bee colonies, patterns and spots on animal coats, heartbeats and arrhythmias, the extinction of living species, the global ecosystem, the origin of life, brain plasticity and memory, the human immune system, patterns in mollusk shells, and stock market catastrophes.[8] This remarkable list clearly ranges from the most mundane to the most abstract of scientific inquiries, attesting to the comprehensiveness of complexity theory. Further evincing the prevalence of complexity theory, the entire April 2, 1999, issue of *Science,* one of the most widely read research journals, was devoted to "complex systems."

Complexity theory is a watchword for a way of thinking about the *collective* behavior of myriad basic, interacting units, be they molecules, neurons, computer bits or some other constituent. The goal is to understand how relatively simple laws of interaction between the basic units lead to a wealth of complex behavior of the system as a whole. The wide range of applicability of complexity theory is a consequence of its emphasis on the laws of interaction of the basic units, or the "rules of engagement," rather than their composition. The essential mathematical ingredients for "complexity" are *irreversibility* and *nonlinearity*.[9] These terms are familiar to those who have studied thermodynamics and statistical mechanics in undergraduate chemistry and physics courses at the university level; however, some illustration for a general audience is warranted here.

Irreversibility is a feature of systems in states far from equilibrium. Such processes cannot be reversed by a slow progression of continuous changes in the external conditions. There is a strong arrow of time in such events that cannot be superseded. The breaking of an egg represents an irreversible process, as dramatized in the popular nursery rhyme about Humpty Dumpty. From a chemical perspective, the hard-boiling of an egg is a better example; there is simply no means of regaining the original fluid state of the egg proteins after they have been denatured. In contrast, the melting of ice in a tray is a reversible process because the liquid phase can be transformed back into the solid phase simply by placing the tray back into a freezer. In reality, reversibility is an idealization that provides thermodynamic reference points in scientific analyses. To some extent virtually all processes we experience have some degree of irreversibility. All of us know intuitively that our life processes themselves tick with ultimate irreversibility.

Nonlinearity is present along with irreversibility in systems scientifically labeled "complex." In linear systems the output from a process is directly proportional to the input. An illustration is provided by a favorite family activity of ours in the fall of each year, picking apples in our orchard and making cider with an old-fashioned mill and press. We know that if one bushel of apples yields one gallon of cider, then two bushels give two gallons for our consumption. Thus the reward of this linear process is directly proportional to the manufacturing effort invested. In contrast, in nonlinear systems there is a more intricate connection between input and output. As university professors, we know that the process of instruction is often a nonlinear enigma. No matter how hard the instructor tries, some students will not grasp the subject; no matter how lazy the instructor is, some students will still master the subject. If the instructor is too demanding, many students will give up in frustration; if the instructor requires too little effort, many students will not be challenged to learn. The technical meaning of nonlinearity ultimately rests in the mathematical structure of the coupled differential equations that describe the time evolution of a complex system. In chemical kinetics, for example, the nonlinearity in complex systems often entails feedback mechanisms whereby the initial products of a chemical reaction may either promote or suppress the further production of the same products. Such nonlinear feedback mechanisms greatly enhance the dynamical possibilities of a system. The self-regulation of the biochemistry in living species is built on this type of nonlinearity.

Additional general properties are found in complex systems on top of the essential ingredients of irreversibility and nonlinearity. These characteristics, which constitute central concepts of complexity theory, include emergent properties, dynamism, adaptiveness, self-organization, chaos, fractals and strange attractors.[10] Some degree of technical discussion is required to explain the last three properties, and we will indeed expound on the mathematical meaning of chaos below, but beforehand we will give a qualitative description of the first four properties.

Emergent properties are those exhibited by a complex system as a whole that are not possessed by the individual components of the system. In thermodynamics, temperature is an emergent property in this sense, because it is defined on the basis of a particular statistical distribution of energy among the individual modes of motion in a large ensemble of molecules. Individual molecules have well-defined energies, but cannot have a temperature, which is a property emerging only in a large collection of molecules. Consciousness is a more striking example of an emergent property. Individual molecules, interacting strictly according to the laws of physics, do not have consciousness; however, consciousness is a characteristic of the human brain, which consists of about 10^{11} neurons, each neuron in turn comprising an enormous number and diversity of molecules.

The dynamism property introduced above implies that the individual units of a complex system are in a state of constant change and are interacting on short time scales. Such a system is able to respond to external influences rapidly, sampling new modes of behavior and potentially achieving effective adaptation to the environment. Dynamism in biochemical systems is made possible by the ultrashort time scale inherent in molecular motion (e.g., rates of atomic vibrations within molecules of more than 10^{13} cycles per second). Thus, myriad dynamical possibilities can be sampled on the microscopic level before the elapse of much time on a macroscopic level.

Self-organization occurs when new and often intricate spatial structure arises after a complex system is placed in dynamic motion from some initial state of more limited order. A simple illustration of self-organization is the consistent development of a vortex once a drain is opened in the bottom of a water reservoir of sufficient height (a filled bathtub, for example). A vortex will remain in a steady state as long as there is a continuing influx of water to maintain the water level (e.g., a hose feeding the other end of the bathtub). No construction by an external agent is necessary to create the vortex; it forms as a consequence of the mathematical structure of the differential equations that govern the fluid dynamics of the system. If the dynamics of the process are externally arrested, either by replacing the plug in the drain or by stopping the influx of water, the vortex structure soon vanishes. A far more involved and mysterious example of self-organization is ontogeny, the biological development of a living being from a single cell to a mature adult. In particular, embryonic development entails the remarkable differentiation of cells and the seemingly miraculous formation of complex physical structures (organs) from a single, formless zygote. This paragon of self-organization is directed by a massive set of elementary physicochemical processes operating in exquisite concert. The preprogramming of the entire system is due to the presence of an enormous amount of information in the finely tuned initial conditions.

In complex systems, matter often exhibits an innate tendency to self-organize and generate new structural features. This property is in both real and perceived tension with the well-known second law of thermodynamics,[11] which holds that in isolation a nonequilibrium system will spontaneously evolve toward a state of higher entropy (disorder). In figure 17.1 appears a pictorial representation of the disparate endpoints into which an initial distribution of a system might evolve, one being a self-organized steady state, and the other a uniform thermodynamic equilibrium of maximum entropy. We can think of these depictions as representing the two-dimensional concentration profile of some chemical species involved in an elaborate system of chemical reactions. The key to resolving the apparent conflict between self-organization and thermodynamic equilibrium is the phrase "in isolation" in the statement of the second law. If we cut off all matter and energy flow to the system, the initial distribution will eventually proceed toward the equilibrium state of uniform concentration. However, if the boundaries of the system are open to matter and energy exchange, then a dynamic steady state of self-organization might develop under certain circumstances. As in the water vortex analogy discussed above, this steady state may be long-lived, but it cannot persist once the influx of material and/

or energy is halted. Even if a system is in a state of self-organization, the second law is not violated, because the total entropy of the system *plus* its surroundings continues to increase with time.

Figure 17.1. Disparate endpoints for the time evolution of an initial distribution, illustrating the tension between self-organization and thermodynamic equilibrium

Qualitative descriptions are ultimately insufficient in complexity theory. Indeed, the proper language of complexity theory is mathematics and its tool is computer technology. This point is no better illustrated than in the history of the modern, scientific understanding of the chaos property, an aspect of classical physics that remained undiscovered for almost three hundred years after the pioneering work of Isaac Newton. The fascinating story starts in 1960 with Edward Lorenz at MIT.[12] Lorenz was pioneering weather simulations on the clumsy early computers of the day. One day in the winter of 1961, he wanted to study a particular computer simulation for a longer time period, and in order to save effort, he restarted the run at its midpoint by typing in initial conditions from the printout of the earlier job. Then he walked down the hall for a cup of coffee and to escape the irritating noise of his Royal McBee computer. An hour later he returned to find something entirely unexpected: the new computer simulation had quickly diverged from the earlier run merely due to the three-digit round off in the input. This serendipitous discovery of hypersensitivity of the differential equations to initial conditions was a seed for the new science of chaos. By the early 1970s, studies of deterministic, nonlinear behavior had given the term "chaos" its current mathematical meaning. To quote Garnett Williams:

> The scattered bits and pieces of chaos began to congeal into a recognizable whole in the early 1970s. It was about then that fast computers started becom-

ing more available and affordable. It was also about then that the fundamental importance of *nonlinearity* began to be appreciated. The improvement in and accessibility to fast and powerful computers was a key development in studying nonlinearity and chaos. . . . Chaos today is intricately, permanently and indispensably welded to computer science.[13]

In common parlance, chaos means a condition of utter confusion, totally lacking in order. However, from a scientific perspective, chaos is sustained, long-term evolution of a system that is disordered only in appearance. Some characteristic aspects of mathematical chaos are as follows:

1. Chaos results from fully deterministic processes, not from any degree of indeterminism.

2. Chaos occurs only in nonlinear systems.

3. Chaotic behavior for the most part looks disorganized and erratic, usually passing all statistical tests for randomness.

4. Chaos happens in feedback systems, where the initial output of the process either suppresses or stimulates further output.

5. Chaos can arise in relatively simple systems. With discrete time, chaos can take place in a system that has only one variable. With continuous time, as few as three degrees of freedom may be necessary.

6. Chaos is entirely self-generated by the mathematical structure of the dynamical equations; no manipulation of external conditions is required.

7. Chaos is not an artifact of data inaccuracies, such as sampling or measurement error.

8. Despite its appearance, chaos invariably includes one or more types of order or structure.

9. The erratic behavior exhibited by chaotic systems is constrained within certain bounds by conservation conditions.

10. Details of chaotic behavior are hypersensitive to changes in initial conditions.

11. Precise forecasts of long-term behavior in chaotic systems are not possible.

12. Short-term predictions in chaotic systems can nonetheless be relatively accurate.

13. The distant history of a chaotic system cannot be practically deter-

mined; precise information about early conditions is irretrievably lost.

14. The trajectories in a chaotic system may have fractal properties (in phase space), exhibiting the same patterns over a broad range of scale; that is, showing the same appearance when viewed under successive levels of magnification.

15. As a control parameter within the dynamical equations of a system changes monotonically, behavior that is initially regular may transform into chaos, according to one of a select few typical scenarios.[14]

Because there is indeed determinism and certain types of structure in mathematical chaos, the term is somewhat a misnomer, but it is too late to change it now! Due to extreme sensitivity to initial conditions, chaotic systems are *unpredictable* in practical terms, even though deterministic. In his *Physical Chemistry* textbook, Peter Atkins clarifies this point:

> The term 'chaos' is in certain respects misleading. . . . In so far as the composition can be specified exactly, then a later composition can be predicted. Our inability to predict the composition of a system that is in a chaotic regime stems from our inability to know exactly the initial experimental conditions or measure the composition at an exact later instant. The unpredictability of a chaotic system lies not in the formulation or solution of the differential equations that describe the rates of processes, but in our ability to relate those solutions to the practical system of interest given the inherent imprecision of experimental observations.[15]

In brief, chaotic systems are different from regular systems in that acceptably approximate input knowledge is *not* sufficient to obtain acceptably approximate output predictions. A simple and highly instructive model of chaos is the widely studied *logistic equation,* whose mathematical form is $x_{n+1} = k\,x_n\,(1 - x_n)$.[16] Among other things, the logistic equation has been used as a model of population development within a species, in which case x_n represents the population in generation n normalized with respect to some maximum possible value. According to the logistic equation, the population in the next generation (x_{n+1}) is determined by a linear birth-rate component (kx_n) and a compensating quadratic death-rate term $(- kx_n^2)$. The proportionality constant k is characteristic of the particular population under investigation, and indeed k serves as the *control parameter* that critically determines the time evolution of the system. For simplicity, time is discretized in the logistic equation; hence, it is a *difference* equation rather than a *differential* equation involving continuous time. Nonetheless, the logistic equation has the rudimentary elements and nonlinear structure necessary for our brief lesson on chaos.

Drastically different behavior in the logistic equation is observed depending on the value of the control parameter k. In figure 17.2, representative population evolution is shown in four distinct plots, all with the same starting population (x_0 = 0.3), but with k in four different regimes (I-IV). In regime I, $0 < k \neq 1$, the population always moves to extinction, regardless of the initial population. In regime II, $1 < k \neq 3$, the population always converges to a steady-state value of $1 - 1/k$, regardless of x_0. In regime III, $3 < k < k_{crit}$, the population evolves into a pattern of oscillation between fixed turning points. For the example in figure 17.2 with $k = 3.4$, there are two turning points (0.842, 0.452) in the cycle. As k approaches the critical value k_{crit}. 3.57, there is a so-called bifurcation of limit cycles, whereby the periodicity (number of turning points) of oscillations steadily increases. Finally, in regime IV, $k_{crit} < k < 4$, mathematical chaos is found. Despite the erratic nature of the time evolution in regime IV, definite patterns are present, although no exact periodicity persists.

The logistic equation nicely illustrates how the onset of chaos affects the sensitivity of time evolution to initial conditions. Consider the case of $k = 2.8$, lying in nonchaotic regime II. If the system is propagated from the initial value x_0 = 0.30000000, then the population after 100 time increments is x_{100} = 0.64285714. If the initial condition is minutely perturbed to x_0 = 0.30000001, then no change results in x_{100} to eight digits of accuracy. Thus, an initial uncertainty of one part in 10^8 leads to no loss of accuracy in the final prediction; no information is lost in this sense. In contrast, consider the case of $k = 3.8$, lying in chaotic regime IV. Starting from x_0 = 0.30000000, the population after 100 time increments is x_{100} = 0.93709460. Now if the starting population is perturbed to x_0 = 0.30000001, an amazing hypersensitivity is witnessed; namely, x_{100} changes to 0.78263771. Therefore, an initial uncertainty of one part in 10^8 leads to an error in the final prediction of 16 percent! Stated another way, seven digits of accuracy have been lost in the process. Considering that in practice we rarely know more than four digits of accuracy in initial conditions, the futility of attempting precise, long-term predictions in chaotic systems becomes obvious. A tremendous epistemic problem thus arises from the hypersensitivity to initial conditions embedded in the deterministic mathematical equations that govern chaotic systems.

Philosophical and Theological Implications of Complexity and Chaos

Some enthusiastic proponents consider complexity and chaos theory as the third scientific revolution of the past century, of equal stature with relativity and quantum mechanics.[17] Coveney and Highfield support this view: "Just as investigating the very small gave us quantum mechanics and investigating the very large led to general relativity, perhaps something new will come from studying the very complex."[18] However, relativity and quantum mechanics ushered in fundamentally new physics, unlike complexity and chaos theory, which simply explore the multifarious behavior of classical, deter-

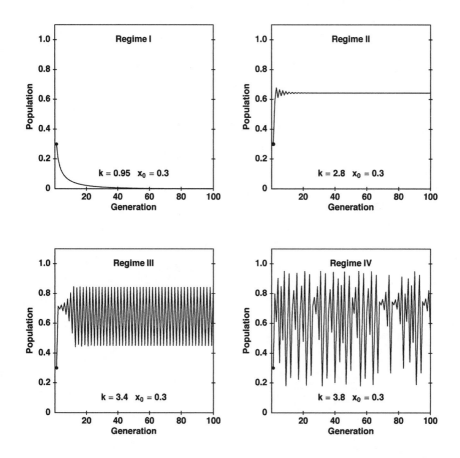

Figure 17.2. Population evolution according to the logistic equation in four regimes of the control parameter *k*

ministic equations. Goldenfeld and Kadanoff offer a balanced assessment: "The complexity of the world is contrasted with the simplicity of the basic laws of physics. In recent years, considerable study has been devoted to systems that exhibit complex outcomes. This experience has not given us any new laws of physics, but has instead given us a set of lessons about appropriate ways of approaching complex systems."[19]

Without question, complexity theory is applicable to a wide range of physical phenomena. However, opinions vary widely as to the prevalence of the chaos property in complex systems, or the general importance of mathematical chaos in the real world.[20] Some hardened skeptics dismiss chaos as a mere mathematical curiosity with little relevance for physical reality. The truth surely lies somewhere between euphoric optimism and radical skepticism about the significance of chaos.

In ecology, over twenty years of searching for chaos has not met with clear success. "There is no unequivocal evidence for the existence of chaotic dynamics in any natural population," according to ecologist David Earn of Oxford University.[21] Nature seems to sit on the edge of chaos without plunging into it. Kevin McCann, a University of California, Davis, ecologist, says, "It seems that for species to persist, nature is biased toward inhibitors and away from oscillators. That's just going to decrease the likelihood of chaos, no matter what."[22]

In contrast, complexity and chaos theory are natural and widely accepted approaches to understanding the human brain.[23] The brain is comprised of 10^{11} neurons, myriad units that are interconnected, dynamic and adaptive, yielding emergent properties such as consciousness. Centers of intelligence and decision making ought to be characterized by high sensitivity to input conditions. In other words, only slight changes in the information flowing into the system should be necessary to alter resulting decisions and the ensuing macroscopic outcomes. For example, when driving on the freeway, a rapid decision to steer clear of a deadly collision might result from picking up a faint noise or a peripheral image from a car positioned in one's blind spot. Therefore, in terms of its physical structure, the human brain is in many respects an ideal "complex" system. Then to what extent can human thinking be understood by theories of complexity and classical chaos embedded in chemical kinetic and electrochemical equations?

Roger Penrose has argued abstractly in a series of popular books that existing physics is fundamentally incapable of explaining the mysteries of the human brain and that some type of nonlocal quantum coherence must be at work.[24] However, counterarguments based on statistical mechanics have disputed the notion that quantum coherence can persist in the brain, because the physical medium is too hot and too disordered.[25] Scientists in this field generally feel that complexity and chaos within classical physics can generate the essential emergent behavior necessary to explain the fundamental processes of the brain. For example, Michael Arbib is quoted by Coveney and Highfield as claiming "Nobody has said they are stuck in their study of the hippocampus because the fundamental laws of physics are restrictive. . . . I would be surprised if Penrose has got the answer in quantum gravity."[26] We are eager to see how this field of inquiry develops. The main point we wish to make here is that current scientific approaches to human thinking and brain function include concepts of complexity and chaos in an integral way. Thus complexity and chaos are not only embedded in many physical phenomena but also appear to be essential elements of how we perceive physical reality itself.

The integrative philosophical agenda of complexity theory is a countercurrent to traditional scientific reductionism as a means of describing the natural world. Coveney and Highfield[23] see complexity theory as an answer to glaring deficiencies of the reductionist approach:

> The reductionist view that anything and everything can be boiled down to atoms and molecules is widely viewed by nonscientists as a philosophy that also erodes our belief in "humanity" and the value we place on it. . . . The vision of the world that naive reductionist science has proclaimed is a cold and solitary one that sets mankind apart from an unseeing and uncaring universe.
>
> . . . Complexity affords a holistic perspective and with it insights into many difficult concepts, such as life, consciousness, and intelligence, that have consistently eluded science and philosophy.[27]

Crutchfield contends that the hope that physics could offer a complete description of physical reality through an increasingly detailed understanding of fundamental particles and forces is unfounded because complex global behavior on a large scale in general cannot be deduced merely from knowledge of the individual components.[28] In this regard, some have proclaimed the discovery of chaos in physical systems as the end of the reductionist program in science. Paul Davies holds that reductionism is not merely inadequate as a global scientific paradigm or metaphysical framework but is actually "nothing more than a vague promise founded on the discredited concept of determinism."[29] As quantum chemists whose daily research involves predicting and explaining chemical phenomena by reducing them to solutions of the Schrödinger equation, we view the Davies characterization as somewhat imbalanced, while not denying the limitations of studying parts in isolation from the whole. We find greater affinity with the comments of Gallagher and Appenzeller concerning reductionism vis-à-vis complexity theory:

> We have the best of reasons for taking this reductionist approach—it works. It has been the key to gaining useful information since the dawn of Western science and is deeply embedded in our culture as scientists and beyond. But shortfalls in reductionism are increasingly apparent. Mostly these arise from information overload. . . . So perhaps there is something to be gained from supplementing the predominantly reductionist approach with an integrative agenda.[30]

Concepts from complexity theory have been used by some scientists to address profound philosophical and theological questions.[31] For example, the essence of life has been described on the basis of emergence and complexity. This approach rejects the age-old concept of vitalism, that some sort of nonphysical dynamism or essence must be added to a material system to make life. Instead, complexity theory views life as an emergent property arising when material systems are organized and interact in certain ways. Coveney and Highfield make rather startling pronouncements about this issue:[27]

> These approaches to complexity are so successful that life itself is now gaining a new meaning. Neither actual nor possible life is determined by the matter that composes it. Life is a *process*, and it is the *form* of this process, not the matter, that is the essence of life. . . . One can ignore the physical medium and concentrate on the *logic* governing this process. . . . Life is fundamentally independent of the medium in which it takes place. The implications of separating living complexity from its medium are stunning.[32]

Some theologians might retort that the realization of Coveney and Highfield is not so stunning, because it has been part of Christian doctrine for two thousand years. Specifically, in 1 Corinthians 15:42-44 and 2 Corinthians 5:1 the apostle Paul speaks about the Christian's ultimate fate—not as a disembodied spirit but in terms of a continuity of existence that proceeds through physical death into a new and glorious medium. In the very least we can applaud the nonreductionist stance of complexity theory in maintaining that a human being is far more than a collection of molecules. The concept of a human soul can be retained in complexity theory as an emergent, nonreducible collection of properties or essences.

The presence of chaos in physical systems has its own far-reaching philosophical ramifications in that it shatters the Newtonian image of a predictable, clockwork universe and the Laplacean dream of a thoroughly predictable and controllable world.[33] In 1986, Professor James Lighthill, president of the International Union of Theoretical and Applied Mechanics, essentially conceded this point in the *Proceedings of the Royal Society of London A*:[29]

> We collectively wish to apologize for having misled the general educated public by spreading ideas about the determinism of systems satisfying Newton's laws of motion that, after 1960, were proved to be incorrect. . . . Modern theories of dynamical systems have clearly demonstrated the unexpected fact that systems governed by the equations of Newtonian dynamics do not necessarily exhibit the "predictability" property. . . . We are able to come to this conclusion without ever having to mention quantum mechanics or Heisenberg's uncertainty principle. A fundamental uncertainty about the future is there, indeed, even on the supposedly solid basis of the good old laws of motion of Newton.[34]

This public apology, made exactly three hundred years after the presentation of Newton's great *Principia Mathematica* to the Royal Society of London, refers to the hypersensitivity to initial conditions exhibited by the differential equations governing the dynamics of chaotic systems. Even though chaos is deterministic, it is fundamentally unpredictable because input information of the greatest possible accuracy will always lead to unacceptable errors in output predictions if enough time has elapsed. While chaos is not the demise of determinism in a technical sense, it does make determinism moot in practical application. The "death" of the dream of unlimited predictability points to fundamental limitations on human knowledge and human control that have been encountered during the last century, arising both from quantum mechanics and classical chaos. John Jefferson Davis points out some of the theological consequences of these limitations:

> From a Christian perspective, such an encounter with the limits inherent in the nature of the physical realm should remind man of the fundamental distinction between an infinite Creator and a finite and limited creation, including man. The new discoveries of chaos theory give man further reason to adopt a stance of "epistemic humility" in the face of a complex and unpredictable world.[35]

One might say that chaos is a chasm separating the limited mind of man from the unbounded mind of God. Because this chasm cannot be bridged, scientific hubris loses justification.

In the view of some, chaos provides a resolution to the vexing problem of human free will versus naturalistic determinism. Although we much prefer the term "human responsibility" to "free will," the latter is too strongly rooted in the lexicon to be replaced. In popular discussions, free will is often only intuitively understood rather than explicitly defined; in fact, our experience is that most people have not thought very carefully about free will and do not know precisely what they mean by this term. It is commonly held that if humankind is ruled by natural forces via deterministic mechanisms, then there is no free will in human actions. Of course, the debate should properly center on the structure and function of the human brain, which is a natural target for the application of complexity theory. If chaos is present in the collective behavior of neurons in the brain, then it is possible that human decisions retain a fundamental lack of predictability, even if thought processes are deterministic. In this case, "free will" arises because the sensitivity to initial conditions of the deterministic laws governing human behavior is so great that alternative outcomes (choices) *seem* equally viable, and it is humanly impossible to discern in advance which choice will be realized. Figure 17.3 is a cartoon of such a model of "free will."

If the hypersensitivity of chaotic neural action to initial conditions (input stimuli and past experiences) extends all the way down to the quantum level, then the causal links of classical determinism are broken, or in the very least muddled, in human decision making. In the words of Crutchfield, because chaotic processes selectively amplify small fluctuations, "chaos provides a mechanism that allows for free will within a world governed by deterministic laws."[36] J. Doyne Farmer echoes this idea: chaos theory might provide "an operational way to define free will, in a way that allowed you to reconcile free will with determinism. The system is deterministic, but you can't say exactly what it's going to do next."[37]

Some Christian scholars hotly object to such ideas. According to John Jefferson Davis:

> The biblical doctrine of the *imago Dei* places a fundamental barrier (from a Christian viewpoint) against all attempts to explain the human person completely or exclusively in terms of scientific laws. . . . [The problem with such implicit identifications of "freedom" with "randomness" or "unpredictability" is

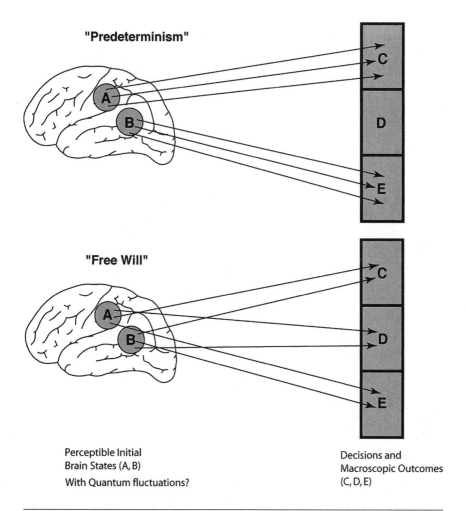

Figure 17.3. A model of "free will" based on chaotic determinism

that it overlooks] the *purposes* of human agents acting for the realization of certain ends. . . . The purposive dimension of human choices, directed toward the realization of certain ends among a number of alternatives, cannot be reduced to the categories of physics—whether or not the physics in question is Newtonian, quantum-mechanical, or "chaotic."[38]

Davis clearly has a restrictive view of the *imago Dei* doctrine. Other thoughtful Christians contend that macroscopic "free will" riding on top of microscopic, chaotic determinism is not necessarily inconsistent with Christianity and indeed provides an avenue for human decisions to be under the control of divine sovereignty. However, both camps would concur with the following broad conclusion from Davis: "What is being suggested here is that the new perspectives arising from chaos research help to make 'cultural and epistemic space' for the human sciences, including religion."[39]

We now understand that chaotic systems provide a deterministic link between imperceptibly small perturbations or fluctuations, of whatever origin, and macroscopic outcomes that can significantly alter human events. Consequently, chaos adds a new perspective on chance events. Metaphysically, chance is "no-thing" insofar as it is not a causative agent. As commonly used, chance describes a situation in which the detailed information necessary to determine an outcome by some rule or physical law is prohibitively voluminous or is fundamentally inaccessible. Lacking knowledge of the necessary input, events are merely ascribed to chance. Chance is not equivalent to indeterminism in classical physics, and even in quantum mechanics this issue is a very subtle matter of interpretation.

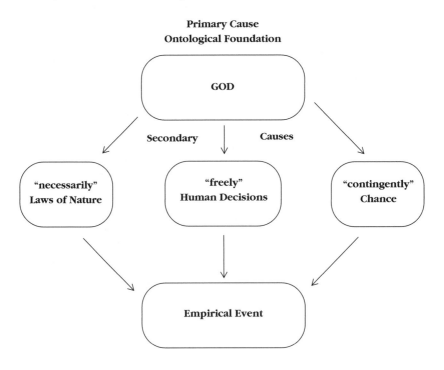

Figure 17.4. The chain of causality in Christian theology, as stated in the Westminster Confession of Faith (1647)

Sequences of events resulting from the intermingling of chance with more identifiable cause and effect relationships generally do not demand explanation that is not naturalistic. But historically many have seen chance situations as avenues through which divine providence can function. According to an amusing aphorism expressing this notion, "chance is the pseudonym of God when he did not want to sign."[40] More seriously, in Christian theology God is the sovereign Creator and Sustainer of the universe, who sometimes uses humanly unpredictable forces of the natural world for his own purposes. In the Westminster Confession of 1647, God is the first cause of all things, but "he ordereth them to fall out, according to the nature of second causes, either necessarily, freely, or contingently." This hierarchy is diagrammed in figure 17.4. In modern parlance, the words *necessarily, freely* and *contingently* mean by laws of nature, acts of human free will, and chance, respectively. Interestingly, the concepts of chaos and complexity provide for a unified view of such secondary causes within the providential workings of God. One aspect of this view is the relationship between chance and law described by Arthur Peacocke, a biochemist and Anglican priest:

> Chaos and complexity provide for a creative interplay of chance and law. Chance operates within a lawlike framework which limits possible outcomes. Chance allows for new forms of life and organization to emerge, while deterministic laws provide the stability for the forms to endure. It is the combination of law and chance which makes possible an ordered universe capable of developing within itself new modes of existence.[41]

Another aspect is the microscopic basis for events whose macroscopic cause is identified as either chance or human decision. Because God is omniscient and omnipotent, all phenomena are known and may be controlled in complete microscopic detail. In complex and chaotic systems, microscopic perturbations of initial conditions are magnified to the level of macroscopic outcomes. Therefore, if human free will or "contingent" causes are chaotic in nature, they can be under complete divine governance (via undetected microscopic control), despite macroscopic appearances to the contrary. In other words, abrupt macroscopic discontinuities in otherwise smooth, naturalistic sequences of events are not necessary in order for divine providence to be at work.

Many proponents of complexity theory strongly adhere to Darwinism and indeed take inspiration from it. Coveney and Highfield are among those whose partisanship pervades their explanation for the origin of living complexity:

> The fabulous complexity of life has been fashioned in the process of evolution by natural selection, a powerful concept that has withstood a great deal of abuse since Charles Darwin unveiled it in 1859. As the foundation stone of modern biology, the theory of evolution has emerged unscathed from more than a century of stringent evaluation by scientists and attacks by creationists. Now it has spread its influence to the inanimate world, where evolutionary metaphors are taking root.[42]

These authors go on to describe developments in genetic algorithms and programming, neural networks, and procedures for evolving answers to intractable optimization problems. Regarding the origin of life, Coveney and Highfield take a questionable Darwinist leap of faith in contending that self-assembly, self-replication and self-organization have the remarkable power necessary to produce vast genetic information and intricate molecular machinery:

> Drawing on the principles of chemical self-organization, we can speculate about the recipe for life to emerge. If a suitable feedback mechanism was operating in the prebiotic soup on the early earth that could account for nonlinearities, then the general conditions were ripe for self-organization. For instance, if a molecule in that soup catalyzed its own replication, then nonlinear feedback, the hallmark of self-organization, would very likely have emerged. This might have led to a breaking of the spatial uniformity of the medium, triggering patterns and rhythms . . . just as a chemical clock can show patterns in space and time. A key ingredient of the prebiotic soup then becomes a self-replicating molecule (or molecules) that catalyzes its own production. . . .
>
> It seems reasonable to speculate that the first primitive life forms were "naked" genes—nucleotide strands with the power to proliferate unaided. The enormous diversity of living matter in nature is made possible by the powers of self-assembly, self-replication, and self-organization implicit within these molecules. . . .
>
> The ever-extending range of RNA chemistry has shown that this giant molecule can hold information *and* carry out functions like a protein. . . . It does not seem too much of a leap of faith to believe that in an unrelated stew of millions of possible polymer species, a reaction network could self-organize to form an ecosystem of molecules, a kind of metabolism.[43]

These are the types of "just-so" stories that Phil Johnson and his colleagues have exposed so clearly in the past fifteen years. We could digress at length on the criticisms of the intelligent design community regarding such origin of life claims. In consideration of our space constraints, we will merely quote from two sources. First, in *Darwin's Black Box*, Michael Behe brings some reality back into RNA self-organization scenarios:

> Unfortunately, the optimism surrounding the RNA world ignores known chemistry. In many ways the RNA-world fad of the 1990s is reminiscent of the Stanley Miller phenomenon during the 1960s: hope struggling valiantly against experimental data. . . . Although a chemist can make nucleotides with ease in a laboratory by synthesizing the components separately, purifying them, and then recombining the components to react with each other, undirected chemical reactions overwhelmingly produce undesired products and shapeless goop on the bottom of the test tube. . . . Origin-of-life chemistry suffers heavily from the problem of road kill.[44]

Second, Dean Overman makes specific arguments against self-organization as the source of information for living complexity:

> All of the different self-organization theories fail because they do not present a plausible method of generating sufficient information content in the time available. . . . The paradigms for the emergence of life are algorithms which must contain at least as much information content as the genetic messages they claim to generate.[45]

Finally, in 2002 Bill Dembski published an entire monograph critiquing "free lunch" mechanisms for generating information and proposed a candidate fourth law of thermodynamics: the complex specified information in an isolated system of natural causes does not increase with time.[46]

One should not deny that complexity and chaos theory are ostensibly reasonable approaches for attempting to construct a purely naturalistic and atheistic solution to the origin of life enigma. However, a proper burden of proof must then be applied as to how complex specified information is generated. Notwithstanding the marvels of self-organization in complex systems, it is still necessary to embed enormous information in the initial conditions of the dynamics in order for processes such as embryonic development to successfully execute. Precisely quantifying and tracking the information flow in such complex systems remains a very important research problem.

Summary

Complex systems with dynamics characterized by nonlinearity and irreversibility are pervasive in nature. Mathematical chaos is also exhibited by some complex systems, but the prominence and practical impact of chaos is less certain. In its proper scientific meaning, chaos is not a state of utter confusion totally lacking in order; rather, it is the occurrence of fully deterministic processes that are hypersensitive to initial conditions. Accordingly, chaos is humanly unpredictable, but not indeterminate and not inconsistent with the notion of cosmos. Contrary to popular misconceptions, complexity and chaos theory do not establish philosophical and moral relativism, insofar as the science of complexity and chaos provides no support for the belief that "nothing is true and everything is permitted."

The existence of chaos does ineffectuate the Newtonian image of a clockwork universe and the Laplacean dream of a completely controllable world. Complexity and chaos invoke no new fundamental physics; instead, they provide new insights into the technical aspects and the philosophical meaning of classical determinism. Nonetheless, in chaotic systems in which quantum mechanical events on the molecular scale are directly connected to macroscopic outcomes, the causal links of classical determinism are broken, or in the very least muddled.

The integrative agenda of complexity theory is a countercurrent to traditional scientific reductionism. Although in tension with one another, both the complexity and reductionist approaches must work in concert to provide a balanced description of nature. The mysteries of the human brain are prime targets of investigation for complexity and chaos theory. Many scientists working in this area believe that human consciousness and decision making can be explained as exotic emergent properties ensuing from the special physical and dynamic structure of the brain. While complexity and chaos are often invoked in Darwinian explanations for the origin of life, they do not effectively address the source of genetic information in living systems.

In theological terms, chaos and complexity confine humanity to inextricable epistemic humility in relation to God. Complexity theory views the essence of life as independent of its particular physical medium, consistent with Christian belief. Chaos and complexity offer approaches for resolving conflicts between free will and determinism in human events. Moreover, chaos and complexity reveal possible mechanisms for divine providence, providing surprising avenues by which classic Christian theism may be at work in physical reality.

Epilogue

Through our collective experiences in times of peace and turmoil, we have become firmly convinced of God's providence in our lives. Fortunately, providence does not mean that Christians receive only special favors in the journey through life. We rest in faith that all events that pass our way are for a greater purpose. We are thankful that the God of Christ's love is also the God of purpose and order who superintends complexity and chaos.

PART VI

EPILOGUE

18

PHILLIP JOHNSON AND THE INTELLIGENT DESIGN MOVEMENT

Looking Back and Looking Forward

WALTER L. BRADLEY

—

I am honored to conclude this tribute to Phillip Johnson and his seminal contributions to the development of the intelligent design (ID) movement by reviewing where we have come from and where we are going.

Looking Back: Intelligent Design—The Early Years (1976-1988)

In 1976 I was invited to write a book that would critique neo-Darwinian evolution and be included in a new series that Probe Ministries (Richardson, Texas) was developing. I felt that the subject of evolution was both too large and too distant from my scientific background, so I offered instead to write on the origin of life, a critical step in the larger metanarrative of natural history that was easily assessible to me as a polymer scientist. As I was in the early stages of this research and writing project, Dr. Charles Thaxton joined Probe Ministries as a staff member and expressed an interest in joining me in this project. His background in biochemistry nicely complemented my background in polymer science and thermodynamics. As we got into the literature, it became clear that key issues in the various origin(s) of life scenarios turned on the atmospheric conditions of the early earth. We were fortunate to have Dr. Roger Olsen, a geochemist, join us to address these questions.

We made four decisions early in this research project that were critical to its success. First, we agreed to be scrupulously fair to the different theories on how life began, without torturing the data to confess what we might prefer it to say. Second, we decided to completely separate our presentation of the results of our scientific investigation from our speculation about philosophical implications, which were addressed in the epilogue. Third, we completely avoided any discussion of biblical implications, believing that this was a topic for another book and another day. Previous scientific works by "creationists" were generally not taken seriously because they were viewed as nothing but a biased polemic to support the Genesis account of creation. We wanted this work to be taken seriously by the broad scientific community. Fourth, we committed ourselves to producing a scientific treatise that would be accepted for publication by a secular publishing company, which would be crucial for us to achieve our goal for our book to be read by secular scientists. We wanted to take our place at the table in the dialogue on the origin of life.

It took us six years to complete our research and the resulting manuscript for *The Mystery of Life's Origin: Reassessing Current Theories*. Unfortunately, the year was 1982. The adverse publicity from the Arkansas trial on teaching about six-day creationism in public schools created a huge skepticism about any work that challenged the naturalistic orthodoxy, which our book certainly did. Our situation was further exacerbated by the fact that the book was clearly too technical to be of interest to Christian publishers, leaving us wondering why God would have had us invest so much time writing a book that would never be published. Fortunately, Jon Buell, who had originally invited me to write *Mystery* while he was working at Probe Ministries, had just started the Foundation for Thought and Ethics. He eagerly took up the challenge of finding us a publisher, contacting an unprecedented 176 publishers over a period of fifteen months before securing a contract with Philosophical Library to publish *Mystery*. We were delighted to get very positive jacket endorsements from Robert Jastrow, founder and former director of NASA's Goddard Institute for Space Studies, and from Robert Shapiro, a Harvard trained DNA chemistry professor at New York University. Dean Kenyon, coauthor of *Biochemical Predestination,* wrote a sterling foreword and we were ready to go, with *Mystery* finally appearing after eight long years of effort.

We were cautiously optimistic that the book would generate some heat and light in the origin-of-life scientific community, gingerly stating as it did that the "emperor had no clothes," so to speak. We were not disappointed.

Mystery was reviewed in at least six major secular journals, including a full two-page review in *Chemical and Engineering News,* one of the most widely read archival journals for chemists and chemical engineers. Happily, these reviews found relatively little of substance in the scientific arguments to criticize, confirming our contention that previously published work on the origin of life had hyped the small successes, often achieved under contrived laboratory circumstances, while ignoring or minimizing fundamental problems with the "soup theory" (the idea that life could originate spontaneously in a primordial chemical soup). The only serious criticism of *Mystery* was directed to the philosophical implications discussed in the appendixes, but these criticisms were rooted in the materialistic presuppositions that *Mystery* challenged.

People who contacted us as a result of *Mystery* included Michael Denton from Australia, who wanted to discuss the book, and Clifford Matthews, one of the most distinguished scientists in the origin of life, who ordered ten copies of *Mystery* for his graduate students. Attending Gordon Research Conferences on the Origin of Life gave us further opportunity to get feedback from researchers in the field, most of whom were generally complimentary of the book, but not necessarily of the epilogue. Several indicated that the ideas in *Mystery* were recognized by many in the scientific community, but had not been openly acknowledged in the literature because funding, which came almost exclusively from NASA, would be in jeopardy if it was generally acknowledged that the prevailing paradigm at the time, the "soup theory," was clearly false. Thus, pragmatic as well as philosophical considerations had driven the study of the origin of life down a cul-de-sac.

The publication in 1986 of Robert Shapiro's *A Skeptics Guide to the Creation of Life in the Universe* and Michael Denton's *Evolution: A Theory in Crisis* continued and greatly expanded some of the conversations begun in *Mystery,* particularly with regard to the question of information and complexity. Denton's book created a much greater controversy than our book or Shapiro's, as the prevailing orthodoxy of neo-Darwinism was and still is widely accepted as a sufficient explanation for the diversity of plant and animal life.

These significant cracks in the prevailing paradigms for the origin of life and neo-Darwinism might have gradually retreated into obscurity like fireworks on a dark evening, dazzling but short lived, had it not been for Phillip Johnson, who helped to turn on the lights permanently.

Phillip Johnson and the Blossoming of the Intelligent Design Movement (1989-2002)

Phillip Johnson brought four critical gifts to this emerging challenge to naturalistic orthodoxy. First, he has the intellectual tools of a master logician, well honed from his long and distinguished career in law. Second, he brought a brilliant rhetorical gift that allowed him to reframe the questions surrounding origins, exposing the hidden metaphysical agendas that often attend such discussions. Third, he brought a winsome personality that allowed him to establish ongoing dialogues, even cordial friendships, with prominent opponents of intelligent design. Fourth, he took a fatherly interest in many young scholars who were interested in plying their intellectual gifts in this field of endeavor, nurturing and directing them in their intellectual pursuits.

When I read an early draft that Phil sent me of *Darwin on Trial,* I was extremely impressed with the tremendous effort that he had put into understanding the many different facets of the neo-Darwinism paradigm, a huge task that I had declined in 1976. Even more impressive was his willingness to seek critiques from ardent neo-Darwinists and to adjust his ideas to those comments that were scientifically based, allowing him to reshape and refine his arguments. At the same time, his background as a logician allowed him to identify those arguments that were materialistic philosophy masquerading as science. By carefully separating the wheat from the chaff, he was able to spotlight the real basis for neo-Darwinian claims, a basis that he found wanting.

With the publication of *Darwin on Trial,* Phillip Johnson immediately became recognized as the intellectual leader of the group of scholars that God had raised up to address the question of biological origins and the troubling issue of naturalism that invariably accompanies this question. He graciously and willingly assumed this role, beginning with his organization of the Pajaro Dunes conference in 1993. His ability to energize and organize all of us who had previously labored more or less independently to begin to function as a larger organic whole was the beginning of the ID movement. Phil saw himself appropriately as the intellectual architect of this movement, sketching in broad strokes the critical questions, both scientifically and philosophically, and then depending on those with various scientific and philosophical expertises to address these questions in depth.

Phillip Johnson also saw very clearly the importance of separating the questions of the possible existence of an intelligent designer from questions associated with harmonizing the Bible and science, a distinction that we also made when we wrote *Mystery.* This was critical because they are distinctly different questions and because any reference to biblical insights, much less biblical authority, will not be well received in the scientific community, the very group with whom we wanted to dialogue. Avoiding exegetical questions about Genesis 1—3 also allowed Phil to involve a wider range of scientists and philosophers, who might disagree on the details of Genesis 1—3 but could unite around the premise that an intelligent designer did indeed create our universe, and a careful examination of the nature of nature should reveal fingerprints that confirm this hypothesis.

With occasional retreats and conferences and an Internet listserv that he supervised for daily conversations, everyone had an opportunity to work together synergistically, with the quality of everyone's individual work in the origins area benefiting greatly from their cyber colleagues. Phillip Johnson was also instrumental in securing financial support for major projects related to intelligent design and helped publicize important advances.

His subsequent books also clarified the larger stakes in the intellectual battle over intelligent design and developed strategies on how to advance the ID movement. With his book *Reason in the Balance,* he showed the pernicious impact of naturalism in many areas of scholarship and public policy, allowing everyone to see the larger cultural erosion that has resulted from naturalistic ideology. He challenged scholars in all areas to adopt a more philosophically neutral playing field that would allow the search for truth to be pursued without blinders. His book *The Wedge of Truth* laid out a simple strategy for the ID community to pursue its proper place in the marketplace of ideas. *The Right Questions* helped us to focus on the fundamental questions rather than getting bogged down on secondary or even tertiary questions. Each of these books has contributed to the intellectual architecture that Phillip Johnson has been constructing for the ID movement and has provided unifying principles both intellectually and strategically for various works in different areas of science and philosophy.

What has been the impact of the intelligent design movement? In 1986, when I first began giving lectures on university campuses on the topic "Is There Scientific Evidence for the Existence of God?" I was often challenged by angry professors in the audience who claimed that any attempt to support belief in a Creator God with scientific evidence was completely "out of bounds," no matter how compelling the evidence might be. Between 1986 and 2002 I gave this talk about 120 times to campus-wide audiences at major college campuses to more than 65,000 students. These presentations were made at Harvard, Yale, Dartmouth, Brown, Cornell, Penn, the University of Michigan, the University of Virginia, the University of Illinois, the University of Texas, Texas A&M University, the University of Wisconsin, the University of Minnesota, the University of Florida, Arizona State, the University of Colorado, the University of Washington, UCLA, USC, Stanford, the University of California, Berkeley, and many other universities. During this time, the objections from the audience about the legitimacy of the argument gradually disappeared, allowing the question and answer period to focus exclusively on the sufficiency of the evidence itself.

In 1992 *Time* magazine ran a cover story titled "What Does Science Tell Us About God?" *Newsweek* ran a similar story in 1998. There have been at least fifteen significant books relevant to intelligent design that have been published between 1986 and 2000, most written by secular authors and published by secular publishers. Each points out the many special features of our universe that seem to suggest a designer. Many of these scientists implicitly or sometimes explicitly acknowledged that the nature of nature could support (or possibly undermine) the paradigm of an intelligent designer. The adoption by the State Board of Education in Ohio of standards that require science classes to consider the weaknesses and well as the strengths of evolution in 2003 is also a milestone of note. The *New York Times* ran a respectful front-page story about intelligent design in 2001, further indicating that the movement was beginning to be taken seriously.

Slowly but surely the ID movement that Phillip Johnson has organized and leads is making an intellectual beachhead, cracking the stranglehold that materialistic and methodological naturalism have had for most of the twentieth century. It is worth noting that the beachhead is greatest in physics and least in biology. Why? Because of the difference in the persuasiveness of the evidence in these two fields and because of the differences in the academic culture in these two disciplines, with physics being the more open to new ideas.

Personal Reflections on Phillip Johnson

At a personal level, I must express to Phil Johnson my great appreciation for his friendship and encouragement over the years. I have been greatly blessed by the "band of brothers and sisters" that I have enjoyed as a result of Phil's leadership in bringing this movement together. The ID movement he has fostered has given legs and longevity to my early work in *Mystery* and has provided a much larger platform and amplifier for my later work. Phil has also given me an even greater appreciation for the importance of work in intelligent design by his framing of the right questions. I am deeply indebted to Phil for enriching my life in these ways.

Looking Forward

I would like to conclude by considering the future prospects of intelligent design. Specifically, I want to consider several challenges of which we must be mindful.

The scope of the intelligent design argument. I think most in the ID movement would agree that ID is not *just* an alternative explanation to evolution concerning the great diversity of flora and fauna. ID addresses the bigger metanarrative of natural history and the sufficient causes that can account for this drama. Yet to many people, ID is narrowly identified as nothing but antievolutionary propaganda. For example, Bill Dembski's most ardent adversaries at Baylor University claimed Kenneth Miller's *Finding Darwin's God* as the proper alternative to Dembski's position. Yet chapter eight in Miller's book ("The Road Back Home") speaks of evidence for an intelligent designer in big bang cosmology and the fine-tuning of the universe. Miller is an ardent critic of intelligent design while warmly affirming what I consider to be some of the most important parts of the ID argument. Obviously, some of our critics do not see intelligent design as this bigger metanarrative. The title of this book, *Darwin's Nemesis,* illustrates the perception, even in the intelligent design community, of intelligent design as an antithtetical alternative to neo-Darwinian evolution.

If we could go back and start over, which of course we can't, we would want to define intelligent design in the broadest possible way. Debate within the theistic community could be used to sort out which parts illustrate God working in his customary way via so-called ordinary processes and which parts of the narrative suggest God working in some extraordinary way. In such a discussion the veracity of ID does not hang on whether macroevolution can be explained by neo-Darwinism. Scientists who study cosmology and the origin of life are much more modest in their claims about the adequacy of completely naturalistic explanations to account for their observations in nature than biologists. By focusing on evolution, we have taken on the biggest challenge first. Since the study of evolution plays a much greater role in science education and in the naturalistic indoctrination that often takes place in public schools, it is understandable that this might naturally be selected as the first and biggest target. But it is also the thickest part of the target to penetrate.

I think it imperative that as a movement we reframe the ID debate to clearly include cosmology and the origin of life along with evolution as critical areas of discussion. Given the current misunderstanding here, it will require a concerted effort to redefine the debate in this way, even as we speak and write to some subset of this bigger story.

Influencing public education. The *O'Reilly Factor* on Fox News (November 7, 2004) noted the widespread attempts of ID advocates to change how evolution is taught in public schools. That such a modest proposal as the one adopted in Ohio, which required that weaknesses as well as strengths in the theory of evolution be taught, would generate such strong opposition indicates the threat that materialist see in intelligent design. As the scientific basis for intelligent design continues to grow and as the question of design is framed in the larger way outlined previously, we can be optimistic that public education will move toward a more balanced treatment of cosmology, the origin of life and evolution than the scientism that has so dominated the past half century.

The challenge of peer reviewed publications. After the publication of *Mystery* in 1984, I decided to submit some of the work that went into that book for peer review. With a friend I prepared a manuscript and submitted it to the most prestigious journal in the origin-of-life research community, *Origins of Life and Evolution of the Biosphere.* This paper showed the errors of an earlier publication from the National Academy of Sciences (NAS) that supported the idea that amino acid sequencing was nonrandom in a way that was determined by dipeptide bonding preferences. Our work, which used twenty-five protein sequences, including the ten used in the NAS article, was declined, we were told, because we did not use a sufficiently large set of proteins. Now ten was sufficiently large if the answer was "right" but twenty-five was too small a set if the answer was "wrong"—even though intelligent design was not mentioned in the paper. We were so annoyed that we went back and used every protein (more than 350) in the *Atlas of Proteins Structures.* The statistical results were now incontrovertible. Indeed, the resubmission was published in 1988 without any further comment from the editor.

A firestorm of protest accompanied the publication of Stephen Meyer's article "The Origin of Biological Information and the Higher Taxonomic Categories" in the September 2004 issue of *Proceedings of the Biological Society of Washington*. In this article, which is reprinted in this volume, Meyer explicitly advocated intelligent design as the best explanation for the advent of novel genes. The journal has promised never again to publish papers that support ID. The editor has stepped down and continues to experience great persecution for breaking the sacred but unwritten rule: namely that no paper that explicitly supports intelligent design can ever be published in a peer-reviewed journal, no matter how cogent the arguments. Notwithstanding, some of our critics have the gall to argue that ID is not credible because it is not supported in the peer-reviewed literature.

One of our great challenges is to find our place at the academic table and to be included in discussions at the frontiers of science. There is no obvious solution to this problem, but we must persist in trying to break this intellectual logjam. Otherwise, it will take a long time to affect the thinking of people who accept the neo-Darwinian paradigm.

Career paths for up-and-coming IDers. Finally, I close with a concern that I have had for the past twenty years, namely, how to counsel young scientists who are interested in ID. How public should they be in their work related to ID, especially before they have a tenured position at a major university? It is clear that some of the most outstanding young scientists in the ID movement have found great difficulty at research universities securing tenure-track positions commensurate with their abilities and training. Here we need to be "as wise as serpents and as gentle as doves." It may be advisable for young scholars who would like a traditional career track in a "Research 1 University" to have a low profile in the ID movement, at least until they receive tenure. If they can contribute in the meantime, it would be best done by publishing in journals without "connecting the dots" by directly drawing ID implications. The significance of this work can then be espoused by ID scientists who are already tenured professors and have a much greater degree of academic freedom.

Summary

It is interesting that the ID movement was born in the dark shadows of the

1982 Arkansas creationists court case and the 1987 Supreme Court decision on the Louisiana creationist case, both resounding victories for advocates of a naturalistic monopoly in the study of and teaching about science. Yet into this stiff wind, the ID movement was born and has sailed with remarkable speed. The rise of the ID movement is in many ways as surprising as the abrupt and surprising fall the Soviet Union, which occurred about the same time (but we agree was not causally connected). Who could have predicted either of these most unlikely turns of events?

The testimony of this book is that much of the credit for this remarkable achievement goes to Phillip Johnson, master logician, visionary leader, fatherly mentor and personal friend to many of us in the ID movement. We hope, Phil, that this book will bring you the intellectual stimulation and personal joy that working with you has brought those of us in the intelligent design movement.

19

THE FINAL WORD

PHILLIP E. JOHNSON

———

Reading the papers collected in this volume has been a great pleasure to me. First, the papers remind me of that wonderful gathering on April 22, 2004, when all my colleagues in the ID movement totally surprised me by luring me into the board of trustees room at Biola University's main administrative building. There I found a festive gathering at which each of the authors represented in this volume orally delivered the gist of his or her paper while I listened, entranced. It was particularly delightful to me that two of the friendly adversaries I have debated frequently during the years since *Darwin on Trial* was published, William Provine and Michael Ruse, were present and participating, although Provine later elected not to include his essay in this volume.

Of course I am pleased by the positive memories of my influence that my colleagues in the intelligent design (ID) movement recount. Even so, the important feature of this book is contained not in anything that anyone says about me but in what the breadth and quality of the papers demonstrate about the excellence of the members of the ID research community in which I am so proud to be included as a founding member. I did not persuade any of these fine men and women to become skeptical of Darwinism; they all had developed their own doubts from a profound understanding of the subject before they encountered my writings or met me in person. What I was able to provide was mainly a legal analyst's instinct for focusing attention on the most important and most vulnerable aspects of the Darwinian system, particularly that its essential foundation is in naturalistic philosophy rather than in observation or experiment. There is also the enormously exaggerated claims for the creative power of the Darwinian mechanism, and the "pea under the shell" deceptiveness with which shifting definitions of *evolution* are employed, so that the stock phrase "evolution has occurred" can mean anything from "disease pathogen populations display adaptive resistance to antibiotic drugs" to the enormous claim that "Darwinian evolution created all living species from a single-celled common ancestor, and ultimately from nonliving chemicals."

The greatest reward I have experienced during the years I have taken a leading role in the ID movement is the close association with the wonderful people represented as authors in the volume. I think particularly of Michael Behe, with his homespun wit and charm, a man who in the Darwinian sense of reproductiveness is clearly the most fit organism currently engaged in the great argument over Darwinism (his family is enormous). I have told many groups, "If you don't like Mike Behe, you just don't like people." Steve Meyer was my earliest comrade, a fast friend and soulmate ever since the fateful conversation at Cambridge University, which he recounts in his essay. Once we had established an institutional home for the movement at the Discovery Institute's Center for Science and Culture, Steve took over as the center's director, and hence as the intellectual supervisor of the work of all the center's research fellows.

In addition to his enormous contribution as supervisor and editor, Steve is the sole author of the landmark paper "The Origin of Biological Information and the Higher Taxonomic Categories," which, I am gratified to say, he has chosen to include in this volume. After passing a stringent peer-review process, Steve's paper was published in the Smithsonian Institution's *Proceedings of the Biological Society of Washington*. That publication attracted sufficient public notice that Darwinist science organizations launched a desperate and vindictive campaign to force the society to discredit the journal editor and to force the society to repudiate the paper. At the same time, Steve was receiving a staggering number of requests from scientists for reprints, thus demonstrating that the Darwinist establishment is able to exclude papers positively discussing intelligent design in biology from the peer-reviewed professional literature only by employing the most heavy-handed methods of coercion, including reckless defamation of the professional standing of any editor who approves for publication any paper deemed subversive of the Darwinist monopoly in scientific discourse.

Bill Dembski is our chief mathematician and theoretician of complex specified information and the discoverer of the explanatory filter. I some-times call him "Bill of the bulging brain and the bulging muscles," because he works out and it shows. Bill's important article in this volume wisely teaches us how to evaluate the furious backlash we are undergoing from Darwinists who are upset by the attention being paid to evidence for an in-telligent cause in biology, especially when, as is the case, many people agree that the evidence is impressive. I admit that the outrage has sometimes come as a shock to me, even though I have predicted for years that some Darwinists would become steadily angrier and more vindictive as they saw their position weakening. Bill wisely tells us not to be impressed by attempts at intimidation. In addition to writing his lovely essay, Bill planned and ed-ited this festschrift.

John Mark Reynolds is my "spiritual adopted son" and my cherished com-panion in the life of the imagination. John Mark is the founding director of Biola's Torrey Honors Institute, which I consider to be the finest undergrad-uate liberal arts program in the United States, bar none. Jonathan Wells has been a tower of strength for our movement, both for his solid scientific knowledge, especially in embryology, and for his bold imagination com-bined with good judgment.

Nancy Pearcey, who interviewed me as a journalist when our movement was young, is now one of our star authors and worldview teachers, with the publication of her monumental book *Total Truth*. Paul Nelson, whom I sometimes call the "walking encyclopedia," has been invaluable not only for the enormous store of information about evolution and embryology at his fingertips, but also for the penetrating acumen with which he directs our at-tention to the right questions. David Berlinski is valuable for similar reasons, and also because he is so independent-minded that he sometimes turns his piercing analytical scrutiny on his comrades, thus making it very difficult for us ever to become complacent.

That's the ID movement as represented in this volume. It's a collection of independent-minded individuals and also a research community tied to-gether by bonds of affection and respect. They all share an interest in freeing science from the thought control of a materialist ideology that forbids scien-tists to follow the evidence if there is a danger that the evidence will lead to knowledge of something that the ideologues don't want the world to know.

NOTES

Chapter 2: From Muttering to Mayhem: How Phillip Johnson Got Me Moving

[1] Here's a question that large families ask. What is the minimum number of children in a family where each child can say "I have brothers and sisters"? In order for a boy to say he has "brothers" (plural), there must be three boys in the family: the one speaking and at least two brothers. For a girl to say the same about sisters there must be three girls. So there must be at least six children.

[2] E. Wasman and H. Muckermann, "Evolution," in *The Catholic Encyclopedia* (New York: Encyclopedia Press, 1909), 5:654.

[3] Grace, Benedict and Clare were then followed by Leo, Rose, Vincent, Dominic, Helen and Gerard.

[4] A few years later in a discussion class I asked students what would be necessary to convince the public that evolution was untrue. One fellow said that to change his own mind, God would have to personally tell him so in a dream. I said, OK, suppose you were told that in a dream and then woke up. Now, how would you go about convincing the public at large. He thought about it for a minute and replied that, well, God should tell *everyone* in a dream. Good strategy.

[5] To *myself!* Never smile at students in a way they think is patronizing or they'll clam up for the rest of the semester!

[6] "Johnson vs. Darwin. A new antievolution book appeared on the scene last month, and from an unlikely source—a University of California law professor. Phillip Johnson, of U.C. Berkeley's Boalt Law School, claims that he is 'not a defender of creation science,' but his book, *Darwin on Trial,* was nevertheless endorsed by the Institute for Creation Research and uses many of the same arguments that its leaders use.

"Johnson admits that religion fuels his personal beef with evolution. 'There is no room [in evolutionary theory] for a life force,' he says, 'for something . . . that cannot be perceived through the tools of science.' But like the creationists he dissociates himself from, Johnson claims to 'examine the scientific evidence [for evolution] on its own terms.' In an interview, Johnson blasted evolutionary theory for being 'constantly reformulated, in the manner of Marxism, on account of the failure of its predictions to come true.' The theory should be abandoned, he argues, and scientists should admit that the origin of species can't be explained without invoking supernatural processes.

"Johnson's arguments demonstrate his misunderstanding of the scientific process, in which theories are continually tested and refined, says Eugenie Scott of the Berkeley-based National Center for Science Education. The problem, says Scott, is that Johnson is a lawyer, not a scientist. 'Theory, proof, and law are very different terms to scientists than to lawyers,' she says.

"Johnson is busy on the talk-show circuit publicizing his book, and Scott worries that his academic position and his approach will win him a wide following. 'I hope scientists find out about this. They really need to know [the book] is out there confusing the public' " (*Science,* July 26, 1991, p. 379).

[7] "Understanding evolution. The briefing about University of California, Berkeley, law professor Phillip E. Johnson's book *Darwin on Trial* (*News & Comment,* July 26, p. 379) is a good illustration of the failure of the scientific community to follow its own advice about the perennial evolution controversy. Instead of simply addressing the skeptical arguments advanced in the

book, the article relies on ad hominem remarks. It is pointed out that Johnson's religious views predispose him against naked materialism (although in his book he states that he finds nothing a priori incredible in God's using Darwinian evolution to produce life), and a science educator is trotted out to opine that Johnson misunderstands the scientific process. Johnson is also found guilty by association because Creationists like his book.

"Well, now. It is also true that fascist governments have embraced Darwinism, that most scientists are not trained logicians, and that many commentators on evolution are predisposed in favor of naked materialism. But all of this is name calling and quite beside the point. In his book Johnson appears to be an interested, open-minded, and very intelligent layman who sees large conclusions drawn from little evidence, notices anomalies in current evolutionary explanations, and will draw his own conclusions, thank you, about the validity of Darwin's theory. A man like that deserves to be argued with, not condescended to.

"The theory of evolution by natural selection is not a difficult concept to grasp, and Charles Darwin addressed *The Origin of Species* itself to a general audience. But neither is it self-evident to many people that natural selection can fully account for the world they observe. Thus when questions about the theory arise in public forums, the scientific community would do much better in the long run to patiently list supporting facts and frankly admit where positive evidence is lacking, rather than paternalistically maintaining that an understanding of the theory of evolution is reserved for the priesthood of professional scientists" (Michael J. Behe, *Science,* August 30, 1991, p. 951).

[8]One incident sticks in my mind. During a break in the symposium I was standing outside having coffee and Phil came over to chat. "So," he asked, "what church do you go to?" I was taken aback and enchanted. "That's not one of the first questions one gets from a new acquaintance in my neck of the woods," I thought to myself. "But, heck, this is Texas." When I told Phil I'm a Roman Catholic, he was a bit surprised. "I thought you were another of our evangelical friends." I commented that in Catholic thinking God could have made life by Darwinian means if he wanted, adding, "But, *did* he?" Phil smiled and said that distinction—between what God could have done and what he in fact actually did do—was one that escaped a surprisingly large number of people.

[9]I remember walking barefoot in the shallow ocean (as I was used to doing at Ocean City, Maryland, where my family vacations) and noticing I was losing feeling in my legs—the water was numbingly cold! Ocean City it wasn't. From then on I stayed either on the beach or in the house.

[10]Since its publication, the argument for intelligent design in *Darwin's Black Box* has received heavy Darwinian criticism. I have replied to those criticisms in a number of places, including the articles cited below. Most of my replies are also reproduced on the Web at <www.crsc.org>. For the criticisms of my argument, all one needs to do (as I tell audiences at lectures) is go to an Internet search engine, type in my name plus any common swear word, and you'll come up with tons of hits.

Chapter 3: How Phil Johnson Changed My Mind

[1]Del Ratzsch, "Design, Chance and Theistic Evolution," in *Mere Creation,* ed. William Dembski (Downers Grove, Ill.: InterVarsity Press, 1998), pp. 289-312; Michael Ruse, *Can a Darwinian Be a Christian? The Relationship Between Science and Religion* (Cambridge: Cambridge University Press, 2004).

[2]Charles Thaxton, Walter Bradley and Roger Olsen, *The Mystery of Life's Origins* (New York: Philosophical Library, 1984).

[3]Phillip E. Johnson, *Darwin on Trial* (Chicago: Regnery Gateway, 1991), p. 14.

[4]Ibid., p. 3.

[5]Steve Meyer and Mike Keas have recently described the various meanings of evolution quite rigorously. See their "The Meanings of Evolution," in *Darwinism, Design, and Public Education,* ed. J. A. Campbell and Stephen Meyer (East Lansing: Michigan State University Press, 2003).

[6]Marvin Lubenow, *Bones of Contention* (Grand Rapids: Baker, 1987).

[7]For a fully developed defense of an interpretation similar to this, see C. John Collins, *Science & Faith: Friends or Foes?* (Wheaton, Ill.: Crossway, 2003).

Chapter 4: Putting Darwin on Trial: Phillip Johnson Transforms the Evolutionary Narrative

[1]For the details of these interactions, see both the text and endnotes of chapter four in my study, *Doubts About Darwin* (Grand Rapids: Baker, 2003).

[2]In fact, the last chapter of his *Evolution: A Theory in Crisis* (Bethesda, Md.: Adler and Adler, 1986), where this quote is found, is peppered with such stigma wording.

[3]Thomas Woodward, *Doubts About Darwin* (Grand Rapids: Baker, 2003), p. 83.

[4]*Pseudoscience* is the strongest stigma word that Johnson employs in *Darwin on Trial,* and not only does it appear in the title and text of his closing chapter (in the first edition of 1991) but the concept is implied throughout the text.

[5]Phillip Johnson, *Darwin on Trial* (Chicago: Regnery Gateway, 1991), p. 14 (emphasis added). His comment on the bias of scientific creationists is one of several distancing statements that rhetorically separate Johnson from creation science.

[6]Even Gould used the *fundamentalist* stigma word on strict Darwinists Dennett and Dawkins (*New York Review of Books* [June 12 and June 26, 1997; part one was titled "Darwinian Fundamentalism"). On this, see Phillip Johnson, "The Gorbachev of Darwinism," in *Objections Sustained* (Downers Grove, Ill.: InterVarsity Press, 1998). Language critic Kenneth Burke viewed this sort of symbolic linkage as a classic case of "perspective by incongruity." Our pious connotations of a symbol are dislodged "by violating the 'proprieties' of the word in its previous linkages" (see Burke's *Permanence and Change* [Indianapolis: Bobbs-Merrill, 1935], pp. 88-92).

[7]To clarify the prepositions: (1) "between" is obvious, as in "between two extremes at the ends of the ideological range of possibilities"; (2) "above" is intended to picture his analytical stance of detachment—above and beyond the inherent constraining effect of commitments or biases of those two kinds of fundamentalism.

[8]This is from Denton's letter to M. Brockey at Regnery Gateway, the publisher. This quote was used verbatim on the back cover of the 1991 hardcover edition. This blurb, with its visibility, possibly did more to spark projection themes of "devastating attack" and "embarrassment of Darwinism" than anything Johnson said in the book.

[9]Stephen Jay Gould, "Impeaching a Self-Appointed Judge," *Scientific American,* July 1992, pp. 118-22.

[10]Ibid.

[11]Ibid. (emphasis added).

[12]This information comes from a series of interviews by phone, from January 1990 through June 1991.

[13]Johnson, *Darwin on Trial,* p. 116.

[14]Norman Macbeth, *Darwin Retried* (Boston: Gambit, 1971); Pierre Grassé, *Evolution of Living Organisms* (Toronto: Academic Press, 1981); Hubert Yockey, articles in *Journal of Theoretical Biology* (1977, 1981); Michael Pitman, *Adam and Evolution* (Grand Rapids: Baker, 1984); Robert Augros and George Stanciu, *The New Story of Science* (Warner, N.H.: Principle Source, 1984), and *The New Biology* (Warner, N.H.: Principle Source, 1987); Thomas Bethell, "Agnostic Evolutionists," *Harper's,* February 1985. Robert Shapiro, while he wrote as an honest evolutionist, nevertheless added to skepticism of current theories of abiogenesis through his book *Origins: A Skeptic's Guide* (New York: Basic Books, 1985).

[15]This oft-quoted assertion is in chapter one of Dawkins's *The Blind Watchmaker* (New York: W.W. Norton, 1985).

[16]In most Western secular universities, the naturalistic origin of the biosphere has been viewed for well over one hundred years as virtually *indisputable fact,* and this belief has had a con-

straining effect on the vast majority of teaching in the sciences that connects in any way to evolutionary issues. This constraining influence is roughly as powerful, though perhaps to a slightly lesser extent, in the arts and humanities. Such Darwinian assumptions marked out the "appropriate directions" of research and teaching for all persons and guilds in the sciences and humanities.

[17]Furthermore, this framework of understanding the drama and the issues debated therein is also clearly part of the emotional and motivational frameworks of the actors and participants themselves.

[18]From Johnson's letter to Alvin Plantinga, December 6, 1990.

[19]Johnson says, "We call our strategy 'the wedge.' A log is a seeming solid object, but a wedge can eventually split it by penetrating a crack and gradually widening the split. In this case the ideology of scientific materialism is the apparently solid log. The widening crack is the important but seldom-recognized difference between the facts revealed by scientific investigation and the materialist philosophy that dominates the scientific culture. What happens when the facts cast doubt on the philosophy? . . . My own books . . . represent the sharp edge of the wedge. I had two goals in writing those books and in pursuing the program of public speaking. . . . First, I wanted to make it possible to question naturalistic assumptions in the secular academic community. Second, I wanted to redefine what is at issue in the creation-evolution controversy so that Christians, and other believers in God, could find common ground in the most fundamental issue—the reality of God as our true Creator." This is from Phillip Johnson, *Testing Darwinism* (London: Inter-Varsity Press, 1997), p. 92.

[20]This goal is made visible throughout Johnson's *Reason in the Balance* (Downers Grove, Ill.: InterVarsity Press, 1994), which forthrightly advocates and defends "theistic realism" as the most robust and fruitful metaphysical basis for research in all fields of the modern university.

[21]Johnson writes, "[Darwinists'] scientific colleagues have allowed them to get away with pseudo-scientific practices primarily because most scientists do not understand that there is a difference between the scientific method of inquiry, as articulated by Popper, and the philosophical program of scientific naturalism. *One reason that they are not inclined to recognize the difference is that they fear the growth of religious fanaticism if the power of naturalistic philosophy is weakened*" (*Darwin on Trial*, p. 156, emphasis added).

[22]The number of such university-based conferences is astonishing. The highlights of this phase of ID's story—rhetorical engagement with opponents in university conferences—would merit an entire book-length study. The most important conference, early on, was the "Darwinism Symposium" at Southern Methodist University in March 1992. For details, see *Darwinism: Science or Philosophy?* ed. Jon Buell and Virginia Hearn (Dallas: Foundation for Thought and Ethics, 1993).

Perhaps the most important occasion of engagement was "The Nature of Nature," a conference at Baylor University, organized by William Dembski in April 2000. The conference subtitle was "An Interdisciplinary Conference on the Role of Naturalism in Science." Speakers included Nobel laureates Steven Weinberg and Christian de Duve, MIT cosmologist Alan Guth, Cambridge paleontologist Simon Conway Morris, historian of science Ronald Numbers, philosophers Alvin Plantinga and John Searle, Michael Behe, Mark Ptashne, Horace Freeland Judson, and Everett Mendelsohn. The overview statement reads:

Is the universe self-contained or does it require something beyond itself to explain its existence and internal function? Philosophical naturalism takes the universe to be self-contained, and it is widely presupposed throughout science. Even so, the idea that nature points beyond itself has recently been reformulated with respect to a number of issues. Consciousness, the origin of life, the unreasonable effectiveness of mathematics at modeling the physical world, and the fine-tuning of universal constants are just a few of the problems that critics have claimed are incapable of purely naturalistic explanation. Do such assertions constitute arguments from incredulity—an unwarranted appeal to ignorance? If not, is the explanation of such phenomena beyond the pale of science?

Is it, perhaps, possible to offer cogent philosophical and even scientific arguments that nature does point beyond itself? The aim of this conference is to examine such questions.

For a set of evaluations of this conference, see <www.arn.org>.

[23]This refers to the major research universities of the United States and Europe, with their related professional associations, technical journals and secular publishers.

[24]See two chapters in Johnson's *Reason in the Balance:* "The Subtext of Contempt" (the Philip Bishop story) and "Is God Unconstitutional?" (the story of Dean Kenyon). These cases describe what happens when a professor lets his "flag of personal faith" be shown or when he or she shares the problems with the naturalistic view on the origin of life. (See Johnson, *Reason in the Balance.*)

[25]The preceding sentences are based on Johnson's remarks as he described the situation in his talk at the February 1990 meeting of the Ad Hoc Origins Committee in Portland. For a further description of this meeting and its historical importance, see chapter four of *Doubts About Darwin.*

[26]For a helpful elaboration on Johnson's view of the secular trends in American universities and how things stand now, see his chapter "Education" in *Reason in the Balance,* the sequel to *Darwin on Trial.*

[27]It may help balance things to share some observations. As a Christian theist I experienced freedom in graduate seminars to offer criticisms from my perspective and to raise questions about philosophical assumptions I saw as dubious and vulnerable. The welcome to the table I experienced in doctoral studies (and an openness to a theistic perspective) may be due to a deep and enduring (historic) resistance within the fields of communication and rhetoric to the impulses of scientific positivism in the modern era. For example, the works of Kenneth Burke typify the critique of two related scientific syndromes—reductionism and the prevalent epistemic hubris of "scientism."

Rarely have I observed rhetoricians marginalize the issue of the existence of God as nonrhetorical. One rare example (with my reactions in parentheses) was in a communication course. We heard a presentation by a famous professor, who explained when a situation is rhetorical. The professor pointed out that "How many are in the room?" was *not rhetorical;* one could count and know instantly. No persuasion needed. (Agreed!) This professor then said that one moves into the truly *rhetorical realm of the uncertain*—an "exigence" or "imperfection marked by urgency"—for example, with the advisability of a proposed bill in Congress. (No problem.) On the far end of the scale the professor then placed the question *"Does God exist?"* as so *radically indeterminate* that the question was inherently "nonrhetorical." (What? That's a highly rhetorical question!) I found this very revealing, and my mind raced. The professor's use of the "existence of God" as a nonrhetorical issue fit with contemporary intellectual fashion in universities, where the truth of monotheism is seen as equivalent in subjectivity to the question, "What is the best flavor of ice cream?" (Incidentally, the flavor question would have been *my chosen example* of a fairly indeterminate, hence "nonrhetorical," situation, although aesthetic questions do enter into the realm of rhetoric as well.) I assume the professor meant that belief in God was nonrhetorical because (1) it is an *inescapably private matter,* rooted in one's personal life narrative, or (2) it is *unconnected* with philosophical, historical and scientific issues. I totally disagree. Note how Dawkins's comment on Darwinism as that which made it possible to be an "intellectually fulfilled atheist" makes clear how closely theism is tied in with other issues that can be rhetorically evaluated. Would not the debates in American universities over the case for or against the existence of God, by philosopher William Lane Craig and others, serve as defeaters of this classification? *Rhetoric is as rhetoric does.*

[28]See Behe's *Darwin's Black Box* (New York: Free Press, 1996). It is one long argument for the intelligent design of irreducibly complex biochemical systems in the cell. Dembski's works provided powerful theoretical support for Behe, especially his book *The Design Inference*

(Cambridge: Cambridge University Press, 1998). Other leading advocates for a new paradigm are Steven Meyer, Paul Nelson and Jonathan Wells.

Johnson did not conceal his belief in design. In chapter 1 of *Darwin on Trial* he identified his theistic bias—and his acceptance of creation in a broad sense. Yet he never built a detailed case for the creation of life. He comes closest when he points out that DNA, RNA and proteins are described with words of human communication like "messages, programmed instructions, languages, information, coding and decoding, libraries" (p. 112). He then asks why Darwinists do not consider the possibility that life is what it "so evidently seems to be—the product of creative intelligence?"

[29]I should add *falsification* to the word *reassessment* (see Johnson, *Darwin on Trial,* pp. 154-55). I see the complete dominance of Popper in chapter 12 as a key both to Johnson's overall rhetorical strategy and to his "bottom line" action that he invites the reader to initiate. In regard to this, just as a speech or sermon ideally issues a challenge to action at the conclusion, Johnson urges the reader to enter a rigorous testing of Darwinism à la Popper.

[30]Within the circle of scientists, evolutionists and those in related fields would be pivotal yet the hardest to influence. In the next circle would be university professors outside the natural sciences. Further out is the university-educated public, especially those in positions of leadership (e.g., publishers, educators, the media) who could exert influence as they are led into Darwinian doubt. The final circle would be laypeople with the educational background to appreciate Johnson's lean but eloquent prose in his moderately college-level vocabulary.

[31]E-mail letter from Johnson, January 2001.

[32]These words have been uttered often by Johnson when asked about his strategic goals in speaking on campuses.

[33]Attitudes and emotions toward creationism are visible in many essays in *Scientists Confront Creationism*. See Richard Lewontin's "Introduction": "The recent massive attack by fundamentalist Christians on the teaching of evolution in the schools has left scientists *indignant and somewhat bewildered.* Creationist arguments have seemed to them a compound of ignorance and *malevolence,* and, indeed, there has been both confusion and *dishonesty* in the creationist attack" (*Scientists Confront Creationism,* ed. Laurie Godfrey [New York: W. W. Norton, 1984], p. xxiii, emphasis added).

[34]Johnson's letter to Alvin Plantinga, December 6, 1990.

[35]Such attitudes and feelings were often linked to noble values. A scientist might bristle with indignation when convinced that his creationist opponent was lying or was willfully ignoring certain scientific evidence.

[36]This is one side of the creationist obstacle that Michael Denton never had to face.

[37]This assessment comes from my survey of a dozen published or recorded interviews with Johnson during the 1990s.

[38]This is the wording used by David Raup in describing Johnson's competence (see earlier in this chapter where I discuss Raup's endorsement of the accuracy of Johnson's factual descriptions).

[39]Typical is a review of Harvard astronomer Owen Gingerich, *Perspectives on Science and Christian Faith* 44 (1992): 140-42. Gingerich was an attendee at Campion, and he noted Johnson's "covetably sharp pen" and his knack at turning a clever phrase. (Gingerich quoted a few of his favorites.) Henry Morris, the creationist author and lecturer, wrote, "Phil Johnson has a remarkable gift of pithy expression, probably unmatched by any other writer in the creation/evolution field, and approached only by Stephen Jay Gould" (personal letter, July 1, 1991). Even Ruse (*The Evolution Wars: A Guide to the Debates* [Piscataway, N.J.: Rutgers University Press, 2000]) compared Johnson's writing flair with Gould's. From my survey of sixty reviews, it seems that only Gould ("Impeaching a Self-Appointed Judge") criticized Johnson's writing—or at least his style of chapter transitions.

[40]His narrative arguments work together in a way somewhat analogous to Denton's approach, and yet in significant ways Johnson's arguments are different from Denton's. One of these

ways is Johnson's narrative emphasis on Gould and the entire development of punctuated equilibrium and the other controversies that took place in the 1980s.

[41]Walter Fisher subjected stories to two tests: *narrative probability* ("what constitutes a coherent story") and *narrative fidelity* ("whether the stories . . . ring true with stories they know to be true in their lives") (see Fisher in Carl R. Burgchardt, *Readings in Rhetorical Criticism* [State College, Penn.: Strata, 1989], pp. 64-66). In Johnson's attacks on Darwinism's genesis story, it is *coherence*—narrative probability—that is held up to scrutiny.

[42]By "act" I do not imply that Johnson favors a "single act" or even a "single literal week" view of creation. I use the word *act* in the collective or complex-unity sense, as a "setting up," in which one may remain agnostic about the details.

[43]Johnson, *Darwin on Trial*, p. 14.

[44]Ibid., p. 112, emphasis added.

[45]Johnson says he has no quarrel with the conventional time scale for the appearance of different types of animals. Creation in scattered events over eons, the implied position of Johnson, is called "progressive creation."

[46]I see two levels of narrative—*factual stories* and *projection themes*. Johnson uses much factual narrative, which must be "true to the facts" to persuade. (For example, my quotations of Gould or Raup at Campion would be stripped of virtually all power if I said that I had invented them, and even more if I said that I had even invented the meeting.) Yet if a rhetor signals that he is extending upward from factual narrative into *quasi-imaginative* projection themes, not only are these stories enjoyed, they can catch fire and "chain out" (be repeated and further developed by others). As complex rhetorical symbols, projection themes enjoy vast power. Thus the factual and projection narratives live in vibrant symbiosis, without melding their distinct function and force or defining traits. My theory is that projection themes in paradigm-crisis discourses (e.g., Phil Johnson's Darwinian battleship that is leaking) have plausibility and rhetorical power to the degree the author can first build an underlying structure of narrative arguments, using factual narratives to convince the reader that the paradigm's problems really are *severe and fundamental*, justifying the projected story or image of the coming revolution.

[47]There is a two-way relationship. Personal/incident stories *power* the three types of narrative argument, but these arguments, in turn, *make sense of* personal narratives. The vesting of meaning goes both directions. Thus the Wistar story would appear to be an anomalous but trivial incident in the Darwinian HEB (and likely would not even be mentioned), but in the new-HOS narrative, it is important and highly meaningful; it makes perfect sense.

[48]Master narratives are either the evolutionary story or the history of Darwinism or any of *their key stories*.

[49]This story could also be described as a "controversy story" that draws macroevolution into the swirl of some substantive controversy. Many of Johnson's personal or incident narratives can be described as controversy stories.

[50]Gould makes exactly this point, including a reference to media coverage of Johnson ("Impeaching a Self-Appointed Judge," opening section).

[51]See Timothy Ferris's *The Whole Shebang* (New York: Touchstone, 1997), filled with marvelous narrative, and Robert Jastrow's *God and the Astronomers*, 2nd ed. (New York: W. W. Norton, 1992), which uses a narrative approach to teach the discoveries leading up to the big bang theory.

[52]Here are some of Johnson's key narratives: In *Reason in the Balance*, three scientific-legal stories are the Dean Kenyon case (pp. 29-30), the Phillip Bishop case (pp. 173-78, 181-82), and the Henry F. Schaeffer incident (p. 179). He tells a scientific-theological story in the Nancey Murphy interaction (pp. 97-102) and educational stories in the Carper-Sears solution and the Provine-Johnson cooperation (pp. 185-91). In *Testing Darwinism*, Johnson devotes a chapter to the story of "Inherit the Wind" (pp. 24-34) and the Danny Phillips incident (pp. 34-36). His last chapter is a comparison of the life stories of three men: John Shelby Spong, Charles Tem-

pleton and Billy Graham. Many essays in *Objections Sustained* (Downers Grove, Ill.: InterVarsity Press, 1998) are biographical or focus on incidents in the life of a key figure, such as T. H. Huxley, Paul Feyerabend and Gould. *The Wedge of Truth* (Downers Grove, Ill.: InterVarsity Press, 1999) is full of anecdotes. For example, chapter one is Johnson's commentary on a true, published autobiographical piece on Phillip Wentworth's losing his faith in the transition from high school to Harvard. The core of chapter two, "The Information Quandary," is a debate between Paul Davies and Christian de Duve at a science conference in Italy. "The Kansas Controversy" (chap. 3) is narrative with commentary.

Chapter 6: It's the Epistemology, Stupid!

[1] Hadley Arkes, *Natural Rights and the Right to Choose* (New York: Cambridge University Press, 2002), pp. 6-7.

[2] Ibid., pp. 6-7, quoting and citing John Nicolay and John Hay, *Abraham Lincoln: A History* (New York: Century, 1886), 7: 278.

[3] Arkes, *Natural Rights,* p. 7.

[4] Scientific materialism, which is sometimes employed interchangeably with the terms *naturalism, philosophical naturalism, ontological materialism* or *materialism,* is the view that the natural universe is all that exists, and all the entities in it can be accounted for by strictly material processes without resorting to any designer, creator or nonmaterial entity as an explanation or cause for either any aspect of the natural universe or the universe as a whole. Thus if science is the paradigm of knowledge (as is widely held in our culture), and it necessarily presupposes methodological naturalism, then ontological materialism is the only worldview for which one can have "knowledge."

[5] *Segraves* v. *California Board of Education,* No. 278978 (Cal. Super. Ct. June 12, 1981). See Peter Gwynne et al., " 'Scopes II' in California," *Newsweek,* March 16, 1981, p. 67; and Kenneth M. Pierce, "Putting Darwin Back in the Dock: 'Scientific' Creationists Challenge The Theory of Evolution," *Time,* March 16, 1981, p. 80. A special thank you to one of my M.A. students at Trinity International University, Daniel J. DeWitt, who was able to track down the text of the Segraves case on the Internet.

[6] Gwynne, "Scopes II," p. 67.

[7] *Segraves* v. *California Board of Education,* n.p.

[8] Bill Carlson, "Antidogmatism Policy: But What About Their Rights?," EducationNews.org, April 13, 2002 <www.educationnews.org/antidogmatism_policy_but_what_ab.htm>.

[9] Alvin Plantinga, "Creation and Evolution: A Modest Proposal," in *Intelligent Design Creationism and Its Critics: Philosophical, Theological, and Scientific Perspectives,* ed. Robert T. Pennock (Cambridge, Mass.: MIT Press, 2001). For a reply to this essay, see Robert T. Pennock, "Reply to Plantinga's 'Modest Proposal,' " in *Intelligent Design Creationism.*

[10] See Alvin Plantinga, "Is Belief in God Rational?" in *Rationality and Religious Belief,* ed. C. F. Delaney (Notre Dame, Ind.: University of Notre Dame Press, 1997), pp. 7-27.

[11] Alvin Plantinga, *Warrant and Proper Function* (New York: Oxford University Press, 1993), pp. 216-37. For a collection of essays specifically dealing with Plantinga's argument, see James K. Beilby, ed., *Naturalism Defeated? Essays on Plantinga's Evolutionary Argument Against Naturalism* (Ithaca, N.Y.: Cornell University Press, 2002). See also Victor Reppert, *C. S. Lewis's Dangerous Idea: In Defense of the Argument From Reason* (Downers Grove, Ill.: InterVarsity Press, 2003).

[12] *Segraves* v. *California Board of Education,* n.p.

[13] Stephen Jay Gould, "Nonoverlapping Magisteria," *Natural History* 106 (1997): 16.

[14] *Segraves* v. *California Board of Education,* n.p.

[15] *Segraves* v. *California Board of Education,* n.p.

[16] State Board of Education, *Science Framework for California Public Schools Kindergarten Through Grade Twelve* (hereafter *CSF*), p. 5 (adopted February 6, 2002).

[17] It does, however, mention "divine creation" (see *CSF,* p. 5).

[18]See, for example, Phillip E. Johnson, *Darwin on Trial* (Chicago: Regnery Gateway, 1991), pp. 140-44.

[19]*CSF,* p. 5

[20]Ibid.

[21]Ibid.

[22]Ibid.

[23]*Freiler* v. *Tangipahoa Parish Board of Education* 185 F.3d 337 (5th Cir. 1999) *(Freiler II), reh'g denied en banc,* 201 F.3d 602 (5th cir. 2000) *(Freiler III), cert. denied,* 530 U.S. 1251 (2000) (mem.) *(Freiler IV).*

[24]*Freiler* v. *Tangipahoa Parish Board of Education* 97 F. Supp. 819 (E.D. La. 1997) *(Freiler I).*

[25]This portion of the sentence is a mistaken quote by the Freiler court. The original resolution reads "to form his/her own opinion *or* maintain beliefs taught by parents" rather than what the Freiler court quoted, "to form his/or opinion *and* maintain beliefs taught by parents" (emphasis added). One could argue that whether the resolution says "or" or "and" makes a significant difference. The court, in a denial of rehearing en banc, argued that "the improper substitution of 'and' for 'or' does not affect the outcome of the case." *(Freiler III,* 201 F.3d, 603). The dissent, however, thought it was important: "The first-purpose-is-a-sham conclusion is unwarranted. As noted, the panel misquoted the following portion of the disclaimer: 'it is the basic right and privilege of each student to form his/her opinion or [not "and," as the panel opinion mistakenly quoted] maintain beliefs taught by parents on [the] . . . matter of the origin of life and matter.' This mistaken reading of the disclaimer as conjunctive, rather than disjunctive, perhaps explains why the panel discounted the disclaimer's clear message that, concerning the origin of life and matter, students are free to either maintain their current beliefs, including those taught by their parents, or to form their own, new, independent opinion" *(Freiler III,* 201 F.3d, 605 [Barksdale, J., dissenting]).

[26]*Freiler II,* 185 F.3d, 346.

[27]In *Lemon* v. *Kurtzman* 403 U.S. 602 (1971), the Supreme Court contended that the Lemon test is based on the history of the Court's decisions on the matter of church and state. Thus if a challenged policy or law passes this test, it is constitutional. However, it need only fail one prong of the test in order to be declared unconstitutional:

> Every analysis in this area [church-state cases] must begin with consideration of the cumulative criteria developed by the Court over many years. Three such tests may be gleaned from our cases. First, the statute must have a secular legislative purpose; second, its principle or primary effect must be one that neither advances nor inhibits religion. Finally, the statute must not foster "an excessive government entanglement with religion" *(Lemon,* 403 U.S. at 612-13 [citations omitted]).

It should be noted that some post-Lemon opinions by Supreme Court justices have criticized and questioned certain aspects of the Lemon test. See, e.g., *Edwards,* 482 U.S. at 636-37 (Scalia dissenting) (criticizing the "purpose" prong of the Lemon test); *Wallace* v. *Jaffree,* 472 U.S. 38, 112 (1985) (Rehnquist dissenting) (arguing that the Lemon test is "a constitutional theory [that] has no basis in the history of the amendment it seeks to interpret, it is difficult to apply, and yields unprincipled results"); *Lynch* v. *Donnelly,* 465 U.S. 668, 687-94 (1984) (O'Connor concurring) (suggesting an "endorsement test."); *Marsh* v. *Chambers,* 463 U.S. 783 (1983) (upholding "the Nebraska Legislature's practice of opening each legislative session with a prayer by a chaplain paid by the State," but not applying the Lemon test); *Mueller* v. *Allen,* 463 U.S. 388 (1983) (upholding Minnesota's policy that allowed taxpayers to deduct from gross income actual expenses incurred for textbooks, tuition and transportation for dependents attending elementary and secondary schools whether public or nonpublic, maintaining that Lemon is settled law but is "no more than a signpost"); and *Meek* v. *Pittenger,* 421 U.S. 349, 374 (1975) (Brennan dissenting) (finding a fourth prong to the Lemon test: "four years ago, the Court, albeit without express recognition of the fact, added a significant fourth factor to the test: 'A broader base of entanglement of yet a different character is presented by the divisive political potential of these state programs'").

[28]*Freiler II,* 185 F.3d, p. 344.

[29]Ibid.

[30]The district court, in contrast, ruled that the resolution does not pass Lemon's first prong. *See Freiler I,* 97 F. Supp., pp. 826-30.

[31]*Freiler II,* 185 F.3d, p. 345.

[32]Ibid., pp. 344-45.

[33]Ibid., p. 345.

[34]Ibid.

[35]Ibid., pp. 344-45.

[36]The following is a brief summary of the three most important cases that have shaped the current law under which school boards must make their judgments. In *Epperson* v. *Arkansas,* 393 US 97 (1968), the Supreme Court struck down an Arkansas law that was similar to the anti-evolution statute upheld in the Scopes Trial. The Court held that the statute "must be stricken because of its conflict with the constitutional prohibition of state laws respecting an establishment of religion or prohibiting the free exercise thereof." The Court concluded that the statute proscribed evolution solely because it is inconsistent with the creation story in the book of Genesis. Thus, the statute had no secular purpose.

Opponents of evolution, in the face of *Epperson,* developed a "balanced treatment" approach, a strategy that resulted in the crafting of statutes that required a balanced treatment in public schools between evolution and "creation science," a particular religious doctrine, transparently derived from a literal reading of the first chapters of the book of Genesis though portrayed by its proponents as a scientific alternative to evolution.

Balanced-treatment acts in Arkansas and Louisiana were struck down as unconstitutional by a federal district court in *McLean* v. *Arkansas* 529 F. Supp. 1255 (1982) and the U.S. Supreme Court in *Edwards* v. *Aguillard* 482 U.S. 578 (1987), respectively. Although the statutes were not identical, they were similar. And the reasoning on which the courts based their judgments were similar as well. The courts held that the real purpose of the acts was to advance a particular religious viewpoint, creation science, and thus they violated the Establishment Clause. Four issues dominated the analysis in both cases: (1) the statute's historical continuity with *Scopes* as well as the creation-evolution battles throughout the twentieth century,(2) how closely the curricular content required by the statute parallels the creation story in Genesis, and/or whether the curricular content prohibited or regulated by the statute is treated as such because it is inconsistent with the creation story in Genesis, (3) the motives of those who supported the statute in either the legislature or the public square; and (4) whether the statute was a legitimate means to achieve appropriate state ends.

[37]*Freiler II,* 185 F.3d, p. 346.

[38]Ibid.

[39]Ibid. (footnote omitted).

[40]*Freiler II,* 185 F.3d, p. 346 n. 4.

[41]The school board did argue that the disclaimer's mention of "Biblical version of Creation" is merely illustrative. But the court did not buy it: "[T]he record does not comport with the School Board's characterization of its reason for including 'Biblical version of Creation' in the disclaimer. When the School Board debated the propriety of the proposed disclaimer, a member suggested deleting the reference to the Biblical version of creation. The Board ultimately rejected that suggestion, apparently not because doing so might confuse students who needed an illustrative reference, but because doing so would, in the words of the disclaimer's sponsor, 'gut . . . the basic message of the [disclaimer]' " (ibid.).

[42]Edward McGlynn Gaffney Jr., "Critical Thinking Prohibited," *First Things* 112 (2001), §9 <www.firstthings.com/ftissues/ft0104/opinion/gaffney.html>.

[43]*Freiler II,* 185 F.3d, pp. 341-42, 345, 346 n. 4

[44]Gaffney, "Critical Thinking," §9.

[45]See, for example, Alvin Plantinga and Nicholas Wolterstorff, eds., *Faith and Rationality: Reason and Belief in God* (New York: Oxford University Press, 1984).

[46]See, for example, M. Drew DeMott, "Freiler v. Tangipahoa Parish Board of Education: Disclaiming the Gospel of Modern Science," *Regent University Law Review* 13.2 (2000-2001); Gaffney, "Critical Thinking."

[47]See Francis J. Beckwith, *Law, Darwinism & Public Education: The Establishment Clause & the Challenge of Intelligent Design* (Lanham, Md.: Rowman & Littlefield, 2003); Francis J. Beckwith, "Science and Religion 20 Years After *McLean v. Arkansas*: Evolution, Public Education, and the Challenge of Intelligent Design," *Harvard Journal of Law & Public Policy* 26, no. 2 (2003): 456-99; and Francis J. Beckwith, "Public Education, Religious Establishment, and the Challenge of Intelligent Design," *Notre Dame Journal of Law, Ethics, and Public Policy* 17, no. 2 (2003): 461-519

[48]See, for example, Stephen L. Carter, "Evolutionism, Creationism, and Teaching Religion as a Hobby," *Duke Law Journal* (1987).

[49]*Freiler II*, 185 F.3d, p. 346.

[50]National Academy of Sciences, *Science and Creationism: A View for the National Academy of Sciences* (1984), pp. 5-6, available at <http://stills.nap.edu/books/030903440X/html>.

[51]"The idea that belief in a creator and acceptance of the scientific theory of evolution are mutually exclusive is a false premise and offensive to the religious views of many. . . . Dr. Francisco Ayala, a geneticist of considerable renown and a former Catholic priest who has the equivalent of a Ph.D. in theology, pointed out that many working scientists who subscribed to the theory of evolution are devoutly religious" (*McLean*, 529 F. Supp., p. 1266 n. 23). "Although the subject of origins of life is within the province of biology, the scientific community does not consider origins of life a part of evolutionary theory. The theory of evolution assumes the existence of life and is directed to an explanation of how life evolved. Evolution does not presuppose the absence of a creator or God and the plain inference conveyed by Section 4 [of the statute] is erroneous" (*McLean*, 529 F. Supp., p. 1266 [note omitted]).

[52]*Peloza* v. *Capistrano Unified Sch. Dist.*, 782 F. Supp. 1412 (C.D. Cal. 1992) *("Peloza I"), aff'd in part, Peloza* v. *Capistrano Unified Sch. Dist.*, 37 F.3d 517 (9th Cir. 1994) *("Peloza II")* (maintaining that a public school does not violate the Establishment Clause if it requires teachers to teach evolution, for it is not a "religious belief" because it is not defined as such in the dictionary or in Establishment Clause case law, and it does not explicitly deny the existence of a Creator). According to the 9th Circuit, evolution "has nothing to do with whether or not there is a divine Creator (who did or did not create the universe or did or did not plan evolution as part of a divine scheme)" (*Peloza II*, 37 F.3d, p. 521).

[53]According to the U.S. Supreme Court, courts may not resolve questions of the truth of religious beliefs. See *U.S.* v. *Ballard*, 322 U.S. 78 (1944).

[54]"It is not within the judicial ken to question the centrality of particular beliefs or practices to a faith or the validity of particular litigants' interpretations of [their] creeds" (*Hernandez* v. *Commissioner*, 490 U.S. 680, 699 [1989]).

Chapter 7: Cutting Both Ways

[1]In a recent fundraising letter for the National Center for Science Education (NCSE), an organization dedicated to "Defending the teaching of evolution in public schools," executive director Eugenie Scott claimed the NCSE "monitored and reacted to antievolutionist *[sic]* activities" in twenty-nine of the fifty states (Eugenie Scott, February 2003 NCSE fundraising letter).

[2]For example, Germany (U. Kutschera, "Designer Scientific Literature," *Nature* 423 (2003): 116 and Great Britain (Tania Branigan, "Dawkins Attacks 'Educational Debauchery' of Creationist Schools," *The Guardian* [London], April 29, 2003, Guardian Home Pages, p. 6).

[3]Bruce Chapman, 2004 quoted in "Ohio Votes 13-5 to Approve Lesson Plan Critical of Evolution," Discovery Institute, March 9, 2004 <www.discovery.org/scripts/viewDB/index.php?command=view&id=1898&program=News&callingPage=discoMainPage>.

[4]Personal communication I had with Jonathan Wells, March 26, 2004.

[5]A summary of Stephen Meyer's position can be found in Stephen C. Meyer, "Teach the Controversy," *Cincinnati Enquirer,* March 30, 2002 <www.discovery.org/scripts/viewDB/index.php?program=CRSC&command=view&id=1134>.

[6]For example, in the journal *Science,* the new lesson plan was called "a creationist-promoted biology lesson plan" and scientists were said to be calling the plan "creationist-inspired." In both cases, the impression given is that creationism is somehow being introduced as an alternative to Darwinism. Not only was this not the case, the only people who seemed to be obsessing about creationism were those who opposed the new standards, not those who were advocating for them (C. Holden, ed., "Disappointing News From Ohio in Random Samples," *Science* 303 [2004]: 1761).

[7]The Establishment Clause of the First Amendment to the United States Constitution reads, "Congress shall make no law respecting an establishment of religion."

[8]The term *Christian* is claimed by many different groups with a surprisingly wide spectrum of beliefs. In this paper *Christian* refers to those who subscribe to the propositional truth of the Bible when it claims in Genesis 1—2, Exodus 20 and 31, John 1 and many other places that the earth was organized to support life and that life was created on it by God.

[9]Phillip E. Johnson, *The Right Questions: Truth, Meaning and Public Debate* (Downers Grove, Ill.: InterVarsity Press, 2002).

[10]Jeffrey Jordan, "Evangelicals and Science Education," in *Science and Religion in the Context of Science Education,* ed. J. Kittleson et al. (National Association for Research in Science Teaching annual meeting symposium, Philadelphia, Penn., 2003), pp. 17-19.

[11]For one example, see William A. Dembski, "Science and Design," *First Things* 86 (1998): 21-27.

[12]Many publications have pointed this out. A long list is provided in Henry M. Morris, *Men of Science—Men of God: Great Scientists Who Believed the Bible* (El Cajon, Calif.: Master Books, 1988), p. 107. *Geoscience Reports* carries a regular column on the faith of prominent scientists.

[13]Johannes Kepler, a letter to Herwart von Hohenburg in 1599 reprinted in *Johannes Kepler: Life and Letters,* ed. Carola Baumgardt (New York: Philosophical Library, 1951), p 50.

[14]Harvard biologist Richard Lewontin put it this way: "We take the side of science *in spite of* the patent absurdity of some of its constructs, *in spite of* its failure to fulfill many of its extravagant promises of health and life, *in spite of* the tolerance of the scientific community for unsubstantiated just-so stories, because we have a prior commitment, a commitment to materialism" (Richard Lewontin, *New York Review of Books,* January 9, 1997, p. 31).

[15]Draft of the 2003 *State of Ohio Academic Content Standards: K-12 Science* Grade 10, Scientific Ways of Knowing (Nature of Science) 3.

[16]2003 *State of Ohio Academic Content Standards: K-12 Science* Grade 10, Scientific Ways of Knowing (Nature of Science) 3, p. 147. Approved December 10, 2002 <http://www.ode.state.oh.us/academic_content_standards/ScienceContentStd/PDF/SCIENCE.pdf>.

[17]Johnson, *Right Questions,* p. 68.

[18]For a detailed discussion of this problem with the presentation of evolution, see Jonathan Wells, *Icons of Evolution: Why Much of What We Teach About Evolution Is Wrong* (Washington, D.C.: Regnery, 2000).

[19]C. Richard Dawkins, a review of *Blueprints: Solving the Mystery of Evolution* by Maitland A. Edey and Donald C. Johanson, in the *New York Times,* April 9, 1989, sec. 7, p. 34.

[20]Ernst Mayer, "Darwin's Influence on Modern Thought," *Scientific American* 28, no. 1 (2000): 78-83.

[21]"Details of Hybrid Clone Revealed" *BBC News,* June 18, 1999 <http://news.bbc.co.uk/1/hi/sci/tech/371378.stm>.

[22]Nicholas Wade, "Stem Cell Mixing May Form a Human-Mouse Hybrid," *New York Times,* November 27, 2002, sec. A17.

[23]"Creation of Human 'She-Males' Sparks Outrage," *Reuters,* July 2, 2003

<http://my.aol.com/news/news_story.psp?type=1&cat=0200&id=030702130833171086>.
[24]Paul R. Ehrlich, "Evolution of an Advocate," *Science* 287 (2000): 2159.
[25]Johnson, *Right Questions,* p. 59.
[26]This has been a consistent theme of Johnson's books and can be found clearly stated in all his publications since *Darwin on Trial* was published. Phillip E. Johnson, *Darwin on Trial* (Chicago: Regnery Gateway, 1991).
[27]Jonathan Wells, *Icons of Evolution.*
[28]See, e.g., Kenneth R. Miller and Joseph Levine, *Biology,* 3rd ed. (Englewood Cliffs, N.J.: Prentice Hall, 1995), p. 283, figs. 13-16. The authors recognized that this illustration does not reflect the reality of embryonic development and discuss it at <www.millerandlevine.com/km/evol/embryos/Haeckel.html>. A more detailed discussion of Haeckel's fraud can be found in M. K. Richardson et al., "Haeckel, Embryos, and Evolution," *Science* 280 (1997): 983-84.
[29]For example, see N. A. Campbell, J. B. Reece, and L. G. Mitchell, *Biology,* 5th ed. (Menlo Park, Calif.: Benjamin Cummings, 1999).
[30]See, e.g., G. M. Price, *Evolutionary Geology and the New Catastrophism* (Portland, Ore.: Pacific Press, 1926).
[31]For an excellent discussion of this, see Stephen C. Meyer et al., "The Cambrian Explosion: Biology's Big Bang," in *Darwinism, Design and Public Education,* ed. J. A. Campbell and S. C. Meyer (East Lansing: Michigan State University Press, 2003), pp. 323-402.
[32]See the video *Evidences II: The Tale of a Trilobite,* IVd Tech and The Geoscience Research Institute, 2002. This video elegantly discusses these issues in understanding the geological column.

Chapter 9: Darwinism and the Problem of Evil

[1]Richard Dawkins, cited in J. Brockman, *The Third Culture: Beyond the Scientific Revolution* (New York: Simon & Schuster, 1995), pp. 85-86.
[2]Richard Dawkins, *A River Out of Eden* (New York: Basic Books, 1995), p. 133.
[3]Richard Dawkins, "Obscurantism to the Rescue," *Quarterly Review of Biology* 72 (1997): 397.
[4]Richard Dawkins, cited in Brockman, *Third Culture,* p. 86.
[5]Ibid., p. 85.
[6]Michael Ruse, *On a Darkling Plain: The Evolution-Creation Struggle* (Cambridge, Mass.: Harvard University Press, 2005).
[7]Stephen Jay Gould, *Wonderful Life: The Burgess Shale and the Nature of History* (New York: W. W. Norton, 1989), p. 318.
[8]Simon Conway Morris, *Life's Solution: Inevitable Humans in a Lonely Universe* (Cambridge: Cambridge University Press, 2003), p. 196.
[9]Plato *Phaedo* 96 d, in *Plato: Complete Works,* ed. J. M. Cooper (Indianapolis: Hackett, 1997), pp. 83-84.
[10]Ibid., 97 b-c.
[11]Richard Dawkins, *The Blind Watchmaker* (New York: Norton, 1986).
[12]David Hume, *Dialogues Concerning Natural Religion,* ed. N. K. Smith (1779; reprint, Indianapolis, Ind.: Bobbs-Merrill, 1947), p. 172.
[13]Ibid., pp. 203-4 (his italics).
[14]Charles Darwin, *On the Origin of Species* (London: John Murray, 1859), p. 63.
[15]Ibid., pp. 80-81.
[16]Jerry Coyne, "Intergalactic Jesus" *London Review of Books,* 2002, available online at <www.lrb.co.uk/v24/n09/coyn01_.html>.
[17]George Orwell, cited in ibid.
[18]Charles Darwin, in a letter to Asa Gray, May 22, 1860.
[19]Dawkins, *River Out of Eden,* pp. 131-33.
[20]Darwin, in a letter to Asa Gray.
[21]Augustine *The City of God Against the Pagans,* ed. and trans. R. W. Dyson (Cambridge: Cambridge University Press, 1998).

[22]David Hume, *A Treatise of Human Nature* (Oxford: Oxford University Press, 1978).

[23]Richard C. Lewontin, *Biology as Ideology: The Doctrine of DNA* (Toronto: Anansi, 1991).

[24]D. Wooldridge, cited in Daniel C. Dennett, *Elbow Room* (Cambridge, Mass.: MIT Press, 1984), p. 82.

[25]Bruce R. Reichenbach, "Natural Evils and Natural Laws: A Theodicy for Natural Evil," *International Philosophical Quarterly* 16 (1976): 185.

[26]Richard Dawkins, "Universal Darwinism," in *Molecules to Men,* ed. D. S. Bendall (Cambridge: University of Cambridge Press, 1983), p. 423.

[27]Michael Ruse, *Taking Darwin Seriously: A Naturalistic Approach to Philosophy,* 2nd ed. (Buffalo, N.Y.: Prometheus, 1998); and *Mystery of Mysteries: Is Evolution a Social Construction?* (Cambridge, Mass.: Harvard University Press, 1999).

[28]Richard Dawkins, *A Devil's Chaplain: Reflections on Hope, Lies, Science and Love* (Boston and New York: Houghton Mifflin, 2003), p. 19.

Chapter 10: The Wedge of Truth Visits the Laboratory

[1]Maniatis is known as the bible of molecular biology techniques.

[2]Y. Inai, Y. Ohta and M. Nishikimi, "The Whole Structure of the Human Nonfunctional L-Gulono-Gamma-Lactone Oxidase Gene—the Gene Responsible for Scurvy—and the Evolution of Repetitive Sequences Thereon," *J Nutr Sci Vitaminol* (Tokyo) 49 (2003):315-19.

[3]Michael Denton, *Evolution: A Theory in Crisis* (Bethesda, Md.: Adler & Adler, 1986); Phillip E. Johnson, *Darwin on Trial* (Downers Grove, Ill.: InterVarsity Press, 1993).

[4]Christopher Macosko, personal correspondence.

[5]William B. Provine, "The Origin of Species Revisited: The Theories of Evolution and of Abrupt Appearance," *Biology & Philosophy* 8 (1993): 111-24.

[6]Ibid.

[7]Tim Stafford, "The Making of a Revolution," *Christianity Today* 41 (1997), available online at <www.arn.org/johnson/revolution.htm>.

[8]Ibid.

[9]Thomas S. Kuhn, *The Structure of Scientific Revolutions* (Chicago: University of Chicago Press, 1962).

[10]Phillip E. Johnson, *Reason in the Balance* (Downers Grove, Ill.: InterVarsity Press, 1995).

[11]Kuhn, *Structure of Scientific Revolutions.*

[12]If Johnson and the intelligent-design community were faced with the latter challenge, we all might as well throw in the towel!

[13]This is taken from Phil Johnson's final address at the 1996 "Mere Creation" conference. See his chapter "How to Sink a Battleship," in *Mere Creation,* ed. William A. Dembski (Downers Grove, Ill.: InterVarsity Press, 1996).

[14]Stephen Jay Gould, a review of *Darwin on Trial* by Phillip E. Johnson, *Scientific American* 118, no. 4 (1992).

[15]Thomas Woodward, *Doubts About Darwin: A History of Intelligent Design* (Grand Rapids: Baker, 2003), p. 83; and Philip Johnson, personal communication (also on videotape from the February 1990 Ad Hoc Origins Committee meeting in Portland, Oregon).

[16]Stephen Jay Gould, "Nonoverlapping Magisteria," *Natural History* 106 (1997).

[17]Stephen Jay Gould, *Rocks of Ages: Science and Religion in the Fullness of Life* (New York: Ballantine, 1999).

[18]The author of this chapter includes himself in this category.

Chapter 11: Common Ancestry on Trial

[1]The fact that I considered most books critical of Darwinism second-rate does not apply to Michael Denton's excellent book *Evolution: A Theory in Crisis* (Bethesda, Md.: Adler & Adler, 1985). In 1991, however, I had not yet read Denton's book. In fact, I did not read it until after I learned how much it had inspired Phil.

[2]I describe the development of my early views on Darwinism in "Why I Went for a Second Ph.D.," 1995 <www.tparents.org/library/unification/talks/wells/DARWIN.htm>.

[3]Many Westerners have the impression that Zoroastrianism is not monotheistic but dualistic, with two ontologically equal deities (one of light and one of darkness). According to living Zoroastrians, however (the Parsis of India), there is only one god, Ahura-Mazda, and the principle of darkness is a subordinate being, like Satan in Christian theology (see Ninian Smart, "Zoroastrianism," in *The Encyclopedia of Philosophy*, ed. Paul Edwards [New York: Macmillan, 1967], 8:380-82).

[4]Jacques Monod, cited in Horace Freeland Judson, *The Eighth Day of Creation* (New York: Simon & Schuster, 1979), p. 217.

[5]A summary of the fruits of my Yale Ph.D. research on the nineteenth-century Darwinian controversies can be found in my 1991 article, "Darwinism and the Argument to Design" <www.discovery.org/scripts/viewDB/index.php?program=CRSC&command=view&id=102>.

[6]Phillip E. Johnson, *Darwin on Trial,* Revised Edition (Downer's Grove, Ill.: InterVarsity Press, 1993), p. 50.

[7]Ibid., p. 53.

[8]Charles Darwin, September 10, 1860, letter to Asa Gray, in *The Life and Letters of Charles Darwin,* ed. Francis Darwin (New York: D. Appleton, 1896), 2:131.

[9]Misia Landau, *Narratives of Human Evolution* (New Haven, Conn.: Yale University Press, 1991), p. 148.

[10]Johnson, *Darwin on Trial,* pp. 117-18.

[11]Ibid., p. 101.

[12]Ibid., p. 152.

[13]Ibid., p. 68

[14]Theodosius Dobzhansky, "Nothing in Biology Makes Sense Except in the Light of Evolution," *The American Biology Teacher* 35 (1973): 125-29.

Chapter 12: The Origin of Biological Information and the Higher Taxonomic Categories

[1]Gerd B. Muller and Stuart A. Newman. "Origination of Organismal Form: The Forgotten Cause in Evolutionary Theory," in *Origination of Organismal Form: Beyond the Gene in Developmental and Evolutionary Biology,* ed. Gerd B. Muller and Stuart A. Newman (Cambridge, Mass.: MIT Press, 2003), pp. 3-10.

[2]Ibid., p. 3.

[3]Ibid.

[4]Ibid., p. 7.

[5]K. S. Thomson, "Macroevolution: The Morphological Problem," *American Zoologist* 32 (1992): 107.

[6]G. L. G. Miklos, "Emergence of Organizational Complexities During Metazoan Evolution: Perspectives from Molecular Biology, Palaeontology and Neo-Darwinism," *Memoirs of the Association of Australasian Palaeontologists* 15 (1993): 29.

[7]S. F. Gilbert, J. M. Opitz and R. A. Raff, "Resynthesizing Evolutionary and Developmental Biology," *Developmental Biology* 173 (1996), p. 361.

[8]Specifically, Gilbert, Opitz and Raff argued that changes in morphogenetic fields might produce large-scale changes in the developmental programs and, ultimately, body plans of organisms (ibid., pp. 357-72). Yet they offered no evidence that such fields—if indeed they exist—can be altered to produce advantageous variations in body plan, though this is a necessary condition of any successful causal theory of macroevolution.

[9]G. Webster and B. Goodwin, *Form and Transformation: Generative and Relational Principles in Biology* (Cambridge: Cambridge University Press, 1996); N. H. Shubin and C. R. Marshall, "Fossils, Genes, and the Origin of Novelty," in *Deep Time,* ed. Douglas H. Erwin and Scott L. Wing (Lawrence, Kans.: Allen, 2000); D. H. Erwin, "Macroevolution Is More than Repeated

Rounds of Microevolution," *Evolution & Development* (2000); S. Conway Morris, "Evolution: Bringing Molecules into the Fold," *Cell* 100 (2000); S. Conway Morris, "Cambrian 'Explosion' of Metazoans and Molecular Biology: Would Darwin Be Satisfied?" *International Journal of Developmental Biology* 47, nos. 7-8 (2003); R. L. Carroll, "Towards a New Evolutionary Synthesis," *Trends in Ecology and Evolution* 15 (2000); G. P. Wagner, "What Is the Promise of Developmental Evolution? Part II: A Causal Explanation of Evolutionary Innovations May Be Impossible," *Journal of Experimental Zoology* (Molecular and Developmental Evolution) 291 (2001); H. Becker and Wolf-Ekkehard Loennig, "Transposons: Eukaryotic," in *Nature Encyclopedia of Life Sciences*, vol. 18 (London: Nature Publishing, 2001); B. M. R. Stadler et al., "The Topology of the Possible: Formal Spaces Underlying Patterns of Evolutionary Change," *Journal of Theoretical Biology* 213 (2001); Wolf-Ekkehard Loennig and H. Saedler, "Chromosome Rearrangements and Transposable Elements," *Annual Review of Genetics* 36 (2002); G. P. Wagner and P. F. Stadler, "Quasi-Independence, Homology and the Unity-C of Type: A Topological Theory of Characters," *Journal of Theoretical Biology* 220 (2003); J. W. Valentine, *On the Origin of Phyla* (Chicago: University of Chicago Press, 2004), pp. 189-94.

[10]Stephen C. Meyer et al. "The Cambrian Explosion: Biology's Big Bang," in *Darwinism, Design and Public Education,* ed. J. A. Campbell and S. C. Meyer (East Lansing: Michigan State University Press, 2003), pp. 323-402.

[11]S. A. Bowring et al., "Calibrating Rates of Early Cambrian Evolution," *Science* 261 (1993); S. A. Bowring and D. H. Erwin, "A New Look at Evolutionary Rates in Deep Time: Uniting Paleontology and High-Precision Geochronology," *GSA Today* 8 (1998): 1; S. A. Bowring, "Geochronology Comes of Age," *Geotimes* 43 (1998): 40; R. A. Kerr, "Evolution's Big Bang Gets Even More Explosive," *Science* 261 (19930); R. Monastersky, "Siberian Rocks Clock Biological Big Bang," *Science News* 144 (1993); S. Aris-Brosou and Z. Yang, "Bayesian Models of Episodic Evolution Support a Late Precambrian Explosive Diversification of the Metazoa," *Molecular Biology and Evolution* 20 (2003).

[12]Meyer, "The Cambrian Explosion," pp. 323-402.

[13]Miklos, "Emergence of Organizational Complexities"; D. H. Erwin, J. Valentine and D. Jablonski, "The Origin of Animal Body Plans," *American Scientist* 85 (1997): 132; M. Steiner and R. Reitner, "Evidence of Organic Structures in Ediacara-Type Fossils and Associated Microbial Mats," *Geology* 29, no. 12 (2001); S. Conway Morris, "Cambrian 'Explosion' of Metazoans and Molecular Biology: Would Darwin Be Satisfied?" *International Journal of Developmental Biology* 47, nos. 7-8 (2003): 510; J. W. Valentine and D. Jablonski, "Morphological and Developmental Macroevolution: A Paleontological Perspective," *International Journal of Developmental Biology* 47 (2003): 519-20.

[14]M. Foote, "Sampling, Taxonomic Description, and Our Evolving Knowledge of Morphological Diversity," *Paleobiology* 23 (1997); M. Foote et al., "Evolutionary and Preservational Constraints on Origins of Biologic Groups: Divergence Times of Eutherian Mammals," *Science* 283 (1999); M. Benton, and F. J. Ayala, "Dating the Tree of Life," *Science* 300 (2003); Meyer, "Cambrian Explosion."

[15]Conway Morris, "Cambrian 'Explosion' of Metazoans and Molecular Biology," p. 505.

[16]S. Conway Morris, "The Question of Metazoan Monophyly and the Fossil Record," *Progress in Molecular and Subcellular Biology* 21 (1998); S. Conway Morris, "The Cambrian 'Explosion' of Metazoans," in *Origination of Organismal Form: Beyond the Gene in Developmental and Evolutionary Biology,* ed. Gerd B. Muller and Stuart A. Newman (Cambridge, Mass.: MIT Press, 2003); Conway Morris, "Cambrian 'Explosion' of Metazoans and Molecular Biology," p. 510; P. Willmer, *Invertebrate Relationships: Patterns in Animal Evolution* (Cambridge: Cambridge University Press, 1990); P. Willmer, "Convergence and Homoplasy in the Evolution of Organismal Form," in *Origination of Organismal Form: Beyond the Gene in Developmental and Evolutionary Biology,* ed. G. B. Muller and S. A. Newman (Cambridge, Mass.: MIT Press, 2003).

[17]Conway Morris, "Cambrian 'Explosion' of Metazoans and Molecular Biology," pp. 505-6; Val-

entine and Jablonski, "Morphological and Developmental Macroevolution."

[18]G. A. Wray, J. S. Levinton and L. H. Shapiro, "Molecular Evidence for Deep Precambrian Divergences Among Metazoan Phyla," *Science* 274 (1996).

[19]F. Ayala, A. Rzhetsky and F. J. Ayala, "Origin of the Metazoan Phyla: Molecular Clocks Confirm Paleontological Estimates," *Proceedings of the National Academy of Sciences USA* 95 (1998).

[20]If one takes the fossil record at face value and assumes that the Cambrian explosion took place within a relatively narrow 5 to 10 million year window, explaining the origin of the information necessary to produce new proteins, for example, becomes more acute in part because mutation rates would not have been sufficient to generate the number of changes in the genome necessary to build the new proteins for more complex Cambrian animals (S. Ohno, "The Notion of the Cambrian Pananimalia Genome," *Proceedings of the National Academy of Sciences, U.S.A.* 93 [1996]: 8475-78). This review will argue that even if we allow several hundred million years for the origin of the metazoan, significant probabilistic and other difficulties remain with the neo-Darwinian explanation of the origin of form and information.

[21]Claude Elwood Shannon, "A Mathematical Theory of Communication," *Bell System Technical Journal* 27 (1948).

[22]Ibid.

[23]H. P. Yockey, *Information Theory and Molecular Biology* (Cambridge: Cambridge University Press, 1992), p. 110.

[24]Francis Crick, "On Protein Synthesis," *Symposium for the Society of Experimental Biology* 12 (1958): 144, 153; S. Sarkar, "Biological Information: A Skeptical Look at Some Central Dogmas of Molecular Biology," in *The Philosophy and History of Molecular Biology: New Perspectives,* ed. S. Sarkar (Dordrecht: Kluwer Academic, 1996), p. 191. As Crick put it, "information means here the *precise* determination of sequence, either of bases in the nucleic acid or on amino acid residues in the protein" (Crick, "On Protein Synthesis," pp. 144, 153).

[25]J. Levinton, *Genetics, Paleontology, and Macroevolution* (Cambridge: Cambridge University Press, 1988), p. 485.

[26]William A. Dembski, *No Free Lunch: Why Specified Complexity Cannot Be Purchased Without Intelligence* (Lanham, Md.: Rowman & Littlefield, 2002).

[27]J. J. Brocks et al., "Archean Molecular Fossils and the Early Rise of Eukaryotes," *Science* 285 (1999).

[28]J. P. Grotzinger et al., "Biostratigraphic and Geochronologic Constraints on Early Animal Evolution," *Science* 270 (1995).

[29]Bowring, "Calibrating Rates."

[30]J. W. Valentine, "Late Precambrian Bilaterians: Grades and Clades," in *Temporal and Mode in Evolution: Genetics and Paleontology 50 Years After Simpson,* ed. W. M. Fitch and F. J. Ayala (Washington, D.C.: National Academy Press, 1995), pp. 91-93.

[31]E. Koonin, "How Many Genes Can Make a Cell? The Minimal Genome Concept," *Annual Review of Genomics and Human Genetics* 1 (2000).

[32]J. Gerhart and M. Kirschner, *Cells, Embryos, and Evolution* (London: Blackwell Science, 1997), p. 121; M. D. Adams et al., "The Genome Sequence of *Drosophila melanogaster,*" *Science* 287 (2000).

[33]Michael Denton, *Evolution: A Theory in Crisis* (London: Adler & Adler, 1986), pp. 309-11.

[34]Ibid., pp. 301-24.

[35]M. Eden, "The Inadequacies of Neo-Darwinian Evolution as a Scientific Theory," in *Mathematical Challenges to the Darwinian Interpretation of Evolution,* ed. P. S. Morehead and M. M. Kaplan, Wistar Institute Symposium Monograph (New York: Allen R. Liss, 1967); M. Schutzenberger, "Algorithms and the Neo-Darwinian Theory of Evolution," in *Mathematical Challenges to the Darwinian Interpretation of Evolution,* ed. P. S. Morehead and M. M. Kaplan, Wistar Institute Symposium Monograph (New York: Allen R. Liss, 1967); S. Lovtrup, "Semantics, Logic and Vulgate Neo-Darwinism," *Evolutionary Theory* 4 (1979).

[36]Eden, "The Inadequacies of Neo-Darwinian Evolution"; Denton, *Evolution: A Theory in Crisis.*

[37]J. Bowie and R. Sauer, "Identifying Determinants of Folding and Activity for a Protein of Unknown Sequences: Tolerance to Amino Acid Substitution," *Proceedings of the National Academy of Sciences, U.S.A.* 86 (1989); J. Reidhaar-Olson and R. Sauer, "Functionally Acceptable Solutions in Two Alpha-Helical Regions of Lambda Repressor," *Proteins, Structure, Function, and Genetics* 7 (1990); S. V. Taylor et al., "Searching Sequence Space for Protein Catalysts," *Proceedings of the National Academy of Sciences, U.S.A.* 98 (2001).

[38]M. F. Perutz and H. Lehmann, "Molecular Pathology of Human Hemoglobin," *Nature* 219 (1968).

[39]Bowie and Sauer, "Identifying Determinants"; Reidhaar-Olson and Sauer, "Functionally Acceptable Solutions"; C. Chothia, I. Gelfland, and A. Kister, "Structural Determinants in the Sequences of Immunoglobulin Variable Domain," *Journal of Molecular Biology* 278 (1998); D. D. Axe, "Extreme Functional Sensitivity to Conservative Amino Acid Changes on Enzyme Exteriors," *Journal of Molecular Biology* 301, no. 3 (2000); Taylor, "Searching Sequence Space."

[40]Axe, "Extreme Functional Sensitivity."

[41]Reidhaar-Olson and Sauer, "Functionally Acceptable Solutions"; Michael Behe, "Experimental Support for Regarding Functional Classes of Proteins to Be Highly Isolated from Each Other," in *Darwinism: Science or Philosophy?* ed. J. Buell and V. Hearn (Richardson, Tex.: Foundation for Thought and Ethics, 1992); S. Kauffman, *At Home in the Universe* (Oxford: Oxford University Press, 1995), p. 44; William A. Dembski, *The Design Inference* (Cambridge: Cambridge University Press, 1998), pp. 175-223; Axe, "Extreme Functional Sensitivity"; Axe, "Estimating the Prevalence of Protein Sequences Adopting Functional Enzyme Folds," *Journal of Molecular Biology* (2004).

[42]Richard Dawkins, *Climbing Mount Improbable* (New York: W. W. Norton, 1996).

[43]For a review, see W. E. Loennig, "Natural Selection," in *The Corsini Encyclopedia of Psychology and Behavioral Sciences,* ed. W. E. Craighead and C. B. Nemeroff, 3rd ed., vol. 3 (New York: John Wiley, 2001).

[44]M. Kimura, *The Neutral Theory of Molecular Evolution* (Cambridge: Cambridge University Press, 1983).

[45]Reidhaar-Olson and Sauer, "Functionally Acceptable Solutions"; Behe, "Experimental Support for Regarding Functional Classes of Proteins," pp. 65-69

[46]H. P. Yockey, "A Calculation of the Probability of Spontaneous Biogenesis by Information Theory," *Journal of Theoretical Biology* 67 (1978).

[47]Axe, "Extreme Functional Sensitivity"; and Axe, "Estimating the Prevalence of Protein Sequences."

[48]Axe, "Estimating the Prevalence of Protein Sequences."

[49]Ohno, "Notion of the Cambrian Pananimalia Genome."

[50]See Dembski's *Design Inference* (pp. 175-223) for a rigorous calculation of this "Universal Probability Bound." See also Axe, "Estimating the Prevalence of Protein Sequences."

[51]Bowring, "Calibrating Rates of Early Cambrian Evolution"; Bowring, "New Look at Evolutionary Rates in Deep Time," p. 1; Bowring, "Geochronology Comes of Age," p. 40; Kerr, "Evolution's Big Bang"; Monatersky, "Siberian Rocks."

[52]Wray, "Molecular Evidence."

[53]S. Conway Morris, "Early Metazoan Evolution: Reconciling Paleontology and Molecular Biology," *American Zoologist* 38 (1998).

[54]Ohno, "Notion of the Cambrian Pananimalia Genome," p. 8475. To solve this problem Ohno himself proposes the existence of a hypothetical ancestral form that possessed virtually all the genetic information necessary to produce the new body plans of the Cambrian animals. He asserts that this ancestor and its "pananimalian genome" might have arisen several hundred million years before the Cambrian explosion. On this view, each of the different Cambrian animals would have possessed virtually identical genomes, albeit with considerable latent and unexpressed capacity in the case of each individual form (ibid., pp. 8475-78). While this pro-

posal might help explain the origin of the Cambrian animal forms by reference to preexisting genetic information, it does not solve, but instead merely displaces, the problem of the origin of the genetic information necessary to produce these new forms.

[55]Richard Dawkins, *The Blind Watchmaker* (London: Penguin, 1986), p. 139.

[56]Axe, "Extreme Functional Sensitivity."

[57]F. Blanco, I. Angrand and L. Serrano, "Exploring the Confirmational Properties of the Sequence Space Between Two Proteins with Different Folds: An Experimental Study," *Journal of Molecular Biology* 285 (1999): 741.

[58]Ibid.; Axe, "Extreme Functional Sensitivity."

[59]W. Arthur, *The Origin of Animal Body Plans* (Cambridge: Cambridge University Press, 1997), p. 21.

[60]K. S. Thomson, "Macroevolution: The Morphological Problem," *American Zoologist* 32 (1992).

[61]B. John and G. Miklos, *The Eukaryote Genome in Development and Evolution* (London: Allen & Unwin, 1988), p. 309.

[62]Arthur, *Origin of Animal Body Plans,* p. 21.

[63]C. Nusslein-Volhard and E. Wieschaus, "Mutations Affecting Segment Number and Polarity in Drosophila," *Nature* 287 (1980); P. A. Lawrence and G. Struhl, "Morphogens, Compartments and Pattern: Lessons from Drosophila?" *Cell* 85 (1996); Gerd B. Muller and Stuart A. Newman, "Origination of Organismal Form: The Forgotten Cause in Evolutionary Theory," in *Origination of Organismal Form: Beyond the Gene in Developmental and Evolutionary Biology,* ed. Gerd B. Muller and Stuart A. Newman (Cambridge, Mass.: MIT Press, 2003). Some have suggested that mutations in "master regulator" Hox genes might provide the raw material for body plan morphogenesis. Yet there are two problems with this proposal. First, Hox gene expression begins only after the foundation of the body plan has been established in early embryogenesis (E. Davidson, *Genomic Regulatory Systems: Development and Evolution* [New York: Academic Press, 2001], p. 66). Second, Hox genes are highly conserved across many disparate phyla and so cannot account for the morphological differences that exist between the phyla (Valentine, *On the Origin of Phyla,* p. 88).

[64]Kauffman, *At Home in the Universe,* p. 200.

[65]J. F. McDonald, "The Molecular Basis of Adaptation: A Critical Review of Relevant Ideas and Observations," *Annual Review of Ecology and Systematics* 14 (1983): 93.

[66]Notable differences in the developmental pathways of similar organisms have been observed. For example, congeneric species of sea urchins (from genus *Heliocidaris*) exhibit striking differences in their developmental pathways (R. Raff, "Larval Homologies and Radical Evolutionary Changes in Early Development," in *Homology,* Novartis Symposium 222 [Chichester, U.K.: John Wiley, 1999], pp. 110-21). Thus it might be argued that such differences show that early developmental programs can in fact be mutated to produce new forms. Nevertheless, there are two problems with this claim. First, there is no direct evidence that existing differences in sea urchin development arose by mutation. Second, the observed differences in the developmental programs of different species of sea urchins do not result in new body plans, but instead in highly conserved structures. Despite differences in developmental patterns, the endpoints are the same. Thus, even if it can be assumed that mutations produced the differences in developmental pathways, it must be acknowledged that such changes did not result in novel form.

[67]Darwin, *Origin of Species,* p. 108.

[68]B. C. Goodwin, "What Are the Causes of Morphogenesis?" *BioEssays* 3 (1985); H. F. Nijhout, "Metaphors and the Role of Genes in Development," *BioEssays* 12 (1990); J. Sapp, *Beyond the Gene* (New York: Oxford University Press, 1987); Muller and Newman, "Origination of Organismal Form."

[69]Of course, many posttranslation processes of modification also play a role in producing a functional protein. Such processes make it impossible to predict a protein's final sequencing from its corresponding gene sequence alone (Sarkar, "Biological Information," pp. 199-202).

[70]F. M. Harold, "From Morphogenes to Morphogenesis," *Microbiology* 141 (1995): 2774; L. Moss,

What Genes Can't Do (Cambridge, Mass.: MIT Press, 2004).

[71]F. M. Harold, *The Way of the Cell: Molecules, Organisms, and the Order of Life* (New York: Oxford University Press, 2001), p. 125.

[72]Ibid.

[73]M. A. McNiven and K. R. Porter, "The Centrosome: Contributions to Cell Form," in *The Centrosome*, ed. V. I. Kalnins (San Diego: Academic Press, 1992), pp. 313-29.

[74]B. M. Lange et al., "Centriole Duplication and Maturation in Animal Cells," in *The Centrosome in Cell Replication and Early Development*, ed. R. E. Palazzo and G. P. Schatten, Current Topics in Developmental Biology 49 (San Diego: Academic Press, 2000), pp. 235-49; W. F. Marshall and J. L. Rosenbaum, "Are There Nucleic Acids in the Centrosome?" in *The Centrosome in Cell Replication and Early Development*, ed. R. E. Palazzo and G. P. Schatten, Current Topics in Developmental Biology 49 (San Diego: Academic Press, 2000), pp. 187-205.

[75]Lange, "Centriole Duplication and Maturation in Animal Cells."

[76]T. M. Sonneborn, "Determination, Development, and Inheritance of the Structure of the Cell Cortex," in *Symposia of the International Society for Cell Biology* 9 (1970): 1-13; J. Frankel, "Propagation of Cortical Differences in *Tetrahymena*," *Genetics* 94 (1980); D. L. Nanney, "The Ciliates and the Cytoplasm," *Journal of Heredity* 74 (1983).

[77]Moss, *What Genes Can't Do*.

[78]Harold, "From Morphogenes to Morphogenesis," p. 2767.

[79]Kauffman, *At Home in the Universe*.

[80]Ibid., pp. 47-92.

[81]Ibid., pp. 86-88.

[82]Ibid., pp. 53, 89, 102.

[83]Ibid., pp. 199-201.

[84]Ibid., p. 201.

[85]D. H. Erwin, J. Valentine and J. J. Sepkoski, "A Comparative Study of Diversification Events: The Early Paleozoic Versus the Mesozoic," *Evolution* 41 (1987); R. Lewin, "A Lopsided Look at Evolution," *Science* 241 (1988); Valentine and Jablonski, "Morphological and Developmental Macroevolution," p. 518.

[86]N. Eldredge and Stephen Jay Gould, "Punctuated Equilibria: An Alternative to Phyletic Gradualism," in *Models in Paleobiology*, ed. T. Schopf (San Francisco: W. H. Freeman, 1972).

[87]J. W. Valentine and D. H. Erwin, "Interpreting Great Developmental Experiments: The Fossil Record," in *Development as an Evolutionary Process*, ed. R. A. Raff and E. C. Raff (New York: Alan R. Liss, 1987), p. 96. Erwin ("One Very Long Argument," *Biology and Philosophy* 19 [2004]: 21), although friendly to the possibility of species selection, argues that Gould provides little evidence for its existence. "The difficulty" writes Erwin of species selection, "is that we must rely on Gould's arguments for theoretical plausibility and sufficient relative frequency. Rarely is a mass of data presented to justify and support Gould's conclusion." Indeed, Gould himself admitted that species selection remains largely a hypothetical construct: "I freely admit that well-documented cases of species selection do not permeate the literature" (Stephen Jay Gould, *The Structure of Evolutionary Theory* [Cambridge, Mass.: Harvard University Press, 2002], p. 710).

[88]Ibid. "I do not deny either the wonder, or the powerful importance, of organized adaptive complexity. I recognize that we know no mechanism for the origin of such organismal features other than conventional natural selection at the organismic level—for the sheer intricacy and elaboration of good biomechanical design surely precludes either random production, or incidental origin as a side consequence of active processes at other levels" (ibid.). "Thus, we do not challenge the efficacy or the cardinal importance of organismal selection. As previously discussed, I fully agree with Dawkins (1986) and others that one cannot invoke a higher-level force like species selection to explain 'things that organisms do'—in particular, the stunning panoply of organismic adaptations that has always motivated our sense of wonder about the natural world, and that Darwin (1859) described, in one" of his most famous

lines (3), as 'that perfection of structure and coadaptation which most justly excites our ad-
miration' " (ibid., p. 886).
[89]Gerry Webster and Brian Goodwin, "A Structuralist Approach to Morphology," *Rivista di Biologia* 77 (1984); Webster and Goodwin, *Form and Transformation*.
[90]D'Arcy W. Thompson, *On Growth and Form*, 2nd ed. (Cambridge: Cambridge University
Press, 1942).
[91]Webster and Goodwin, "A Structuralist Approach to Morphology," pp. 510-11.
[92]J. Maynard Smith, "Structuralism Versus Selection: Is Darwinism Enough?" in *Science and Beyond*, ed. S. Rose and L. Appignanesi (London: Basil Blackwell, 1986).
[93]Yockey, *Information Theory and Molecular Biology;* Michael Polanyi, "Life Transcending
Physics and Chemistry," *Chemical and Engineering News* 45, no. 35 (1967); Michael Polanyi,
"Life's Irreducible Structure," *Science* 160 (1968); Stephen C. Meyer, "DNA and the Origin of
Life: Information, Specification and Explanation," in *Darwinism, Design and Public Education*, ed. J. A. Campbell and Stephen C. Meyer (East Lansing: Michigan State University Press,
2003).
[94]Yockey, *Information Theory and Molecular Biology*, pp. 77-83.
[95]Polanyi, "Life Transcending Physics and Chemistry"; Polanyi, "Life's Irreducible Structure";
Yockey, *Information Theory and Molecular Biology*, p. 290.
[96]Yockey, *Information Theory and Molecular Biology*, p. 290.
[97]G. E. Budd and S. E. Jensen, "A Critical Reappraisal of the Fossil Record of the Bilaterian
Phyla," *Biological Reviews of the Cambridge Philosophical Society* 75 (2000): 253.
[98]Ibid., p. 288.
[99]Conway Morris, "Evolution: Bringing Molecules into the Fold," p. 8; Conway Morris, "Cambrian 'Explosion' of Metazoans and Molecular Biology," p. 511.
[100]Denton, *Evolution: A Theory in Crisis;* Michael Denton, *Nature's Density* (New York: Free
Press, 1998); Charles B. Thaxton, Walter L. Bradley and Roger L. Olsen, *The Mystery of Life's
Origin: Reassessing Current Theories* (Dallas: Lewis & Stanley, 1992); D. Kenyon and G. Mills,
"The RNA World: A Critique," *Origins & Design* 17, no. 1 (1996); Michael Behe, *Darwin's
Black Box* (New York: Free Press, 1996); Michael Behe, "Irreducible Complexity: Obstacle to
Darwinian Evolution," in *Debating Design: From Darwin to DNA*, ed. William A. Dembski
and Michael Ruse (Cambridge: Cambridge University Press, 2004); Dembski, *The Design In-
ference;* Dembski, *No Free Lunch;* William A. Dembski, "The Logical Underpinnings of Intel-
ligent Design," in *Debating Design: From Darwin to DNA*, ed. William A. Dembski and
Michael Ruse (Cambridge: Cambridge University Press, 2004); Conway Morris, "Evolution:
Bringing Molecules into the Fold"; Conway Morris, "The Cambrian 'Explosion' of Metazoans
and Molecular Biology"; Loennig, "Natural Selection"; Loennig and Saedler, "Chromosome
Rearrangements and Transposable Elements"; P. Nelson and J. Wells, "Homology in Biology:
Problem for Naturalistic Science and Prospect or Intelligent Design," in *Darwinism, Design
and Public Education*, ed. J. A. Campbell and Stephen C. Meyer (East Lansing: Michigan State
University Press, 2003); Stephen C. Meyer, "DNA and the Origin of Life: Information, Speci-
fication and Explanation," in *Darwinism, Design and Public Education*, ed. J. A. Campbell
and Stephen C. Meyer (East Lansing: Michigan State University Press, 2003); Stephen C.
Meyer, "The Cambrian Information Explosion: Evidence for Intelligent Design," in *Debating
Design: From Darwin to DNA*, ed. William A. Dembski and Michael Ruse (Cambridge: Cambridge University Press, 2004); W. Bradley, "Information, Entropy and the Origin of Life," in
Debating Design: From Darwin to DNA, ed. William A. Dembski and Michael Ruse (Cambridge: Cambridge University Press, 2004).
[101]N. C. Gillespie, *Charles Darwin and the Problem of Creation* (Chicago: University of Chicago
Press, 1979); T. Lenior, *The Strategy of Life* (Chicago: University of Chicago Press, 1982), p. 4.
[102]F. Ayala, "Darwin's Revolution," in *Creative evolution?!* ed. J. Campbell and J. Schopf (Boston: Jones & Bartlett, 1994), p. 5; Dawkins, *Blind Watchmaker*, p. 1; E. Mayr, foreword to
M. Ruse, *Darwinism Defended* (Boston: Pearson Addison Wesley, 1982), pp. xi-xii;

R. Lewontin, "Adaptation," in *Evolution: A Scientific American Book* (San Francisco: W. H. Freeman, 1978).

[103]Ayala, "Darwin's Revolution," p. 5.

[104]J. Weiner, *The Beak of the Finch* (New York: Vintage, 1994); P. R. Grant, *Ecology and Evolution of Darwin's Finches* (Princeton, N.J.: Princeton University Press, 1999).

[105]Richard Dawkins, *River Out of Eden* (New York: Basic Books, 1995), p. 11; Bill Gates, *The Road Ahead* (Boulder, Colo.: Blue Penguin, 1996), p. 228.

[106]Gillespie, *Charles Darwin and the Problem of Creation*.

[107]P. Lipton, *Inference to the Best Explanation* (New York: Routledge, 1991), pp. 32-88; S. G. Brush, "Prediction and Theory Evaluation: The Case of Light Bending," *Science* 246 (1989): 1124-29; E. Sober, *The Philosophy of Biology*, 2nd ed. (San Francisco: Westview Press, 2000), p. 44.

[108]Stephen C. Meyer, *Of Clues and Causes: A Methodological Interpretation of Origin of Life Studies*, doctoral dissertation, Cambridge: University of Cambridge, 1991; Stephen C. Meyer, "The Scientific Status of Intelligent Design: The Methodological Equivalence of Naturalistic and Non-Naturalistic Origins Theories," in *Science and Evidence for Design in the Universe*, Proceedings of the Wethersfield Institute (San Francisco: Ignatius Press, 2002); C. Cleland, "Historical Science, Experimental Science, and the Scientific Method," *Geology* 29 (2001): 987-89; C. Cleland, "Methodological and Epistemic Differences Between Historical Science and Experimental Science," *Philosophy of Science* 69 (2002): 474-96. Theories in the historical sciences typically make claims about what happened in the past, or what happened in the past to cause particular events to occur (Meyer, *Of Clues and Causes*, pp. 57-72). For this reason, historical scientific theories are rarely tested by making predictions about what will occur under controlled laboratory conditions (Cleland, "Historical Science," p. 987; Cleland, "Methodological and Epistemic Differences," pp. 474-96). Instead, such theories are usually tested by comparing their explanatory power against that of their competitors with respect to already known facts. Even in the case in which historical theories make claims about past causes, they usually do so on the basis of preexisting knowledge of cause and effect relationships. Nevertheless, prediction may play a limited role in testing historical scientific theories since such theories may have implications as to what kind of evidence is likely to emerge in the future. For example, neo-Darwinism affirms that new functional sections of the genome arise by a trial-and-error process of mutation and subsequent selection. For this reason, historically, many neo-Darwinists expected or predicted that the large noncoding regions of the genome—so-called junk DNA—would lack function altogether (L. E. Orgel and Francis H. Crick, "Selfish DNA: The Ultimate Parasite," *Nature* 284 [1980]). On this line of thinking, the nonfunctional sections of the genome represent nature's failed experiments that remain in the genome as a kind of artifact of the past activity of the mutation and selection process. Advocates of the design hypotheses, on the other hand, would have predicted that noncoding regions of the genome might well reveal hidden functions, not only because design theorists do not think that new genetic information arises by a trial and error process of mutation and selection but also because designed systems are often functionally polyvalent. Even so, as new studies reveal more about the functions performed by the noncoding regions of the genome (W. W. Gibbs, "The Unseen Genome: Gems Among the Junk," *Scientific American* 289 [2003]), the design hypothesis can no longer be said to make this claim in the form of a specifically future-oriented prediction. Instead, the design hypothesis might be said to gain confirmation or support from its ability to explain this now-known evidence, albeit after the fact. Of course, neo-Darwinists might also amend their original prediction using various auxiliary hypotheses to explain away the presence of newly discovered functions in the noncoding regions of DNA. In both cases, considerations of ex post facto explanatory power reemerge as central to assessing and testing competing historical theories.

[109]Charles Darwin, "Letter to Asa Gray," in *Life and Letters of Charles Darwin*, ed. F. Darwin,

vol. 1 (London: D. Appleton, 1896), p. 437.

[110]Lipton, *Inference to the Best Explanation,* pp. 32-88.

[111]Stephen Jay Gould, "Is Uniformitarianism Necessary?" *American Journal of Science* 263 (1965); G. Simpson, "Uniformitarianism: An Inquiry into Principle, Theory, and Method in Geohistory and Biohistory," in *Essays in Evolution and Genetics in Honor of Theodosius Dobzhansky,* ed. M. K. Hecht and W. C. Steered (New York: Appleton-Century-Crofts, 1970); M. G. Rutten, *The Origin of Life by Natural Causes* (Amsterdam: Elsevier, 1971); R. Hooykaas, "Catastrophism in Geology, Its Scientific Character in Relation to Actualism and Uniformitarianism," in *Philosophy of Geohistory (1785-1970),* ed. C. Albritton (Stroudsburg, Penn: Dowden, Hutchinson & Ross, 1975).

[112]Henry Quastler, *The Emergence of Biological Organization* (New Haven, Conn.: Yale University Press, 1964), p. 16.

[113]Dawkins, *Blind Watchmaker,* pp. 47-49; B. O. Kuppers, "On the Prior Probability of the Existence of Life," in *The Probabilistic Revolution,* ed. L. Kruger et al. (Cambridge, Mass.: MIT Press, 1987), pp. 355-69.

[114]Stephen C. Meyer, "DNA by Design: An Inference to the Best Explanation for the Origin of Biological Information," *Rhetoric & Public Affairs* 1, no. 4 (1998): 127-28; Meyer, "The Cambrian Information Explosion," pp. 247-48.

[115]David Berlinski, "On Assessing Genetic Algorithms," public lecture at the "Science and Evidence of Design in the Universe" conference, Yale University, New Haven, Connecticut, November 4, 2000.

[116]Denton, *Evolution: A Theory in Crisis,* pp. 309-11.

[117]Polanyi, "Life Transcending Physics and Chemistry"; and "Life's Irreducible Structure."

Chapter 13: Genetic Analysis of Coordinate Flagellar and Type III Regulatory Circuits in Pathogenic Bacteria

[1]For a review of flagellum synthesis see R. M. Macnab, "How Bacteria Assemble Flagella," *Annual Review of Microbiology* 57 (2003): 77-100.

[2]R. M. Macnab, "The Bacterial Flagellum: Reversible Rotary Propellor and Type III Export Apparatus," *Journal of Bacteriology* 181, no. 23 (1999):7149-53.

[3]G. R. Cornelis et al. "The Virulence Plasmid of *Yersinia:* An Antihost Genome" *Microbiology and Molecular Biology Reviews* 62 (1998): 1315-27.

[4]R. R. Rohde, J. M. Fox and Scott A. Minnich, "Thermoregulation in *Yersinia enterocolitica* Is Coincident with Changes in DNA Supercoiling." *Molecular Microbiology* 12, no. 2 (1994): 187-99.

[5]G. Ramakrishnan, J. L. Zhao, and A. Newton, "The Cell Cycle-Regulated Flagellar Gene *flbF* of *Caulobacter crescentus* Is Homologous to a Virulence Locus (*lcrD*) of *Yersinia pestis,*" *Journal of Bacteriology* 173, no. 22 (1991): 7283-92; and L. A. Sanders, S. Van Way, D. A. Mullin, "Characterization of the *Caulobacter crescentus flbF* Promoter and Identification of the Inferred FlbF Product as a Homolog of the LcrD Protein from a *Yersinia enterocolitica* Virulence Plasmid," *Journal of Bacteriology* 174, no. 3 (1992): 857-66.

[6]W. Y. Zhuang and L. Lucy, "*Caulobacter* FliQ and FliR Membrane Proteins, Required for Flagellar Biogenesis and Cell Division, Belong to a Family of Virulence Factor Export Proteins," *Journal of Bacteriology* 177, no. 2 (1995): 343-56.

[7]J. C. Haller et al., "A Chromosomally Encoded Type III Secretion Pathway in *Yersinia enterocolitica* Is Important in Virulence," *Molecular Microbiology* 36 (2000): 1436-46.

[8]V. Kapatral and Scott A. Minnich, "Co-ordinate, Temperature-Sensitive Regulation of the Three *Yersinia enterocolitica* Flagellin Genes," *Molecular Microbiology* 17 (1995): 49-56.

[9]Michael J. Smith and Scott Minnich, "Genetic Regulation of Type III Secretion Systems in *Yersinia enterocolitica,*" Ph.D. diss., University of Idaho, Moscow, Idaho, 2000.

[10]T. Kubori et al., "Supramolecular Structure of the *Salmonella typhimurium* Type III Protein Secretion System," *Science* 280 (1998): 602-5.

[11]E. Hoiczyk and G. Blobel, "Polymerization of a Single Protein of the Pathogen *Yersinia en-*

terocolitica into Needles Punctures Eukaryotic Cells," *Proceedings of the National Academy of Sciences, U.S.A.* 98, no. 8 (2001): 4669-74.

[12]D. M. Anderson, D. E. Fouts, A. Collmer and O. Schneewind, "Reciprocal Secretion of Proteins by the Bacterial Type III Machines of Plant and Animal Pathogens Suggests Universal Recognition of mRNA Targeting Signals," *Proceedings of the National Academy of Sciences, U.S.A.* 96 (1999): 12839-43.

[13]P. F. McDermott et al., "High-Affinity Interaction Between Gram-Negative Flagellin and a Cell Surface Polypeptide Results in Human Monocyte Activation," *Infection and Immunity* 68, no. 10 (2000): 5525-29.

[14]M. J. Smith and S. A. Minnich, unpublished observation, University of Idaho, 2000.

[15]S. R. Monday, S. A. Minnich, and P. C. F. Feng, "A 12-Base-Pair Deletion in the Flagellar Master Control Gene *flhc* Causes Nonmotility of the Pathogenic German Sorbitol-Fermenting *Escherichia coli* O157:H⁻ Strains," *Journal of Bacteriology* 186 (2004): 2319-27.

[16]B. M. Pruss, X. Liu, W. Hendrickson and P Matsumura. "FlhD/FlhC Regulated Promoters Analyzed by Gene Array and *lacZ* Fusions," *FEMS Microbiology Letters* 197 (2001): 91-97.

[17]U.S. Department of Energy's "Genomes to Life: Accelerating Biological Discovery" (April 2001) <www.doegenomestolife.org/roadmap/GTLcomplete_web.pdf>.

[18]Michael J. Behe, *Darwin's Black Box: The Biochemical Challenge to Evolution* (New York: Simon & Schuster, 1996).

[19]K. R. Miller, *Finding Darwin's God: A Scientist's Search for Common Ground Between God and Evolution* (New York: Cliff Street Books, 1999); and K. R. Miller, "The Bacterial Flagellum Unspun," in *Debating Design: From Darwin to DNA*, ed. William A. Dembski and Michael Ruse (Cambridge: Cambridge University Press, 2004), pp. 81-97.

[20]L. Nguyen, I. T. Paulsen, J. Tchieu, C. J. Hueck and M. H. Saier Jr., "Phylogenetic Analyses of the Constituents of the Type II Protein Secretion Systems," *Journal of Microbiology and Biotechnology* 2 (2000): 125-44.

[21]Stephen C. Meyer, "The Cambrian Information Explosion: Evidence for Intelligent Design," in *Debating Design: From Darwin to DNA*, ed. William A. Dembski and Michael Ruse (Cambridge: Cambridge University Press, 2004), pp. 371-91; Stephen C. Meyer, "DNA and the Origin of Life: Information, Specification and Explanation," in *Darwinism, Design and Public Education*, ed. J. A. Campbell and Stephen C. Meyer (East Lansing: Michigan State University Press, 2003), pp. 223-85.

Chapter 14: Intelligent Design and the Defense of Reason

[1]James M. Kushiner, "Berkeley's Radical: An Interview with Phillip E. Johnson," *Touchstone*, June 2002 <www.touchstonemag.com/docs/_issues/15.5docs/15-5pg40.html>. For a fuller discussion of the impact of evolutionary thought on a broad range of fields such as philosophy, law, education and theology, see Nancy Pearcey, *Total Truth: Liberating Christianity from Its Cultural Captivity* (Wheaton, Ill.: Crossway, 2004).

[2]Phillip Johnson, *The Wedge of Truth: Splitting the Foundations of Naturalism* (Downers Grove, Ill.: InterVarsity Press, 2000), p. 166. See also my review: Nancy Pearcey, "A New Foundation for Positive Cultural Change: Science and God in the Public Square," *Human Events*, September 15, 2000 <www.arn.org/docs/pearcey/np_hewedgereview091200.htm>.

[3]C. P. Snow, *The Two Cultures and the Scientific Revolution* (New York: Cambridge University, 1959).

[4]The history of the division of the university curriculum into these two broad categories is recounted in Jon Roberts and James Turner, *The Sacred and the Secular University* (Princeton, N.J.: Princeton University Press, 2000). Part one of the book deals with the sciences, part two with the humanities.

[5]Richard Rorty, "Untruth and Consequences," a review of *Killing Time* by Paul Feyerabend, *The New Republic,* July 31, 1995, pp. 32-36.

[6]Rorty, *Contingency, Irony, and Solidarity* (New York: Cambridge University Press, 1989), p. 17.

[7]For a more detailed discussion of Dewey and his evolutionary epistemology, along with the other classical pragmatists, see chap. 8 of Pearcey, *Total Truth*.

[8]Allan Bloom in *The Republic of Plato*, translated with notes and an interpretative essay by Allan Bloom (New York: Basic Books, 1968), p. x.

[9]Peter Kreeft, in a book interview on the Christian Book Distributor's website <www.christian book.com/Christian/Books/cms_content/_92165368?page=364779&in sert=7843899&event= ESRC>100f743.jpg>.

[10]*Economics for Decision Making* (Boston: D.C. Heath, 1988), p. 5. This textbook and others are discussed in Warren A. Nord, *Religion and American Education: Rethinking a National Dilemma* (Chapel Hill: University of North Carolina Press, 1995), chap. 4.

[11]Francis Schaeffer deals with the divided concept of truth in *Escape from Reason* (Downers Grove, Ill.: InterVarsity Press, 1968), and *The God Who Is There*, 2nd ed. (Downers Grove, Ill.: InterVarsity Press, 1998).

[12]Elizabeth Flower and Murray G. Murphey, *A History of Philosophy in America* (New York: Putnam, 1977), 2:553.

[13]Neal Gillespie, *Charles Darwin and the Problem of Creation* (Chicago: University of Chicago Press, 1979), p. 16 (emphasis added).

[14]Daniel Dennett, *New York Times*, July 12, 2003, sec. A, p. 11, column 1.

[15]Peter Berger, *Facing Up to Modernity: Excursions in Society, Politics, and Religion* (New York: Basic Books, 1977), p. 133.

[16]Much of what follows is adapted from the study guide in the 2005 revised edition of *Total Truth: Liberating Christianity from Its Cultural Captivity*. The material was originaly developed for a C-SPAN lecture, "Total Truth: Liberating Christianity from Its Cultural Captivity" <www.booktv.org/General/index.asp?segID=5201&schedID=314>.

[17]Thomas Byrne Edsall, "Blue Movie," *Atlantic Monthly* (January-February 2003) <www .theatlantic.com/doc/200301/edsall>.

[18]Christopher Reeve, cited in "Reeve: Keep Religious Groups out of Public Policy," Associated Press, April 3, 2003 (emphasis added).

[19]Johnson, *Wedge of Truth*, p. 148 (emphasis added).

[20]Eleanor Clift, "Faith Versus Reason," *Newsweek*, August 13, 2004 (emphasis added).

[21]Transcript of a television interview with John Searle from a program titled "Thinking Allowed: Conversations on the Leading Edge of Knowledge and Discovery," with Dr. Jeffrey Mishlove <www.williamjames.com/transcripts/searle.htm>.

[22]Steven Pinker, *How the Mind Works* (New York: W. W. Norton, 1997), pp. 55-56 (emphasis added). See Phillip Johnson's discussion of Pinker in *Wedge of Truth*.

[23]Steven Pinker, *The Blank Slate: The Modern Denial of Human Nature* (New York: Viking, 2002), p. 240.

[24]Rodney Brooks, *Flesh and Machines* (New York: Pantheon, 2002), p. 174.

[25]Marvin Minsky, *The Society of Mind* (New York: Simon & Schuster, 1985), p. 307 (emphasis added).

[26]Paul Ramsey, *The Patient as Person* (New Haven, Conn.: Yale University Press, 1970).

[27]John Kerry, *ABC News* interview with Peter Jennings, July 22, 2004, reported in *USA Today* <www.usatoday.com/printedition/news/20040920/a_kerryreligion20.art.htm>.

[28]Bill Allen, *Court TV*, March 24, 2005 <www.courttv.com/talk/chat_transcripts/2005/ 0324schiavo-debate.html>.

[29]See Robert Johansen, "Starving for a Fair Diagnosis," *National Review Online*, March 16, 2005 <www.nationalreview.com/comment/johansen200503160848.asp>.

[30]"What Kids Want to Know About Sex and Growing Up," Children's Television Workshop, 1998.

[31]Bret Johnson, "Coming Out Healthy," *In the Family*, July 1998, quoted in Laura Markowitz, "A Different Kind of Queer Marriage: Suddenly Gays and Lesbians Are Wedding Partners of the Opposite Sex," *Utne Reader*, September/October 2000 <www.yvonnesplace.net/news/ differentkindofqueer.htm>.

[32]Gene Edward Veith, "Identity Crisis," *World*, March 27, 2004 <www.yvonnesplace.net/news/differentkindofqueer.htm>.

[33]Charles Darwin, cited in *Life and Letters of Charles Darwin*, ed. Francis Darwin, vol. 1 (New York: D. Appleton, 1898), p. 285. For several additional quotations from Darwin, see Nancy Pearcey, "The Influence of Evolution on Philosophy and Ethics," in *Science at the Crossroads: Observation or Speculation?* papers of the 1983 National Creation Conference (Richfield, Minn.: Onesimus, 1985), pp. 166-71.

[34]In recent years, this argument has been developed by Al Plantinga. He explains: "What evolution guarantees is (at most) that we behave in certain ways—in such ways as to promote survival. It does not guarantee mostly true or verisimilitudinous beliefs" (*Warrant and Proper Function* [New York: Oxford University Press, 1993], p. 218).

[35]Rorty, *Contingency, Irony, and Solidarity*, p. 5.

[36]Ibid., p. 21.

[37]Roberts and Turner, *Sacred and the Secular University*, p. 90, emphasis added.

Chapter 15: Phillip Johnson Was Right

[1]C. S. Lewis, "The Funeral of a Great Myth," in *Christian Reflections*, ed. Walter Hooper (Grand Rapids: Eerdman's, 1967), pp. 86-88.

Aldous Huxley, "Confessions of a Professed Atheist," *Report: Perspective on the News* 3 (1966), p. 19.

[3]The Bloodhound Gang, "The Bad Touch," *Hooray for Boobies*, Interscope Records.

[4]Phillip E. Johnson, *Reason in the Balance: The Case Against Naturalism in Science, Law, and Education* (Downers Grove, Ill.: InterVarsity Press, 1995).

[5]The idea of a divine authority behind the natural law is often misunderstood. Some people imagine that if God had ordained that we rape instead of marry, murder instead of cherish, hate him instead of love him, then such things would be right. The absurdity of this idea is considered an objection to God's authority. What the objection overlooks is that a being capable of commanding such things would not be God. God is neither constrained by nor indifferent to the good; he *is* the good, the uncreated good in which created goods are grounded.

[6]For further discussion, see J. Budziszewski, *What We Can't Not Know: A Guide* (Dallas: Spence, 2003), chaps. 4-5.

[7]William B. Provine, "Scientists, Face It! Science and Religion Are Incompatible," *The Scientist*, September 5, 1988, pp. 10-11. See also William Provine, "Evolution and the Foundation of Ethics," *MBL Science* 3, no. 1 (1988): 25-29. The article is conveniently reprinted in Steven L. Goldman, *Science, Technology, and Social Progress* (Bethlehem, Penn.: Lehigh University Press, 1989).

[8]Richard Dawkins, *River Out of Eden* (New York: HarperCollins, 1995), pp. 132-33.

[9]Edward O. Wilson, *On Human Nature* (Cambridge, Mass.: Harvard University Press, 1978), p. 176.

[10]Michael Ruse and E. O. Wilson, "The Evolution of Ethics," *New Scientist* 108, no. 1478 (1985): 51-52.

[11]Robert Wright, *The Moral Animal: The New Science of Evolutionary Psychology* (New York: Random House, 1994), p. 212.

[12]Richard Dawkins, *The Selfish Gene* (Oxford: Oxford University Press, 1989), p. 3.

[13]Edward O. Wilson, "What Is Nature Worth?" *San Francisco Chronicle*, May 5, 2002, opinion sec. Wilson's book *The Future of Life* develops this argument further and was published by Alfred Knopf (New York: 2002).

[14]It seems likely that imagination is not a property of matter either, but we will leave this question for another time.

[15]There is a danger of circularity: Unless they had already developed the tendency to mutual aid, why *would* they have lived in family groups?

[16]William D. Hamilton, "The Evolution of Altruistic Behavior," *American Naturalist* 97 (1963):

354-56; and "The Genetical Evolution of Social Behavior," *Journal of Theoretical Biology* 7 (1964): 1-52.

[17]Wright, *Moral Animal,* pp. 313-14.

[18]Ibid., pp. 211-12.

[19]Ibid., 332-33.

[20]For more detailed discussion of the significance of these four steps for utilitarianism, see J. Budziszewski, *Written on the Heart: The Case for Natural Law* (Downers Grove, Ill.: Inter-Varsity Press, 1997), chaps. 10-12.

[21]See John Stuart Mill's *Utilitarianism.*

[22]Larry Arnhart, *Darwinian Natural Right: The Biological Ethics of Human Nature* (Albany: State University of New York Press, 1998). See also Larry Arnhart, Michael J. Behe and William A. Dembski, "Conservatives, Darwin and Design: An Exchange," *First Things* 107 (2000): 23-31; and Larry Arnhart, "Evolution and Ethics: E. O. Wilson Has More in Common with Thomas Aquinas Than He Realizes," *Books & Culture* 5, no. 6 (1999): 36.

[23]John Hare pursues a similar line of reasoning in his paper "Evolutionary Naturalism and the Reduction of the Ethical Demand," which was presented at "The Nature of Nature: An Inter-disciplinary Conference on the Role of Naturalism in Science," Baylor University, April 2000. Because Hare's purpose in writing is somewhat different than mine, he does not comment on the confusion between naturalism and natural law, nor does he draw out the parallel between Arnhart's theory and utilitarianism. However, he vigorously criticizes Arnhart for what he calls the "double identity" of equating the good with the desirable and the desirable with what in fact is desired (my steps two and three), and our arguments coincide at several points.

[24]C. S. Lewis, *The Pilgrim's Regress,* 3rd ed. (New York: Bantam, 1986), p. xii. Lewis's analysis of the experience is illuminating.

[25]Thomas Aquinas *Summa Theologica* I-II, Q. 100.

[26]Not all exceptionless precepts are "first" principles. The prohibition of murder, for example, is not a first principle because it rests on the still more basic precept that we must never gratuitously harm our neighbors; nevertheless it binds without exception.

[27]Aquinas *Summa Theologica* I-II, Q. 94, art. 4.

[28]Arnhart, *Darwinian Natural Right,* p. 147 (emphasis added). Page numbers taken from the *NetLibrary* edition <www.netlibrary.com>.

[29]Ibid., pp. 30-31.

[30]Ibid., p. 149, section on "The Moral Complementarity of Male and Female Norms."

[31]Ibid., p. 143.

[32]Ibid., p. 149.

[33]Ibid., p. 124.

[34]Ibid., p. 275.

Chapter 16: A Taxonomy of Teleology

[1]We will use *naturalism* throughout this essay to refer to what Joel Cracraft calls "the fundamental precept of science," namely, "that phenomena in the natural world should be interpreted through naturalistic explanations that are accepted (always tentatively) or rejected by reference to observation" ("The New Creationism and Its Threat to Science Literacy and Education," *American Institute of Biological Sciences,* January 2004 <www.aibs.org/bioscience editorials/editorial_2004_01.html>). Defenders of naturalism usually add the adjective "methodological" (MN), to distinguish this precept from *philosophical* naturalism (i.e., the doctrine that nature is the whole of reality, and God does not exist; abbreviated as PN). For the practice of science, however, MN versus PN is a distinction without a difference. Ask oneself a simple question. *Suppose life actually were designed by a nonhuman intelligence—would MN allow us to discover that?* If the answer is no, then MN hinders scientific discovery and dictates the shape of reality as thoroughly as PN. If the answer is yes, then MN is superfluous and says nothing more than that science should be empirical and testable.

[2]Who uses a name, how, and where—i.e., in what contexts—are, of course, social realities of more than passing significance. For example, many evolutionary biologists strongly resent being called "Darwinists." As they see it, they are *not* followers of Charles Darwin, who died in 1882 and erred in much of his thinking (e.g., in his Lamarckian understanding of heredity). To give the question some pungency, the slang words *white trash* and *nigger* may be humorous and contextually appropriate when used self-referentially by white or African American comedians (for instance, Jeff Foxworthy or Bernie Mac). The same words can be profoundly offensive when used by others, with the intent not to describe but to humiliate. When Stephen Jay Gould, for instance, writes that "creationist-bashing is a noble and necessary pursuit these days," we may assume that he intends *creationist* to be a term of reproach (Stephen Jay Gould, *Dinosaur in a Haystack: Reflections in Natural History* [New York: Three Rivers Press, 1996], p. 422).

[3]Phillip E. Johnson, *Darwin on Trial,* 2nd ed. (Downers Grove, Ill.: InterVarsity Press, 1993), p. 4.

[4]Ibid.

[5]Koalas are marsupials; dolphins and shrews are placentals; all are mammals.

[6]Much of the discussion here is modified from M. R. Ross, "Who Believes What? Clearing Up Confusion About Intelligent Design and Young-Earth Creationism," *Journal of Geoscience Education* 23, no. 3 (2004): 319-23.

[7]"Questions About Intelligent Design," *Discovery Institute Center for Science and Culture* <www.discovery.org/csc/topQuestions.php>.

[8]"What Is Intelligent Design?" *Access Research Network* <www.arn.org/idfaq/What%20is%20 intelligent%20design.htm>.

[9]Kurt P. Wise, *Faith, Form, and Time: What the Bible Teaches and Science Confirms About Creation and the Age of the Universe* (Nashville: Broadman & Holman, 2002), p. 287.

[10]Paul Nelson and John Mark Reynolds, "Young Earth Creationism," in *Three Views on Creation and Evolution,* ed. J. P. Moreland and John Mark Reynolds (Grand Rapids: Zondervan, 1999), p. 42.

[11]William A. Dembski, *Intelligent Design: The Bridge Between Science and Theology* (Downers Grove, Ill.: InterVarsity Press, 1999), p. 312.

[12]Henry R. Morris, "Design Is Not Enough!" *Impact* 127 (1999). This article is also available at the Institute for Creation Research website at <www.icr.org/index.php?module=articles& action=view&ID=859>.

[13]Carl Weiland, "AiG's Views on the Intelligent Design Movement," August 30, 2002 <http:// answersingenesis.org/docs2002/0830_IDM.asp>.

[14]Wise, *Faith, Form, and Time,* p. 281.

[15]Eugenie C. Scott, "The Creation/Evolution Continuum," *NCSE Reports* 19, no. 4 (1999): p. 22. This article is on the web at <www.ncseweb.org/resources/articles/5232_the_creationevolu tion_continu_12_7_2000.asp>.

[16]Donald U. Wise, "Creationism's Propaganda Assault on Deep Time and Evolution," *Journal of Geoscience Education* 49, n. 1 (2001), p. 33.

[17]Scott, "The Creation/Evolution Continuum," p. 22.

[18]See L. Laudan, "Science at the Bar—Causes for Concern," in *But Is It Science?* ed. Michael Ruse (Buffalo, N.Y.: Prometheus, 1982), pp. 351-55; and P. L. Quinn, 1984, "The Philosopher of Science as Expert Witness," in *But Is It Science?* ed. Michael Ruse (Buffalo, N.Y.: Prometheus, 1984), pp. 367-85.

[19]Steve A. Austin et al., "Catastrophic Plate Tectonics: A Global Flood Model of Earth History," *Proceedings of the Third International Conference on Creationism,* ed. R. E. Walsh (Pittsburgh: Creation Science Fellowship, 1994), pp. 609-21.

[20]Scott, "The Creation/Evolution Continuum," p. 16.

[21]Michael J. Denton, *Nature's Destiny: How the Laws of Biology Reveal Purpose in the Universe* (New York: Free Press, 1998), p. 476; and Michael J. Denton, C. G. Marshall and M. Legge,

"The Protein Folds as Platonic Forms: New Support for the Pre-Darwinian Conception of Evolution by Natural Law," *Journal of Theoretical Biology* 219 (2002): 325-42.

[22]"The Raelian Movement Supports ID Theory" <http://www.prweb.com/releases/2002/11/prweb50443.php>.

[23]Johnson, *Darwin on Trial*, p. 4.

Chapter 17: Complexity, Chaos and God

[1]David F. Wells, *God in the Wasteland: The Reality of Truth in a World of Fading Dreams* (Grand Rapids: Eerdmans, 1995), p. 14.

[2]Wesley D. Allen, in "Modern Science: Charting a Course for the Future," *Science: Christian Perspectives for the New Millennium*, ed. S. B. Luley, P. Copan and S. W. Wallace (Addison, Tex.: CLM & RZIM Publishers, 2003), 2:229-48; Nancey R. Pearcey and Charles B. Thaxton, *The Soul of Science: Christian Faith and Natural Philosophy* (Wheaton, Ill.: Crossway, 1994).

[3]Peter Coveney and Roger Highfield, *Frontiers of Complexity: The Search for Order in a Chaotic World* (New York: Fawcett Columbine, 1995).

[4]Ibid., p. 5.

[5]Immanuel Kant, *Universal Natural History and Theory of the Heavens*, trans. S. Jaki (Edinburgh: Scottish Academic Press, 1981), p. 87.

[6]Carl Sagan, *Cosmos* (New York: Ballantine, 1980).

[7]See a reproduction of an illustration from Thomas Burnet's *Telluris Theoria Sacra* (1681), in I. Stewart, *Does God Play Dice? The Mathematics of Chaos* (Oxford: Basil Blackwell, 1989), p. 6.

[8]See Coveney and Highfield, *Frontiers of Complexity*.

[9]Ibid.

[10]Stewart, *Does God Play Dice?* James Gleick, *Chaos: Making a New Science* (New York: Penguin, 1987); Garnett P. Williams, *Chaos Theory Tamed* (Washington, D.C.: Joseph Henry Press, 1997); and E. N. Lorenz, *The Essence of Chaos* (Seattle: University of Washington Press, 1993).

[11]Peter W. Atkins, *The 2nd Law: Energy, Chaos, and Form* (New York: Scientific American Books, 1994).

[12]See Gleick, *Chaos*, pp. 9-31.

[13]Williams, *Chaos Theory Tamed*, pp. 18-19.

[14]Ibid., pp. 209-10.

[15]Peter W. Atkins and Julio de Paula, *Physical Chemistry*, 7th ed. (New York: W. H. Freeman, 2001), p. 920.

[16]Stewart, *Does God Play Dice?* pp. 155-64; Gleick, *Chaos*, pp. 59-80; Williams, *Chaos Theory Tamed*, chapter 10; Wesley J. Wildman and Robert John Russell in *Chaos and Complexity: Scientific Perspectives on Divine Action*, ed. R. J. Russell, N. Murphy and A. R. Peacocke (Vatican City: Vatican Observatory Publications, 1995), pp. 49-74.

[17]Williams, *Chaos Theory Tamed*, p. 13.

[18]Coveney and Highfield, *Frontiers of Complexity*, p. 325.

[19]N. Goldenfeld and L. P. Kadanoff, "Simple Lessons from Complexity," *Science* 284 (1999): 87.

[20]See Gleick, *Chaos*, p. 13.

[21]C. Zimmer, "Life After Chaos," *Science* 284 (1999): 85.

[22]Ibid.

[23]See Coveney and Highfield, *Frontiers of Complexity*, chapter 9.

[24]Roger Penrose, *The Emperor's New Mind* (Oxford: Oxford University Press, 1989), and *Shadows of the Mind* (New York: Oxford University Press, 1994), and *The Large, the Small, and the Human Mind* (Cambridge: Cambridge University Press, 1997).

[25]C. Seife, "Cold Numbers Unmake the Quantum Mind," *Science* 287 (2000): 791.

[26]Michael Arbib, cited in Coveney and Highfield, *Frontiers of Complexity*, pp. 324-25.

[27]Coveney and Highfield, *Frontiers of Complexity*, p. 13.

[28]J. P. Crutchfield et al., "Chaos," *Scientific American* 255, no. 12 (1986).

[29]Paul Davies, *The Cosmic Blueprint* (London: Unwin Hyman, 1989), p. 56.

[30]R. Gallagher and T. Appenzeller, *Science* 284 (1999): 79.

[31]Coveney and Highfield, *Frontiers of Complexity;* and Russell, Murphy and Peacocke, *Chaos and Complexity.*

[32]Coveney and Highfield, *Frontiers of Complexity,* pp. 16-17.

[33]Stewart, *Does God Play Dice?* John Jefferson Davis, "Theological Reflections on Chaos Theory," *Perspectives on Science and Christian Faith* 49 (1997): 75-84.

[34]James Lighthill, *Proceedings of the Royal Society of London A* 407(1986): 35, 38, 47.

[35]Davis, *Perspectives on Science and Christian Faith,* p. 80.

[36]Crutchfield, "Chaos," p. 57.

[37]J. Doyne Farmer, cited in Gleick, *Chaos,* p. 251.

[38]Davis, *Perspectives on Science and Christian Faith,* p. 79.

[39]Ibid., p. 81.

[40]Stewart, *Does God Play Dice?* p. 302.

[41]Arthur Peacocke, *Theology for a Scientific Age* (London: SCM Press, 1993), p. 65.

[42]Coveney and Highfield, *Frontiers of Complexity,* p. 117.

[43]Ibid., pp. 198, 199, 209.

[44]Michael J. Behe, *Darwin's Black Box: The Biochemical Challenge to Evolution* (London: Free Press, 1996), pp. 171-72.

[45]Dean L. Overman, *A Case Against Accident and Self-Organization* (Lanham, Md.: Rowman & Littlefield, 1997), pp. 74-75.

[46]William A. Dembski, *No Free Lunch: Why Complexity Cannot Be Purchased Without Intelligence* (Lanham, Md.: Rowman & Littlefield, 2002), p. 166-73.

CONTRIBUTORS

Wesley D. Allen is a senior research scientist and adjunct professor at the Center for Computational Chemistry at the University of Georgia. Dr. Allen received a B.A. in chemistry and physics from Vanderbilt University (1983) and a Ph.D. in theoretical chemistry from the University of California, Berkeley (1987). He served as an assistant professor in the department of chemistry at Stanford University from 1988-1994, before moving to his current position at the University of Georgia in 1995. Dr. Allen is an expert in fundamental theories and large-scale computational applications of molecular quantum mechanics. He has authored over seventy scientific research publications, most appearing in the *Journal of Chemical Physics,* as well as a recent book chapter in *Science: Christian Perspectives for the New Millennium.*

Francis J. Beckwith is associate director of the J. M. Dawson Institute of Church-State Studies, and associate professor of church-state studies at Baylor University. A 2002-2003 Madison Research Fellow in Politics at Princeton University, he has served since 2003 as a member of Princeton's James Madison Society. He is a graduate of Fordham University (Ph.D., philosophy), and holds the Master of Juridical Studies (M.J.S.) degree from the Washington University School of Law in St. Louis. His books include *To Everyone An Answer: A Case for the Christian Worldview* (InterVarsity Press, 2004); *Law, Darwinism & Public Education: The Establishment Clause and the Challenge of Intelligent Design* (Rowman & Littlefield, 2003); and *Do the Right Thing: Readings in Applied Ethics and Social Philosophy* (Wadsworth, 2002). His articles have been published in a number of academic journals across a variety of disciplines, including *Harvard Journal of Law & Public Policy, International Philosophical Quarterly, Public Affairs Quarterly, Notre Dame Journal of Law, Ethics & Public Policy, Social Theory & Practice, American Journal of Jurisprudence, Journal of Medical Ethics, San Diego Law Review, Nevada Law Journal, Journal of Social Philosophy* and *Philosophia Christi.* His website is http://francisbeckwith.com.

Michael J. Behe is professor of biological sciences at Lehigh University in Pennsylvania. He received his Ph.D. in biochemistry from the University of Pennsylvania in 1978. In addition to publishing over thirty-five articles in refereed biochemical journals, he has also written editorial features in *Boston Review, American Spectator* and the *New York Times.* His book, *Darwin's Black Box,* discusses the implications for neo-Darwinism of what he calls "irreducibly complex" biochemical systems. The book was internationally reviewed in over one hundred publications and recently named by *National Review* and *World* magazine as one of the one hundred most important books of the twentieth century. He is a senior fellow with Discovery Institute's Center for Science and Culture.

Walter L. Bradley (Ph.D., P.E.) is currently Distinguished Professor of Engineering at Baylor University. Previously, he was professor of mechanical engineering at Texas A&M University. There he served as department head for sixty-seven professors and fifteen hundred students from 1989-1993. He also served as the director of the Texas A&M University Polymer Technology Center from 1986-1990 and from 1994-2000. He received more than five million dollars in research contracts from government agencies such as NSF, NASA, DOE and AFOSR, and from major corporations such as Dupont, Exxon, Shell, Phillips, Equistar, Texas Eastman, Union Carbide and 3M. He has published more than 145 technical articles in archival journals, conference proceedings and as book chapters. He was honored by being elected a Fellow of the American Society for Materials in 1992. He has received one national and five local research awards and two local teaching awards. He coauthored a seminal work on the origin of life titled *The Mystery of Life's Origin: Reassessing Current Theories* in 1984 and has published several book chapters and journal articles related to the origin of life as well and has spoken on over sixty university campuses on this topic in the past ten years. He is a fellow of Discovery Institute's Center for Science and Culture.

David Berlinski received his Ph.D. in philosophy from Princeton University and was later a postdoctoral fellow in mathematics and molecular biology at Columbia University. He has authored works on systems analysis, differential topology, theoretical biology, analytic philosophy and the philosophy of mathematics as well as three novels. He has also taught philosophy, mathematics and English at such universities as Stanford, Rutgers, the City

University of New York and the Université Paris. In addition, he has held research fellowships at the International Institute for Applied Systems Analysis (IIASA) in Austria and the Institut des Hautes Études Scientifiques (IHES) in France. Recent articles by Dr. Berlinski have been featured in *Commentary, Forbes ASAP* and the *Boston Review*. He is author of numerous books, including *A Tour of the Calculus, The Advent of the Algorithm, Newton's Gift, The Secrets of the Vaulted Sky* and, most recently, *Infinite Ascent: A Short History of Mathematics*. He is a senior fellow with Discovery Institute's Center for Science and Culture.

J. Budziszewski (Ph.D. Yale, 1981) is professor of government and philosophy at the University of Texas at Austin. He is a political theorist and philosopher of natural law. His recent work focuses on the repression of moral knowledge—what we really know, how we tell ourselves that we don't know what we do, and what happens to the structures of conscience and moral judgment when we try. A fellow of the Wilberforce Forum as well as Discovery Institute's Center for Science and Culture, he is also chair of the board of directors of the Institute for Religion and Democracy. His articles have appeared in journals of law, ethics, theology, public policy and political theory, and his academic books include *The Resurrection of Nature: Political Theory and the Human Character* (Cornell University Press, 1986), *The Nearest Coast of Darkness: A Vindication of the Politics of Virtues* (Cornell University Press, 1988), *True Tolerance: Liberalism and the Necessity of Judgment* (Transaction, 1992), *Written on the Heart: The Case for Natural Law* (InterVarsity Press, 1997), *The Revenge of Conscience: Politics and the Fall of Man* (Spence, 1999), and *What We Can't Not Know: A Guide* (Spence, 2003).

William A. Dembski is the Carl F. H. Henry Professor of Theology and Science at Southern Baptist Theological Seminary in Louisville where he directs the seminary's Center for Science and Theology. He is also a senior fellow with Discovery Institute's Center for Science and Culture in Seattle. A mathematician and philosopher, Dembski previously directed Baylor University's Michael Polanyi Center. He has done postdoctoral work in mathematics at MIT, in physics at the University of Chicago, and in computer science at Princeton University. He has been awarded Templeton and NSF research grants. A graduate of the University of Illinois at Chicago, where he earned a B.A.

in psychology, an M.S. in statistics and a Ph.D. in philosophy, he also received a doctorate in mathematics from the University of Chicago in 1988 and a master of divinity degree from Princeton Theological Seminary in 1996. Dembski has published articles in mathematics, philosophy and theology journals and is the author or editor of ten books. In *The Design Inference: Eliminating Chance Through Small Probabilities* (Cambridge University Press, 1998), he examines the design argument in a post-Darwinian context and analyzes the connections linking chance, probability and intelligent causation.

Phillip E. Johnson is the Jefferson Peyser Emeritus Professor of Law at the University of California, Berkeley. Professor Johnson is a well-known speaker and writer on the philosophical significance of Darwinism. His books on this topic include *Darwin on Trial, Reason in the Balance, Defeating Darwinism by Opening Minds, The Wedge of Truth* and *The Right Questions* (all by InterVarsity Press). After completing his law degree at the University of Chicago, Johnson was a law clerk for Chief Justice Earl Warren of the United States Supreme Court. Johnson taught law for over thirty years at the University of California, Berkeley. He is the author of two widely used textbooks on criminal law: *Criminal Law: Cases, Materials, and Text,* 6th ed. (West, 2000) and *Cases and Materials on Criminal Procedure,* 3rd ed. (West, 2000). Johnson entered the evolution controversy because he found the books defending Darwinism dogmatic and unconvincing. Professor Johnson is an adviser to Discovery Institute's Center for Science and Culture.

David J. Keller received his Ph.D. in (bio)physical chemistry at the University of California, Berkeley, for research on the physical chemistry of DNA and the theory of circularly polarized light scattering. He was awarded an NIH postdoctoral fellowship at Stanford University for work on the theory of phase transitions in lipid membranes. He is currently associate professor of chemistry at the University of New Mexico, where his research focuses on understanding biological molecular machines, especially DNA polymerases and reverse transcriptases. He is the author of over fifty scientific articles and papers on topics ranging from the theory of molecular machines to atomic force microscopy to quantum light scattering.

Stephen C. Meyer is senior research fellow at the Discovery Institute (Seattle) and director of Discovery Institute's Center for Science and Culture. Formerly a geophysicist with the Atlantic Richfield Company, Meyer completed a Ph.D. dissertation in history and philosophy of Science at Cambridge University on origin-of-life biology and the methodology of the historical sciences. He has contributed to several scholarly books and anthologies, including *The History of Science and Religion in the Western Tradition: An Encyclopedia; Darwinism: Science or Philosophy; Of Pandas and People: The Central Question in Biological Origins; The Creation Hypothesis: Scientific Evidence for an Intelligent Designer; Mere Creation: Science, Faith & Intelligent Design;* and *Facets of Faith and Science: Interpreting God's Action in the World.* Dr. Meyer has cowritten or edited two books: *Darwinism, Design, and Public Education* with Michigan State University Press and *Science and Evidence of Design in the Universe* (Ignatius, 2002). In addition to his technical articles on design, Meyer has written many editorial features in newspapers and magazines, including the *Wall Street Journal,* the *Los Angeles Times* and the *Chicago Tribune.* He has also appeared on national TV and radio programs such as *Technopolitics,* PBS's *Freedom Speaks,* CNBC's *Hardball with Chris Matthews* and National Public Radio's *Talk of the Nation.* He was the main content person for the video *Unlocking the Mystery of Life.* His article on the information problem in the Cambrian Explosion that appeared in the *Proceedings of the Biological Society of Washington* in 2004 created an international sensation.

Scott A. Minnich is a microbiology professor at the University of Idaho. He received his Ph.D. in microbiology at Iowa State University, conducting studies on the rapid diagnostic methods for detection of bacteria. He pursued postdoctoral studies at Purdue and Princeton universities in molecular genetics and molecular biology. Dr. Minnich's research is centered on the genetics of flagellar biosynthesis and type III protein secretion in Gram-negative pathogens and novel strategies for vaccine production. Over the years he has published numerous peer-reviewed publications on these topics in such journals as the *Proceedings of the National Academy of Sciences, Journal of Molecular Biology, Journal of Bacteriology* and *Molecular Microbiology.* From October 2003 to May 2004, Scott Minnich served in Iraq as a subject matter expert with the Iraq Survey Group WMD Inspection Team.

Paul Nelson is a philosopher of biology who received his Ph.D. from the

University of Chicago (1998), where he specialized in the philosophy of biology and evolutionary theory. He is currently a Fellow of the Discovery Institute, and adjunct professor in the M.A. program in science and religion, Biola University. Paul's articles have appeared in *Biology & Philosophy, Zygon, Rhetoric and Public Affairs* and *Touchstone,* and chapters in the anthologies *Mere Creation, Signs of Intelligence, Intelligent Design Creationism and Its Critics,* and *Darwin, Design, and Public Education.* His forthcoming monograph, *On Common Descent,* critically evaluates the theory of common descent. Paul is a member of the Society for Developmental Biology (SDB) and the International Society for the History, Philosophy and Social Studies of Biology (ISHPSSB).

Nancy R. Pearcey is the Francis A. Schaeffer Scholar at the World Journalism Institute, a visiting scholar at Biola University's Torrey Honors Institute and a senior fellow at the Discovery Institute. She studied under Schaeffer at L'Abri Fellowship in Switzerland and went on to earn a master's degree from Covenant Theological Seminary, followed by graduate work in history of philosophy at the Institute for Christian Studies in Toronto. A frequent lecturer, Pearcey has spoken to actors and screenwriters in Hollywood; at universities such as Dartmouth, Stanford, USC and Princeton; at national labs like Sandia and Los Alamos; in Congress and the White House; and for political organizations. She has been writing on science and Christian worldview since 1977. In 1991, she became the founding editor of *BreakPoint,* a daily syndicated radio commentary program hosted and voiced by Chuck Colson, and was executive editor of the program for nearly nine years. During the same period, she served as policy director and senior fellow of the Wilberforce Forum, and coauthored a monthly column in *Christianity Today.* Her articles have appeared in *The Washington Times, Human Events, First Things, Books & Culture, World, The Human Life Review, Christianity Today* and the *Regent University Law Review.* She has authored or contributed to several books, including the ECPA Gold Medallion winner *How Now Shall We Live?* and, most recently, *Total Truth: Liberating Christianity from Its Cultural Captivity,* which was featured on C-SPAN, received an award of merit in the Christianity Today 2005 Book Awards and won an ECPA Gold Medallion Award.

John Mark Reynolds is the founder and director of the Torrey Honors Institute, and associate professor of philosophy, at Biola University. (He has also taught philosophy at Roberts Wesleyan College, Whitworth College and Saint John Fisher College.) In 1996 he received his Ph.D. in philosophy from the University of Rochester, where he wrote a dissertation analyzing the cosmology and psychology found in Plato's *Timaeus*. A revised version of his dissertation "Toward a Unified Platonic Psychology" is published by the University Press of America. His first book, *Maker of Heaven and Earth: Three Views on the Creation and Evolution Debate,* was coedited with J. P. Moreland. He has written several technical articles in philosophy of religion and several semipopular articles in journals such as the *New Oxford Review* and *Touchstone.* Dr. Reynolds lectures frequently on ancient philosophy, philosophy of science, home schooling and cultural trends. He is a fellow of the Discovery Institute's Center for Science and Culture.

Jay Wesley Richards holds a Ph.D. in philosophy and theology from Princeton Theological Seminary, where he was formerly a teaching fellow. His masters thesis (Th.M., Calvin Theological Seminary) treated the philosopher of science Michael Polanyi. From 1996-1998, he was executive editor of *The Princeton Theological Review* and president of the Charles Hodge Society at Princeton Theological Seminary. He has published in academic journals such as *Religious Studies, Christian Scholars' Review, The Heythrop Journal, Encounter, The Princeton Theological Review, Perspectives on Science and the Christian Faith* as well as editorial features in *The Washington Post, Seattle Post-Intelligencer* and IntellectualCapital.com. He is coeditor, with William Dembski, of *Unapologetic Apologetics: Meeting the Challenges of Theological Studies* (InterVarsity Press, 2001), and coeditor, with George Gilder, of *Are We Spiritual Machines? Ray Kurzweil vs. the Critics of Strong AI* (Discovery Institute Press, 2002). He is also author of *The Untamed God: A Philosophical Exploration of Divine Perfection, Immutability, and Simplicity* (InterVarsity Press, 2003) and coauthor, with astronomer Guillermo Gonzalez, of *The Privileged Planet: How Our Place in the Cosmos Is Designed for Discovery* (Regnery, 2004). He is vice president of Discovery Institute and senior fellow with its Center for Science and Culture.

Marcus R. Ross is assistant professor of geology and assistant director of creation studies at Liberty University. He is currently completing a Ph.D. in geoscience from the University of Rhode Island, where his research focuses on analyzing the biostratigraphy and extinction of mosasaurs, a group of Late Cretaceous marine reptiles. He received an M.S. in paleontology at the South Dakota School of Mines and Technology, and a B.S. in earth science from Pennsylvania State University. His work has been published in *Northeastern Geology and Environmental Science, Journal of Geoscience Education* and the anthology *Darwinism, Design, and Public Education.* He has given presentations at the Geological Society of America, the Society of Vertebrate Paleontology and the American Scientific Affliation. He and his wife, Corrina, live in Lynchburg, Virginia.

Michael Ruse is Lucyle T. Werkmeister Professor of Philosophy at Florida State University. He received his B.A. in philosophy and mathematics from Bristol University, an M.A. in philosophy from McMaster University and his Ph.D. from Bristol University. He was full professor of philosophy at Guelph from 1974 to 2000. Dr. Ruse is a Fellow of the Royal Society of Canada and of the American Association for the Advancement of Science. He has received numerous visiting professorships, fellowships and grants. Ruse's many publications include *The Philosophy of Biology, Sociobiology: Sense or Nonsense? The Darwinian Revolution: Science Red in Tooth and Claw, Darwinism Defended: A Guide to the Evolution Controversies, Taking Darwin Seriously: A Naturalistic Approach to Philosophy, But Is It Science? The Philosophical Question in the Evolution/Creation Controversy,* and he edited *The Philosophy of Biology.* His most recent works include *Can a Darwinian Be a Christian? The Relationship Between Science and Religion* (2000) and *Darwin and Design: Does Evolution Have a Purpose?* (2003). Dr. Ruse is also on the editorial board for major journals such as *Zygon, Philosophy of Science* and *Quarterly Review of Biology.* On a more public level, Ruse has appeared on television programs, including *Firing Line,* and was a witness for the ACLU in the 1981 Arkansas hearings which overturned a creation science law.

Henry F. Schaefer III, was professor of chemistry at the University of California, Berkeley, for eighteen years. Since 1987 he has been the Graham Perdue Professor of Chemistry and director of the Center for Computational Quantum Chemistry at the University of Georgia. Schaefer is the author of more than one thousand scientific publications, the majority appearing in the *Journal of Chemical Physics* or the *Journal of the American Chemical*

Society. He has been the research director of seventy-three successful doctoral students and has presented plenary lectures at more than 125 national or international scientific conferences. He is the editor in chief of *Molecular Physics* and president of the World Association of Theoretically Oriented Chemists. His major awards include the American Chemical Society Award in Pure Chemistry (1979), the American Chemical Society Leo Hendrik Baekeland Award (1983), and the Centenary Medal of the Royal Society of Chemistry (London 1992). The science citation index shows that Professor Schaefer is one of the most widely cited chemists in the world. His research involves the use of state-of-the-art computational hardware and theoretical methods to solve important problems in molecular quantum mechanics. The *U.S. News and World Report* cover story of December 23, 1991, describes Schaefer as a "five-time nominee for the Nobel Prize." Professor Schaefer is a fellow of the Discovery Institute's Center for Science and Culture.

Timothy G. Standish earned a Ph.D. in environmental biology and public policy from George Mason University, where he studied application of molecular techniques to the identification and classification of nematode worms. As an educator, Standish has taught in colleges and universities in the United States, South Africa, Austria and the Philippines. He has also taught public high school science classes in the United States and currently works closely with science teachers in Christian secondary and tertiary schools. His publications cover topics ranging from the molecular basis of behavior in crickets to intelligent design and public policy. In addition, he has authored many academic and popular papers dealing with faith and science. Interactions between science, faith and public policy are his primary interest at present. Dr. Standish is an active member of his church and holds a research appointment at the Geoscience Research Institute in Loma Linda, California.

Jonathan Wells has a Ph.D. in theology from Yale University (1986) and a Ph.D. in biology from the University of California, Berkeley (1994). He has worked as a postdoctoral research biologist at the University of California, Berkeley, and the supervisor of a medical laboratory in Fairfield, California, and he has taught biology at California State University, Hayward. In 1998 he moved with his family to Seattle to be a full-time senior fellow at the Discovery Institute's Center for Science and Culture. Dr. Wells has published articles in *Development, Proceedings of the National Academy of Sciences USA, BioSystems, The Scientist* and *The American Biology Teacher.* He is also author of *Charles Hodge's Critique of Darwinism* (Edwin Mellen, 1988) and *Icons of Evolution: Why Much of What We Teach about Evolution Is Wrong* (Regnery, 2000). Dr. Wells is currently working on a book criticizing the overemphasis on genes in biology and medicine.

Thomas Woodward is an associate professor of communication, science and theology at Trinity College of Florida in Tampa, and is the founder and president of the C. S. Lewis Society, housed at Trinity College. The C. S. Lewis Society organizes university-based lectures, debates and symposia on the interface between science, philosophy, history and theology. He received his Ph.D. in communication at the University of South Florida, with an emphasis in the rhetoric of science. His dissertation, a rhetorical history of design and its critics, titled *Aroused from Dogmatic Slumber,* was reworked and published as *Doubts About Darwin: A History of Intelligent Design* in 2003 by Baker. This book received *Christianity Today*'s 2004 Book Award for Christianity and Culture. He is also coeditor, with James Gills, M.D., of *Darwinism Under the Microscope* (Strang Communications, 2002). He has published a number of articles on Darwinism and the challenge of ID theory and is a frequent lecturer in universities on the topic. He also has an A.B. from Princeton University in history and Latin American studies, and a Th.M. from Dallas Theological Seminary where he received the Lorraine Chafer Award.